Intermediate Classical Dynamics with Applications to Beam Physics

WILEY SERIES IN BEAM PHYSICS AND ACCELERATOR TECHNOLOGY

Series Editor MEL MONTH

BROWN –	FUNDAMENTALS OF ION SOURCES*
CHAO –	PHYSICS OF COLLECTIVE BEAM INSTABILITIES IN HIGH ENERGY ACCELERATORS
EDWARDS AND SYPHERS –	AN INTRODUCTION TO THE PHYSICS OF HIGH ENERGY ACCELERATORS
MICHELOTTI –	INTERMEDIATE CLASSICAL DYNAMICS WITH APPLICATIONS TO BEAM PHYSICS

*Forthcoming titles

Intermediate Classical Dynamics with Applications to Beam Physics

LEO MICHELOTTI

Fermi National Accelerator Laboratory
Batavia, Illinois

A Wiley-Interscience Publication

JOHN WILEY & SONS, INC.

New York / Chichester / Brisbane / Toronto / Singapore

To my parents, Carlo and Anna Michelotti,
who together taught me all that was really necessary,
and my wife, Kathy, who kept our family going strong
while I took the time to write this.

This text is printed on acid-free paper.

Library of Congress Cataloging-in-Publication Data

Michelotti, Leo, 1945–
Intermediate classical dynamics with applications to beam physics / Leo Michelotti.
p. cm.
Includes bibliographical references.
ISBN 0-471-55384-0 (cloth : acid-free paper)
1. Beam dynamics. 2. Mechanics. I. Title.
QC793.3.B4M5 1993
539.7′3—dc20 93-9418

Printed in the United States of America

10 9 8 7 6 5 4 3 2 1

Contents

Preface

> This is not a work of scholarship. . . . I write for the unlearned about things in which I am unlearned myself. If an excuse is needed . . . for writing such a book, my excuse would be something like this. It often happens that two [students] can solve difficulties in their work for one another better than [a] master can. The fellow-pupil can help more than the master because he knows less. . . . I write as one amateur to another, talking about difficulties I have met . . . with the hope that this might at least interest, and sometimes even help, other inexpert readers.
>
> — C. S. Lewis
> *Reflections on the Psalms*

As there are available now a number of excellent modern books on dynamics, someone who produces yet another, perhaps not as excellent, had better be ready with an apology. Two superb examples of the more comprehensive and scholarly offerings currently in print are *Foundations of Mechanics*, by Abraham and Marsden [1] and *Analysis, Manifolds and Physics*, by Choquet-Bruhat and DeWitt-Morette [22]. If you were to purchase either of these two, say, instead of the one you are holding, and commit yourself to absorbing at least one page each day for a year or two, the (non-monetary) rewards would be incalculable. Unfortunately, many of us who seriously need to know this material are either unfamiliar with or intimidated by advanced formalisms. The present volume is, in some sense, a mathematical halfway house between such works and more introductory material, such as Symon's *Mechanics* [87] or Goldstein's *Classical Mechanics* [41]. It may be considered a supplement to these excellent texts and is written at a level suitable for first-year graduate or senior undergraduate physics students.

Other books, such as Rasband's *Dynamics* [79], attempt this as well. One feature that may distinguish this book from it, and others, is the emphasis placed, via examples and problems, on beam physics, more particularly, on the orbits of particles in synchrotrons and storage rings. Partly, this is due to accidental, personal history. It is also true, however, that accelerator physics is one of the areas in which classical mechanics plays more than an academic role. To people designing accelerators and studying their behavior, it is the bread and butter of everyday existence. The historic successes of high energy physics has been due in no small measure to the ability of these people — from the designers and theorists to the operators in control rooms — to persuade large numbers of like-charged particles to travel billions of kilometers while staying within centimeters of each other, increasing their energy continually, and eventually either directing them to a target or colliding them at a focus narrow enough to make the entire enterprise worth the effort. Now, while several texts on dynamics contain numerous examples from fields like celestial mechanics or mechanical engineering, beam physics has been generally underrepresented. Given its impact on physics as a whole, it is fitting that at least one dynamics text draw from accelerator issues for its motivations and examples.

Much has been written recently about a revolution taking place in the asymptotic analysis of dynamical systems, usually illustrated with beautiful color photographs of objects like chaotic attractors, fractal basin boundaries, and other such Julia-Fatou sets. The first shots of this revolution were fired, however, not in the 1980s or 1970s but toward the close of the nineteenth century. By 1892, Henri Poincaré already had published his landmark work *Les Méthods Nouvelles de la Mécanique Céleste* in which he advanced the theses that differential equations should be viewed as *geometric* objects, in particular, as vector fields on manifolds, and that questions concerning the long-term stability of a dynamical system might be attacked by studying the topological properties of these objects as revealed by maps. In particular, he formulated the goal of finding the dynamically invariant regions of different dimensions and determining how they connect with each other to impose structure on dynamical systems. His work even led him to recognize the extraordinarily complicated behavior of orbits in the vicinity of a separatrix, what today we call "chaotic orbits" and identify as invariant regions of fractal dimension. Much like his predecessor Newton, Poincaré found that ideas and language which he needed did not yet exist and that he had to create entirely new mathematics in order to progress. In time the seeds that he planted grew into branches of modern topology, with all its trappings of tangent and cotangent bundles, differential forms, exterior algebra and calculus, homology and cohomology — all of which are frequently associated with advanced topics, such as general relativity or string theory, but are *almost never mentioned in connection with their primary source, good old classical mechanics.*

The benign neglect accorded to Poincaré in this country[1] was only partially shared by Sophus Lie, whose constructs are all too frequently not taught to physics students until the study of quantum angular momentum or, even worse, quark models. Because Lie's ideas are first presented in such contexts, their initial connection with dynamical systems tends to be forgotten, and it is easy for a student to think of Lie groups only in terms of finite representations and Dynkin diagrams. *The appropriate place to introduce Lie groups is in a course on classical mechanics.* We shall attempt to do that in as natural a manner as possible, following the track of his algebraic constructs

[1] A recent dismissal occurs in a very enjoyable article [19], in the beginning of which a great physicist, while paying homage to a late colleague, parenthetically laments, "I regret to say that he preferred Poincaré . . . to Einstein." The irony is that you are likely to understand Einstein more easily — or, more precisely, a modern exposition, such as Misner, Thorne, and Wheeler's *Gravitation* — by paying some attention to Poincaré early in life.

from their setting in vector fields, through their use in Hamiltonian systems, and in the construction of normal forms.

In 1969 Deprit [28] discovered an algorithm, built on the Lie algebra of Poisson brackets, for performing perturbation theoretic calculations in celestial mechanics. In 1981 Dragt independently introduced another Lie algebraic method, based on discrete maps rather than continuous flows, into accelerator theory. (I was fortunate enough to be assigned the task of editing his lecture notes [2, 30], little suspecting that they would bifurcate my life.) To some these "new" formalisms at first appeared mysterious and somewhat contrived, computational tricks that seemed to have little physics motivation behind them. That this misperception arose may not be totally disconnected from the fact that the key ideas frequently do not appear in most core physics curricula, and practically never in connection with classical mechanics. (This neglect does not extend to generating function techniques, for example, which do possess a strong academic tradition.) We shall introduce the tools naturally in a classical setting so basic that it can be used even at the undergraduate level. We shall also attempt to demonstrate the relations between these "new" methods and the more "traditional" ones used in accelerator physics since the 1950's. What makes the "new" methods so exciting is their imbedding within an effective computing environment. Nonetheless, the underlying ideas go back more than a century. Those who earn their bread, sausage, and beer by using classical dynamics have a glorious heritage, one that is frequently disconnected from its roots.

It is ironic that as we approach the end of the twentieth century, classical mechanics is undergoing a renaissance; it remains a lively subject of active research both in its own domain and in its connections with quantum theory. This is largely due to advances in computation, both symbolic and numeric. Advanced graphics has played a special role, giving us a new "lens" that has opened all our eyes to behavior that previously could barely be imagined only by the most brilliant. In contrast to the impression acquired by too many during their formal education that "classical mechanics" is a dull, closed subject with no mysteries left to explore, the work has just begun, and its prospects are exciting.

This volume grew from courses and lectures given at Northwestern University, at Harvard University, and at Accelerator Schools sponsored by the DoE and organized by Melvin Month. The Accelerator Schools especially have produced much needed documentation on the art and science of accelerators in the form of A.I.P. Conference Proceedings, and they have been a source of fine over-the-shoulder bags as well. My thanks to Melvin Month for giving me a forum for lecturing and for encouraging me to write this book. Additionally, I would like to acknowledge the good fortune of having worked closely with Don Edwards as my first supervisor at Fermilab; I continue to benefit from this early exposure to his physical insights, knowledge, and wisdom. Additionally, I owe a debt of gratitude to Shoroku Ohnuma, who would continually ask the most disturbing and tantalizing questions, only a few of which we ever got around to answering; some of the others are addressed here. As already implied, to Alex Dragt lies the credit for starting the whole business of using Lie operators to solve accelerator problems, and his influence has been pivotal. And I have benefitted from several recent interactions with Etienne Forest, whose insight into the perturbation theories of accelerator orbits is unmatched.

I want to thank one unknown reviewer who read the first rough draft of this book and made several suggestions, many of which I have tried to incorporate into the text. Most of all, however, I thank him for his generous, positive response, without which I probably would have dropped the project.

Regarding the technical errors, the manuscript suffers from not having had a horde of students critically reading the pages for several years, checking the proofs, repeating the calculations, and solving the problems. I regret that very much. However, unlike my colleagues in the universities, I do not have this marvelous resource of undervalued labor at my disposal. So, I must apologize for whatever errors remain — typographical, algebraic, and conceptual — and I urgently request that you communicate them to me as they are unearthed. If possible, I shall archive errata, and make them electronically available to interested reader(s).

A final word to the heroes who actually make accelerators work, most of whom already know it well, though it tends to get submerged: Your field deals with more than environmental impact reports, data bases, impedance measurements, or even the "kick turn turn kick turn . . ."[2] of tracking programs. As was written somewhat earlier, "[t]hose who earn their bread, sausage, and beer by using classical dynamics have a glorious [physics and mathematics] heritage" that should be enjoyed and, more importantly, used.

LEO MICHELOTTI

St. Charles, Illinois

[2]The beginning of a line taken from *The Producers,* by Mel Brooks, the end of which is also appropriate.

CHAPTER 1

Prologue: The Pendulum

> If we want to know where Jupiter will be so as to plan properly the Jupiter shot, then we may proceed in one mathematical direction. If we are interested in whether the solar system is dynamically stable or unstable, we will have to proceed in another. In view of the inherent difficulties of the mathematics, the art of modelling is that of adopting the proper strategy.
>
> —Philip J. Davis and Reuben Hersh
> *The Mathematical Experience*

It is in high school that we are first taught to write Newton's Law as, "$F = ma$." This pedagogical blunder is usually compounded some time later by a teacher of Freshman Physics who one day draws a simple pendulum on a blackboard, more or less as in Figure 1.1, writes the equation of motion, from $F = ma$,

$$-mg\sin\theta = ml\ddot{\theta}, \tag{1.1}$$

and, barely pausing for breath, notes that $\sin\theta \approx \theta$, provided θ is "small," thus transforming this equation into that of a harmonic oscillator,

$$\ddot{\theta} \approx -(g/l)\theta,$$

with a basis of two linearly independent solutions,

$$\theta = \begin{cases} \sin\omega_o t, \\ \cos\omega_o t, \end{cases} \qquad \omega_o = \sqrt{g/l}.$$

All of which is correct, as far as it goes. Regrettably, the teacher of Freshman Physics probably stops the analysis at that point and never returns to consider the motion in

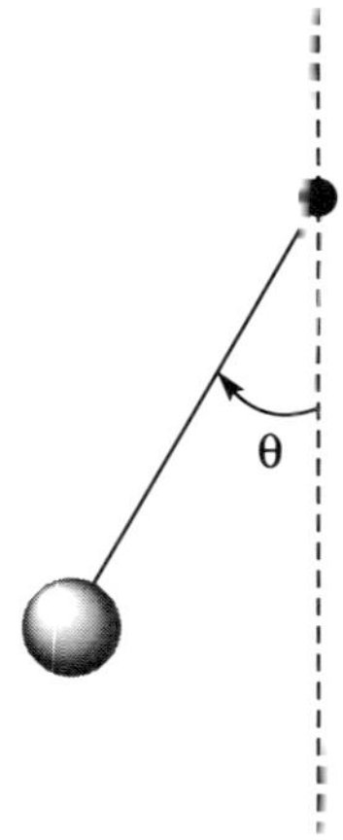

FIGURE 1.1 Simple pendulum: the archetypal nonlinear dynamical system.

more detail. The reason for abandoning the problem is obvious. If you ask the "orbit history" question—"Given initial position and velocity at time $t = 0$, what is the position at time $t \neq 0$?"—then the answer is complicated: in particular, it involves elliptic functions, and freshpersons are not expected to know about such things. This question was the dominant paradigm of physics during the eighteenth and most of the nineteenth centuries: "solving" a dynamical system meant determining its history.

However, there is another sort of question that is easier to answer and, in some ways, more interesting. For the purposes of this chapter, it can be motivated with the observation that the total energy of the pendulum,

$$\begin{aligned} E &= \tfrac{1}{2}mv^2 + V(\theta) \\ &= \tfrac{1}{2}m(l\dot{\theta})^2 - mgl\cos\theta \\ &= \frac{L^2}{2I} - I\omega_o^2\cos\theta, \end{aligned} \tag{1.2}$$

is conserved. In the final line, we have written E in terms of the moment of inertia, $I \equiv ml^2$, and angular momentum, $L \equiv I\dot{\theta}$. If we plot the level sets of E—that is, sets of phase space points for which the value of E is constant—

$$P_W \equiv \{(\theta, L) \mid E(\theta, L) = W\},$$

the resulting curves are paths constraining the system's orbits, as sketched in Figure 1.2. The direction of the orbits is obtained from the definition of L: $L\dot{\theta} > 0$. This picture of the orbit space, sometimes called a "flow diagram," does not contain enough "orbit history" information to answer the original question, but it does contain an enormous amount of qualitative information about the *topological classification of possible orbits* and the relative likelihood that an orbit will belong to one class or

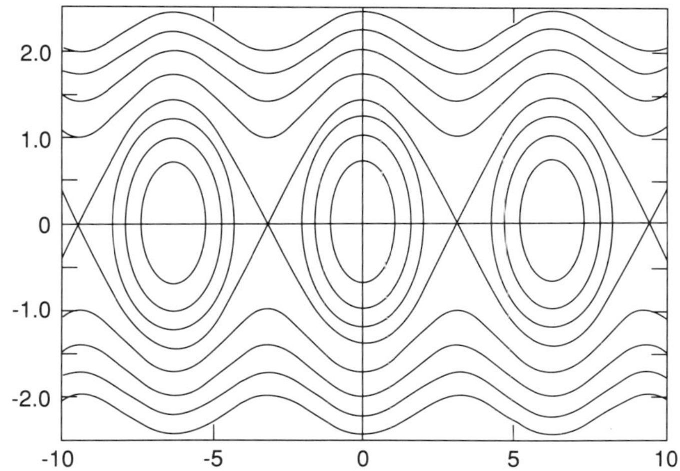

FIGURE 1.2 Level sets of E are paths for the pendulum.

another, the kind of information more appropriate for estimating long-term stability. What it shows is a view of *all possible* paths that the system can traverse, corresponding to all possible sets of initial conditions. Obtaining and interpreting such information form much of the subject matter of "qualitative dynamics."

The remainder of this opening chapter will be devoted to motivating some formal concepts and useful jargon as quickly and painlessly as possible. Before continuing, let me issue a small warning: those who are totally unfamiliar with the more abstract branches of mathematics should simply skim over parts that make them feel uncomfortable. On the other hand, those who are familiar with them to an unhealthy extent may be offended by the cavalier, heuristic manner in which objects such as "charts" or "bundles" are introduced. In either case, please read what follows in the spirit in which it is intended; try to get through it once quickly, getting a feel for the material without bogging down in details. There will be sufficient opportunities for that later.

Critical Orbits and Separatrix The principal feature that appears in Figure 1.2 is the existence of two fixed points — orbits for which $\dot{L} = \dot{\theta} = 0$ — positioned at $L = 0$ and $\theta = 0$ or π. Physically, these correspond to a pendulum at rest, either in the "down" or "up" position. The first fixed point, which is clearly the more stable one, is called an "elliptic" fixed point for the obvious reason that the paths of orbits in its neighborhood look like ellipses. For a similar reason, the second is called "hyperbolic." Loosely stated, the chief difference between them is this: orbits that pass close to an elliptic fixed point remain close to it; in contrast, almost all orbits that pass close to a hyperbolic fixed point do *not* remain close but pull away.

The disclaimer "almost all" is required in the previous sentence.[1] If we focus attention on the neighborhood of the hyperbolic fixed point, we find special orbits

[1]This phrase "almost all" almost always means "all but a set of measure zero," which in turn means "all but a set that has no volume."

that close in on it. All orbits that approach a hyperbolic fixed point as $t \longrightarrow +\infty$ (resp., $t \longrightarrow -\infty$) comprise its "stable (resp., unstable) manifold."[2] In this case, there is only one hyperbolic fixed point and two such orbits, one for $L > 0$, the other for $L < 0$, both of which belong to either kind of manifold. Therefore, in our example, the stable and unstable manifolds are, in fact, identical. That is not generally, or "generically," the case.

Taken together, the union of hyperbolic fixed points and their stable and unstable manifolds form an object called a "separatrix." Its name derives from the observation that this collection of orbits organizes the (θ, L) "phase space" by partitioning it into regions. To complete the terminology, the "center manifold"[3] of this system lies within the "island" that is bounded by the separatrix. The orbits in this island are characterized by bounded excursions in the angle coordinate, θ; speaking physically, their energies are too small to overcome the potential barrier associated with going over the top. Those "outside" the separatrix have unbounded angle excursions.

The pendulum is a very simple system, but this pattern, or its generalization,[4] appears again and again throughout dynamics. The search for separatrices and the study of their properties comprise much of "nonlinear dynamics." It is a fundamental problem, closely related to the search for the smallest invariant subspaces of different dimensions — a task akin to the group theoretical problem of finding irreducible representations.

Amplitude Dependent Frequencies An important feature, not apparent in the static diagram of Figure 1.2, is that the frequency of a pendulum's orbit depends on its amplitude. This is the first and simplest attribute of nonlinearity. The academic folklore of pendula, which overly emphasizes their connection with harmonic oscillators, contains a charming, apocryphal story about Galileo daydreaming in church, timing the swings of a chandelier with his pulse, and noting that their period remained the same as the motion wound down to rest.[5] In fact the period cannot be constant. By continuity, orbits must move slowly in the vicinity of the hyperbolic fixed point. Correspondingly, those orbits that pass closely to this fixed point will take longer to complete their cycle. Therefore if we plot the frequency of an orbit as a function of its energy, we shall get a curve, like that in Figure 1.3, which approaches ω_o for small E and vanishes at $E = I\omega_o^2$, the energy associated with the hyperbolic fixed point. Asymptotically, when the energy is large, the potential energy becomes negligible and we have $E \approx I\omega^2/2$. Therefore the frequency grows as $\sqrt{E}$ as $E \longrightarrow \infty$.

As there is a one-to-one correspondence between the energy of a pendulum and the amplitude of its motion, for $E < I\omega_0^2$, suitably rescaling the abscissa of Figure 1.3 would change it into a plot of frequency versus amplitude.

[2]Simply calling it this does not establish that it *is* a manifold. Indeed, at this point we have not even discussed what constitutes a manifold.

[3]More precise definitions will be given in the next chapter.

[4]We shall see that separatrices possess more complicated structure in higher dimensions.

[5]If the story is true, I suspect that Galileo's pulse rate must have increased as he became more excited by his discovery.

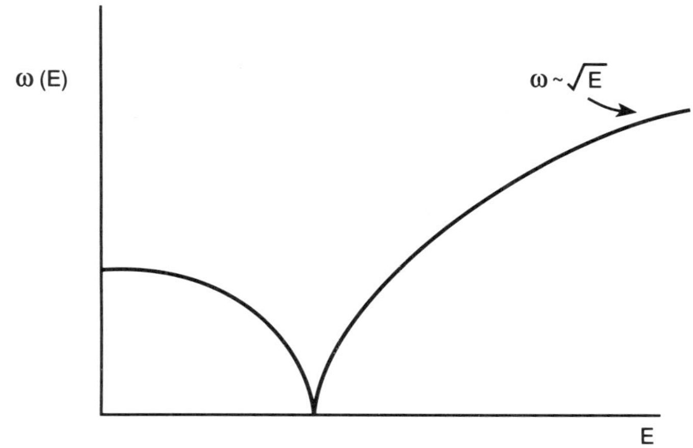

FIGURE 1.3 Rough sketch of frequency versus energy for a simple pendulum.

Phase Space Consider now the very space into which orbits are imbedded, the space that is, in fact, the union of all the paths followed by orbits. Figure 1.2 is drawn as though this were two-dimensional Cartesian space, $R^2 = R \times R$.[6] It is not: any two ordered pairs of the form (θ, L) and $(\theta + 2\pi n, L)$, where $n \in Z$ is an integer, represent the same physical state and therefore must be considered "equivalent." We express this formally by setting up an equivalence relation.[7]

$$(\theta, L) \cong (\theta', L') \quad \textbf{means} \quad L = L' \text{ and } \exists n \in Z : \theta' = \theta + 2\pi n.$$

The appropriate mathematical object that corresponds to a state of the pendulum is not a single ordered pair but a set of equivalent ordered pairs,

$$\lfloor \theta, L \rceil \equiv \{(\theta', L') \mid (\theta', L') \cong (\theta, L)\}, \tag{1.3}$$

which is called the equivalence class of the ordered pair. The true phase space of the system, the space of physically inequivalent states, is technically a "quotient space," which means it is a set of equivalence classes.

$$\mathcal{P} \equiv \{\lfloor \theta, L \rceil \mid (\theta, L) \in R^2\}.$$

A moment's reflection is sufficient to convince one that $\mathcal{P}$ is a cylinder (see Figure 1.4), equivalent to the cross product of a circle with the reals, $\mathcal{P} = S^1 \times R$.

What this means is something more profound than is apparent at first: the stage on which a system *even as simple as a pendulum* plays out its history is not Cartesian space but a manifold, more particularly, a differentiable manifold. These objects are

[6]Notation for a few famous sets: R = reals; Z = integers; S^n = n-dimensional sphere; T^n = n-dimensional torus.

[7]Common set-theoretic notation: $\forall$ = "for all," $\exists$ = "there exists," and $\in$ = "is in the set."

(a) (b)

FIGURE 1.4 The phase space of a simple pendulum is a cylinder.

the subject matter of the most sophisticated, advanced mathematical disciplines ever devised. We shall not even pretend to attempt to begin to do them justice. Fortunately for our purposes, most physicists have an intuitive and generally sufficient understanding of "manifolds" as an extension of "surfaces," such as spheres, cylinders, or tori. Their important feature is that they are *locally Cartesian*. This means that each point on the manifold belongs to a local region on which we can introduce a system of Cartesian coordinates, formally called a chart. More correctly, localized, open sets of an n-dimensional manifold are coordinatized with open subsets of R^n, usually containing the origin. A collection of such charts which covers the entire manifold — with some overlap between charts so as to patch them together smoothly — is called, appropriately enough, an "atlas." (One need only think of real atlases used to map the Earth to get the idea.) Of course, we are glossing over innumerable technical details. For example, the business of patching charts together smoothly means that if two charts overlap, there must exist, as part of the atlas itself, a "smooth" transformation connecting the coordinates associated with those points that belong to both of them. If $\underline{z} \in R^n$ and $\underline{z}' \in R^n$ are the n-tuple representatives of the same point, $\mathbf{p} \in \mathcal{P}$, on two different charts, then the transformation is a one-to-one, highly (usually infinitely) differentiable function, $T : \underline{z} \mapsto \underline{z}'$. Furthermore, if two atlases cover the same manifold, then there are smooth transformations connecting pieces of charts corresponding to the same open sets on the manifold, what physicists think of as "coordinate transformations." The key idea that these and other elaborate constructions are meant to express is that a manifold is an object which can be represented, or coordinatized, in a large number of ways but which we want to think of as an entity superceding all these representations.

A dynamical system — or a "dynamic" — is actually associated with a number of related manifolds. First, there is the configuration manifold, which labels its instantaneous "position" without information as to how that position is changing. For example, the configuration manifold of a harmonic oscillator marks its position along

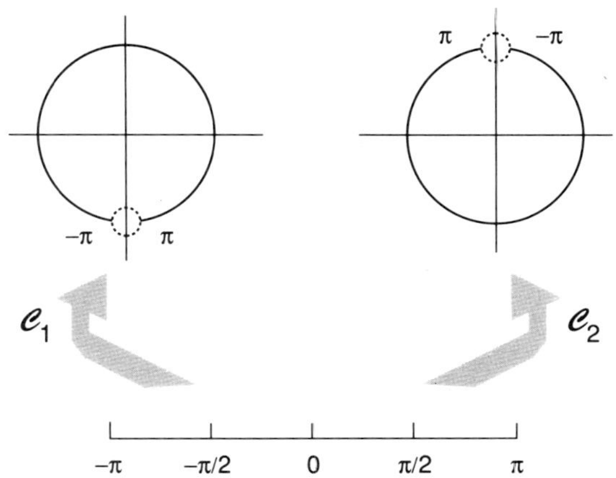

FIGURE 1.5 Charts for S^1, the configuration manifold of a pendulum.

a line and is (topologically equivalent to) R, the set of reals, while that of a pendulum would be (topologically equivalent to) a circle, S^1.[8]

$$S^1 = \{(x, y) \in R^2 \mid x^2 + y^2 = 1\}.$$

Now, although (x, y) Cartesian pairs are used here to define S^1, they cannot serve as coordinates: S^1 is a one-, not two-dimensional object. We have already used an angle coordinate, θ, to denote the pendulum's position, but in fact, $S^1 \neq \{\theta \mid \theta \in (-\pi, \pi)\}$, as this would neglect the upper position of the pendulum, $(0, 1)$. Similarly, using a half-closed interval, $S^1 = \{\theta \mid \theta \in [-\pi, \pi)\}$, would not work, as it would produce a discontinuity in the coordinate.[9] An atlas containing at least two charts, say C_1 and C_2, is required to specify coordinates properly on S^1.One such atlas is sketched in Figure 1.5.

$$\begin{aligned}
S^1 &= S^1_1 \cup S^1_2, \\
S^1_1 &= \{(x, y) \in R^2 \mid x^2 + y^2 = 1 \text{ and } y \neq -1\} = S^1 - \{(0, -1)\}, \\
S^1_2 &= \{(x, y) \in R^2 \mid x^2 + y^2 = 1 \text{ and } y \neq 1\} = S^1 - \{(0, 1)\}, \\
C_1 &: (-\pi, \pi) \longrightarrow S^1_1,\ \theta_1 \longmapsto (\sin\theta_1, \cos\theta_1), \\
C_2 &: (-\pi, \pi) \longrightarrow S^1_2,\ \theta_2 \longmapsto (-\sin\theta_2, -\cos\theta_2).
\end{aligned}$$

We have made the overlap between these two charts rather large, almost all of S^1, in fact. That was not necessary; nothing would be lost by restricting the domains. It also

[8]We shall use the phrase "topologically equivalent to" only sparingly, as it tends to burden presentations without adding much information. It should be considered present but suppressed in sentences like, "This surface is a torus."

[9]Usually, charts are open sets; exceptions are made for manifolds with boundaries.

was not necessary that the domains of C_1 and C_2 be identical. The important thing is that there be some overlap so that the two charts can be related.

Those regions of overlap are the connected components of the intersection of S_1^1 with S_2^1.

$$S_1^1 \cap S_2^1 = S_>^1 \cup S_<^1,$$
$$S_>^1 \equiv \{(x,y) \in R^2 \mid x^2 + y^2 = 1 \text{ and } x > 0\},$$
$$S_<^1 \equiv \{(x,y) \in R^2 \mid x^2 + y^2 = 1 \text{ and } x < 0\}.$$

We mark how C_1 and C_2 map different subsets of $(-\pi, \pi)$ onto $S_>^1$ and $S_<^1$.

$$C_1 : (-\pi, 0) \to S_<^1, \ (0, \pi) \to S_>^1;$$
$$C_2 : (-\pi, 0) \to S_>^1, \ (0, \pi) \to S_<^1.$$

The equations that transform one coordinate into the other within these regions are given below.

$$T_< : (-\pi, 0) \to (0, \pi), \ \theta_1 \longmapsto \theta_2 = \theta_1 + \pi;$$
$$T_> : (0, \pi) \to (-\pi, 0), \ \theta_1 \longmapsto \theta_2 = \theta_1 - \pi.$$

We have gone through these manipulations in a little too much detail to emphasize their involved nature, even in the simplest case. We shall not go through such constructions explicitly again, but one should keep in mind that it must be possible to do so in order for a mathematical object to be a manifold. More frequently, when θ appears only in the arguments of 2π periodic functions, we can effectively replace S^1 with R and let the 2π periodicity absorb the details of working within a quotient space. The coordinate θ then is (mis)treated as though it were a global coordinate; in fact, it is not. Mathematicians like this trick so much they have given it a name, calling R the "universal covering space" of S^1.

The configuration manifold contains information about only the "position" of the pendulum, or general system. For dynamics of interest to physicists, this is not sufficient to propagate an orbit: we also need to know its instantaneous "velocity," a datum contained in the angular momentum coordinate. The full state of the pendulum is a point in the phase space, $\lfloor \theta, L \rceil \in \mathcal{P} \cong S^1 \times R$. Formally, this phase space is a special manifold, called a tangent bundle, built from the configuration manifold as its base. Naively, it arises from attaching a vector space — in this case, R — to each point on the configuration manifold: the vector space of all possible "velocities" that the system can have at that "position." We used $\lfloor \theta, L \rceil$, but this was nothing more than rescaling velocity, as $L = I\dot{\theta}$. Technically, however, the tangent bundle is associated with (*position, velocity*) coordinates — the coordinates that go into Lagrangians — while (*position, momentum*) coordinates — the coordinates that go into Hamiltonians — comprise something called the "cotangent bundle," obtained by attaching to each point of the configuration manifold the *dual* of the vector space attached in the tangent bundle. For our purposes in this book, such distinctions will not matter. We have passed over them quickly here only to point out that these advanced mathematical objects came into play in the natural course of developing classical mechanics.

Mappings In addition to its vector field, a dynamical system must be associated with a family of mappings that formally model its orbits. The terms "mapping" and "map" refer to functions, ϕ, that take the phase space manifold onto itself: $\phi : \mathcal{P} \longrightarrow \mathcal{P}$. We shall require that ϕ be one-to-one (and therefore invertible) and onto (which means that its image is the entire space, $\mathcal{P}$). Of course, we tacitly assume that the mappings under consideration are nicely behaved: in particular, we avoid discontinuities or other behavior not sufficiently "smooth." The collection of interest to us represents the time-evolution of a dynamical system. It is doubly indexed by R^2, and we interpret this "time-evolution," "orbit-evolution," or "flow" map, $\phi_{t_2t_1}$, as taking a dynamical system from its state at time t_1 to its state at time t_2. That is, if $\mathbf{p}_1 \in \mathcal{P}$ is a state of the dynamical system at time t_1, and under the evolution of the system from time t_1 to time t_2 the state moves from $\mathbf{p}_1$ to $\mathbf{p}_2$, then $\phi_{t_2t_1}$ is defined so that it maps $\mathbf{p}_1$ into $\mathbf{p}_2$. This is written formally as $\phi_{t_2t_1} : \mathbf{p}_1 \longmapsto \mathbf{p}_2$, or $\phi_{t_2t_1}(\mathbf{p}_1) = \mathbf{p}_2$.

That the maps in the collection $\{\phi_{t_2t_1} \mid (t_2, t_1) \in R^2\}$ can be defined at all means that each point, $\mathbf{p}_1 \in \mathcal{P}$, must contain all the information necessary to propagate an orbit, given t_1 and t_2. *It is this property that defines what one means by the "phase space" of a dynamical system.* Because of it the collection obeys a group-like property known as the Chapman-Kolmogorov equation,[10]

$$\forall t_1, t_2, t_3 \in R, \forall \mathbf{p} \in \mathcal{P} : \phi_{t_3t_2}(\phi_{t_2t_1}(\mathbf{p})) = \phi_{t_3t_1}(\mathbf{p}),$$

which formally reexpresses the notion that *the orbit launched from a particular point in phase space at a particular time is unique*. As the argument that appears on both sides of this equation is the same, and as the equation holds for all $\mathbf{p}$, it can be written functionally in terms of composition (or catenation) of maps, without reference to its argument.

$$\begin{aligned} (\phi_{t_3t_2} \circ \phi_{t_2t_1})_{(\mathbf{p})} &\equiv \phi_{t_3t_2}(\phi_{t_2t_1}(\mathbf{p})) \\ \phi_{t_3t_2} \circ \phi_{t_2t_1} &= \phi_{t_3t_1}. \end{aligned} \tag{1.4}$$

This clearly can be applied recursively to generate arbitrarily long chains.

$$\phi_{t_nt_{n-1}} \circ \phi_{t_{n-1}t_{n-2}} \circ \cdots \circ \phi_{t_3t_2} \circ \phi_{t_2t_1} = \phi_{t_nt_1}. \tag{1.5}$$

From the Chapman-Kolmogorov equation we see easily that, for all t, ϕ_{tt} is the identity mapping, $\mathbf{id} : \mathbf{p} \longmapsto \mathbf{p}$. Unless we restrict ourselves with a special rule stating

[10]In statistical mechanics the equivalent relation is called the "master equation" but expresses a probabilistic, Markovian dynamic.

that it holds only for $t_1 < t_2 < t_3$, the equation also implies reversibility[11] and the existence of the inverse of each map in the collection — since we have explicitly that $\phi_{t_1t_2} = \phi^{-1}_{t_2t_1}$. This does not quite make the family a group: the composition appearing in the Chapman-Kolmogorov equation requires at least one "time" in common, t_2 in its expression above. However, if the system is periodic, so that for some τ and all t_1, t_2 we have $\phi_{t_2+\tau t_1+\tau} = \phi_{t_2t_1}$, then we can construct a discrete group of mappings by taking as generator a period advance map, $\Phi \equiv \phi_{t+\tau t}$, for some t. The Chapman-Kolmogorov equation then implies that the subset $\{\Phi^n \mid n \in Z\}$ forms a discrete, Abelian group.[12] Furthermore, the different groups obtained by varying the base time, t, are all equivalent (isomorphic).

A system is "autonomous," or time-independent, when $\phi_{t_2+\tau t_1+\tau} = \phi_{t_2t_1}$ for *all* τ. This means that orbits are invariant with respect to shifts in the "time" parameter: the system's evolution depends only on the elapsed interval, not the values of the end-points. Nonautonomy normally implies the influence of external, varying forces. For autonomous dynamics we can build a continuous, one-parameter, Abelian group of mappings by defining $\phi_\tau \equiv \phi_{t+\tau t}$. The Chapman-Kolmogorov equation then implies the group properties,

$$\phi_0 = \mathbf{id}, \quad \phi_\tau \circ \phi_{\tau'} = \phi_{\tau+\tau'}, \quad \text{and} \quad \phi_\tau^{-1} = \phi_{-\tau}. \tag{1.6}$$

Two minor comments: (a) In the Chapman-Kolmogorov equation, Eq. (1.4), the maps $\phi_{t_3t_2}$ and $\phi_{t_2t_1}$ need not be evaluated on the same chart; indeed, it may be impossible to do so using a given atlas. Correspondingly, $\phi_{t_3t_1}$ need not be represented on a single chart; points represented on one chart may be imaged onto those of another; two or more charts may be necessary, or useful. The flow maps map phase space into itself. We are free to implement them in whatever way is convenient in terms of charts and representations. (b) The Chapman-Kolmogorov equation has provided the backbone of numerical analysis in accelerator physics for decades. It is incorporated automatically into "tracking programs." Modern synchrotrons and linacs are chains ("lattices") of small, local electromagnetic environments — magnets, RF cavities, electrostatic separators, septa, and so forth — through which particles must pass. Tracking programs basically catenate maps representing passage through individual elements à la Chapman-Kolmogorov, although it is probable that the connection is usually made unconsciously.

In addition to flow maps, another type of map is important to the analysis of autonomous dynamical systems: the Poincaré map, named after its creator. If one knows on physical or mathematical grounds that a large class of orbits repeatedly pass through a particular subset of the phase space — known as the Poincaré surface of section — then this region can be mapped onto itself via the orbits. The classic example of this is the motion of a proton in Earth's magnetic field, whose orbit repeatedly cuts across the equatorial plane; another, to be used in later chapters, is a

[11] If the indices of the map $\phi_{t_2t_1}$ are constrained to satisfy $t_1 < t_2$, then it is called a "semi-flow" and need not be reversible.

[12] The exponent means repeated composition: $\Phi^0 = \mathbf{id}$, $\Phi^1 = \Phi$, and $\Phi^{n+1} = \Phi^n \circ \Phi$.

storage ring in which particles repeatedly pass through a half-plane slicing through the ring at an arbitrary point. Let $\mathcal{P}' \subset \mathcal{P}$ represent the subset, and let $\mathbf{p}_0 \in \mathcal{P}'$ be an arbirary point in $\mathcal{P}'$. We follow the unique orbit passing through $\mathbf{p}_0$ until it intersects $\mathcal{P}'$ again. If that point of intersection be denoted $\mathbf{p}_1$, then the Poincaré map is defined as $\mathcal{M} : \mathbf{p}_0 \longmapsto \mathbf{p}_1$.

The importance of Poincaré and period advance maps is that they reveal structure that would be much more difficult to observe by following orbits continuously, much as a stroboscope does. We shall illustrate this point in the next chapter using the FODO model of a storage ring.

Vector Fields Some hundreds of years ago the "natural" philosophers of Europe began to pay less attention to "being" and more to "becoming," studying how things *change* rather than what they *are*. As a consequence, things started changing somewhat more rapidly than previously. For better or worse, life has never been the same since.

Modern physics began when people started to use differential equations to model change mathematically. One such for the pendulum was given in Eq. (1.1), but it is not the model we want here, as it does not fit into the "phase space" paradigm. Knowing the initial angle of the pendulum is not sufficient to determine its orbit; we need both position and angular momentum (equiv., velocity). The phase space of the pendulum is a cylinder, $S^1 \times R$, on which we establish coordinates by extending the charts C_1 and C_2 to include L. The ordinary differential equations describing pendulum dynamics take on the following forms on these two charts.

$$\begin{aligned} C_2 : \quad & \begin{pmatrix} \dot{\theta}_2 \\ \dot{L}_2 \end{pmatrix} = \begin{pmatrix} L_2/I \\ -I\omega_o^2 \sin\theta_2 \end{pmatrix}, \\ C_1 : \quad & \begin{pmatrix} \dot{\theta}_1 \\ \dot{L}_1 \end{pmatrix} = \begin{pmatrix} L_1/I \\ I\omega_o^2 \sin\theta_1 \end{pmatrix} \end{aligned} \tag{1.7}$$

These two equations, although slightly different in form, represent the same object: a vector field defined over phase space. As advertised, each point contains all the information necessary to construct the vector propagating the orbit forward in time; by itself, the configuration manifold lacks this property (at least for the pendulum, and most dynamical systems). Analytically, we have substituted coupled sets of first order differential equations, as in Eq. (1.7), for second order ones, as in Eq. (1.1).

Flow maps are clearly related to this vector field: the former are essentially the integrated form of the latter. We shall make this more precise at the cost of increased abstraction. Let $f : \mathcal{P} \longrightarrow R$ be a function which associates a real number with each point in phase space. We shall call such functions "observables," and when necessary, we shall assume they are "smooth" in whatever sense and at whatever level is required.[13] We extend the scope of flow maps by inducing their action on observables:

[13]Later we shall enlarge the definition to include functions whose range (or codomain) is the complex plane. This will be done without further comment.

given a flow map ϕ and observable f, we define the observable ϕf as follows:[14]

$$(\phi f)_{(\mathbf{p})} = f(\phi(\mathbf{p})).$$

The first advantage of doing this is that ϕ is a linear operator when acting on the space of observables, *regardless of how complicated the dynamics.*

$$\begin{aligned}(\phi(f+g))_{(\mathbf{p})} &= (f+g)_{(\phi(\mathbf{p}))}\\ &= f(\phi(\mathbf{p})) + g(\phi(\mathbf{p}))\\ &= (\phi f)_{(\mathbf{p})} + (\phi g)_{(\mathbf{p})}\\ &= (\phi f + \phi g)_{(\mathbf{p})}, \text{ for all } \mathbf{p},\\ \phi(f+g) &= \phi f + \phi g.\end{aligned}$$

The second is that observables, being real-valued functions, possess a natural arithmetic, whereas points on an arbitrary manifold do not. That arithmetic allows us to develop the derivative of a flow with respect to its "time" index. It is itself a time-indexed operator, symbolized here as $\mathbf{V}_t$:

$$\phi_{t+\Delta t,t} f \equiv f + \Delta t\, \mathbf{V}_t f + O((\Delta t)^2),$$

which equation we also write — after blithely brushing aside all the critical issues of topology and limits — in the operator form

$$\mathbf{V}_t \equiv \lim_{\Delta t \to 0} (\phi_{t+\Delta t,t} - \mathbf{id})/\Delta t,$$

where **id** here represents the identity operator, $\mathbf{id} : f \longmapsto f$.

The Chapman-Kolmogorov equation must be modified if we think of flow maps as acting on observables rather than the original manifold.

$$\begin{aligned}(\phi_{t_3t_1} f)_{(\mathbf{p})} &= f(\phi_{t_3t_1}(\mathbf{p}))\\ &= f(\phi_{t_3t_2}(\phi_{t_2t_1}(\mathbf{p})))\\ &= (\phi_{t_3t_2} f)_{(\phi_{t_2t_1}(\mathbf{p}))} \qquad (1.8)\\ &= ((\phi_{t_2t_1} \circ \phi_{t_3t_2}) f)_{(\mathbf{p})},\\ \phi_{t_3t_1} &= \phi_{t_2t_1} \circ \phi_{t_3t_2}.\end{aligned}$$

[14]Looking ahead, see Figure 2.15 in connection with this discussion. Regarding notation: the value of a function f at its argument $\mathbf{p}$ is written $f(\mathbf{p})$, as usual. If f is obtained by manipulating other functions, I will usually put its argument in a subscript, as in composition, $f(g(\mathbf{p})) = (f \circ g)_{(\mathbf{p})}$, or as done here.

The last line follows from the fact the f is arbitrary. It should be contrasted to Eq. (1.4).[15] Using this regrettable result, we can write the "master" differential equation obeyed by the flow, or time-evolution map.

$$\begin{aligned}\phi_{t_2+\Delta t,t_1} f - \phi_{t_2 t_1} f &= \phi_{t_2 t_1} \circ \phi_{t_2+\Delta t,t_2} f - \phi_{t_2 t_1} f \\ &= \left(\phi_{t_2 t_1} \circ (\phi_{t_2+\Delta t,t_2} - \mathbf{id})\right) f \\ &= \Delta t\, \phi_{t_2 t_1}(\mathbf{V}_{t_2} f) + O((\Delta t)^2).\end{aligned}$$

Again, since f is arbitrary, we write this as an operator equation, after formally taking the limit.

$$\frac{d\phi_{t_2 t_1}}{dt_2} = \phi_{t_2 t_1} \circ \mathbf{V}_{t_2}. \tag{1.9}$$

This suggests that the map is obtained from the operator $\mathbf{V}_t$ by some process of exponentiation, and we shall see later that this is exactly correct.

The most important observables serve as the coordinates of a chart. Let $\phi_{t_2 t_1}$ here be the *local representatives* of the time-evolution maps within a particular chart, whose coordinates we shall symbolize as the array, $\underline{z}(\mathbf{p}) = (z_1, z_2, \ldots, z_N)$. The local components of vector field associated with the maps are collected in an array, $\underline{v}$.

$$\underline{v}(\underline{z}, t) \equiv \lim_{\Delta t \to 0} \left(\phi_{t+\Delta t,t}\, \underline{z} - \underline{z}\right)/\Delta t = \left(\mathbf{V}_t \underline{z}\right)_{(\mathbf{p})}.$$

This abstractly defines what is meant by a vector field associated with a dynamical system of orbits. If the system is autonomous, then the right hand side is independent of the value of t, in which case we would write the vector field as $\underline{v}(\underline{z})$. In either case, the shortcut method of writing this relationship is

$$\begin{aligned}\underline{z}(t) &= \phi_{t\,t_0}(\underline{z}(t_0)), \\ \dot{\underline{z}} &= \underline{v}.\end{aligned}$$

In a sense, the programme of physics is to observe the action of ϕ so as to infer $\underline{v}$, subsequently using it to extend ϕ beyond the original observations, or at least to describe its properties (invariant domains, basins of attraction, and so forth).

Notice that we must work within some representation, that is, on some chart, because we have no way of a priori imposing arithmetic, much less analytic, operations to points on an arbitrary manifold. Nonetheless, the term "vector field" is not used lightly. We shall show in the next chapter that it is indeed a *geometric*,

[15]This disaster could have been averted by defining $(\phi f)_{(\mathbf{p})} = f(\phi^{-1}(\mathbf{p}))$. I also could have used another symbol, say ψ, and written $(\psi f)_{(\mathbf{p})} = f(\phi(\mathbf{p}))$. That the ordering gets reversed would then be a statement of anti-morphism between ψ and ϕ. Instead, we shall simply accept the reordering when ϕ acts on observables rather than points.

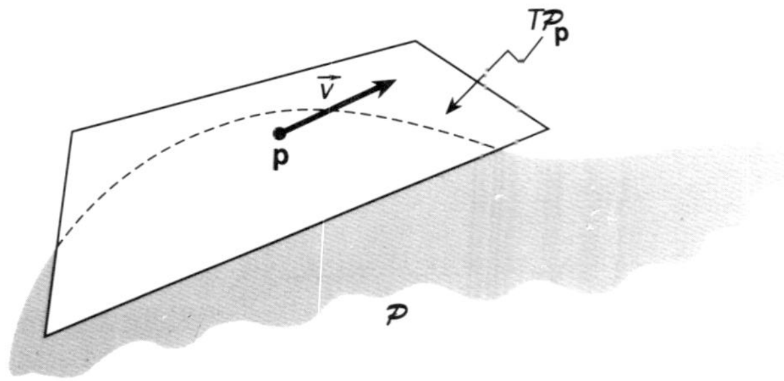

FIGURE 1.6 The tangent space, $T\mathcal{P}_{\mathbf{p}}$, is a vector space attached to the manifold $\mathcal{P}$ at the point **p**. The union of all these spaces is the tangent bundle of the manifold.

chart-transcendent object and has all the properties we intuitively expect vectors to possess. For now, a good heuristic image of a "vector field defined over a manifold" is conveyed by picturing a surface with a tangent vector attached to each point: the "velocity" of the orbit passing through that point. We define it by giving its representation, that is, its components, on an atlas and making sure that everything is consistent and gets patched together smoothly as you jump from one chart to another. To ensure that this is a geometric object requires a consistent set of instructions for transforming the components when you change charts.

The set of all possible tangent vectors that can be attached to a point $\mathbf{p} \in \mathcal{P}$ is itself a vector space, which is typically labeled $T\mathcal{P}_{\mathbf{p}}$, the "tangent space" at **p**; this concept is motivated by the sketch in Figure 1.6. A tangent vector is identified not by $\vec{v}$ alone but also by the point, **p**, to which it is attached, that is, by the doublet $(\mathbf{p}, \vec{v})$. The tangent space, $T\mathcal{P}_{\mathbf{p}}$, is identified as the set $T\mathcal{P}_{\mathbf{p}} = \{(\mathbf{p}, \vec{v}) \mid \text{fixed } \mathbf{p}\}$. On a local chart covering **p**, an element of $T\mathcal{P}_{\mathbf{p}}$ is represented by a multiplet, $(\underline{z}, \underline{v}), \underline{z}, \underline{v} \in R^n$, with twice the number of components of $\underline{z}$.

The union of all these vector spaces is the "tangent bundle" associated with the manifold, loosely $T\mathcal{P} = \bigcup_{\mathbf{p}} T\mathcal{P}_{\mathbf{p}}$. It is a manifold whose dimension is twice that of $\mathcal{P}$ itself. A dynamic, or vector field, is called a "cross section" of the tangent bundle by mathematicians, for reasons that will become obvious upon working Problem 1.9. It is a subset of the tangent bundle, $T\mathcal{P}$, obtained by choosing, in a "continuous" manner, one and only one vector from each tangent space, $T\mathcal{P}_{\mathbf{p}}$. The dynamical "laws of physics" are in this way modeled as cross sections of the tangent bundle of phase space, which itself is *usually* the tangent bundle of a configuration manifold. However, such an observation, while mathematically intriguing, is not important for us, and we shall not return to it. We have touched on these concepts here only to introduce them in their natural setting: dynamics.[16]

Time We have been using the word "time" rather loosely, but it should be made clear that throughout this book "time" has nothing to do with clocks. It simply refers to the parameter used to imbed an orbit into phase space; it is the variable used to

[16]No great mathematical idea was woven out of thin air, independent of preceding problems or applications, most of which were physical.

parametrize orbits, or the index set for the family of maps, $\phi_{t_2 t_1}$. When it is desirable to emphasize this, we shall refer to it as the "orbit parameter" rather than "time." In fact, as much of this book deals with models from accelerator physics, the physical interpretation of the orbit parameter is frequently not time at all but an arc length, or an "angle," marking the azimuthal location of a particle in a synchrotron.

The orbit parameter is not a dynamical coordinate, although one can make it seem so by the trick of "extending" phase space. This is done sometimes to change a nonautonomous vector field into one that is formally autonomous. Suppose that the array of phase space coordinates is $\underline{z}$ and the local representation of the vector field is $\underline{v}(\underline{z}, t)$. If $\mathcal{P}$ is the original phase space, then the extended phase space is $\mathcal{P}^e = \mathcal{P} \times R$ with extra coordinate $s \in R$. In the simplest, most natural scheme, the vector field is extended as follows:

$$\underline{z}^e = \begin{pmatrix} \underline{z} \\ s \end{pmatrix} \quad \text{and} \quad \underline{\dot{z}}^e = \begin{pmatrix} \underline{\dot{z}} \\ 1 \end{pmatrix} = \begin{pmatrix} \underline{v}(\underline{z}, s) \\ 1 \end{pmatrix} \equiv \underline{v}^e(\underline{z}^e).$$

This equates t with the new dynamical coordinate s and forces an autonomous vector field. It is not necessary to choose $\dot{s} = 1$. If we let $\rho : R \rightarrow R$ be any nonvanishing function, with interpretation $\rho(s) = \dot{s}(s)$, then we may just as well have

$$\underline{v}^e(\underline{z}^e) = \begin{pmatrix} \underline{v}(\underline{z}, \int ds/\rho(s)) \\ \rho(s) \end{pmatrix}.$$

However, we shall not need such general constructions.

Indeed, in a triumph of abstraction, it is not even necessary that "time" be a real variable. For example, Eq. (1.6) requires only that τ be a member of an (additive) Abelian group: this could be the integers, a module, the reals, rationals, p-adics, or whatever. Eq. (1.6) then essentially describes a morphism between the index group and the group of mappings. In the "global theory" of linear ordinary differential equations — or linear dynamics — the "time" domain is extended to the complex plane. The orbits of such systems exist on complicated Riemann surfaces, and their analysis involves "hypergeometric functions," "monodromy groups," and associated "monodromy matrices."[17] However, we shall not concern ourselves with these matters.

A Few Words on Notation Before beginning in earnest we should explain a small notational point that is not standard. The single underline was introduced, without comment, to indicate singly indexed arrays, such as $\underline{z}$ and $\underline{v}$. This was done without regard to whether they be "row" or "column" arrays; I see no advantage to making the distinction and shall not write something like "$\underline{z}^T$" unless absolutely necessary for clarity. In later chapters we shall use the double underline to indicate doubly indexed arrays (matrices), as in "$\underline{\underline{A}}$." When a "dot product" is taken between

[17] We shall not engage the "global theory," but it is curious to note the possibility of connecting monodromy matrices and the tunes of an accelerator.

two such symbols it will be understood that adjacent indices are to be contracted. For example, the ith component of $\underline{\underline{A}} \cdot \underline{z}$ is

$$\left(\underline{\underline{A}} \cdot \underline{z}\right)_i = \sum_j A_{ij} z_j,$$

while that of $\underline{z} \cdot \underline{\underline{A}}$ is

$$\left(\underline{z} \cdot \underline{\underline{A}}\right)_i = \sum_j z_j A_{ji}.$$

That is, $\underline{z} \cdot \underline{\underline{A}} = \underline{\underline{A}}^T \cdot \underline{z}$, or if one really wants to be picky, $\underline{z}^T \cdot \underline{\underline{A}} = \underline{\underline{A}}^T \cdot \underline{z}$. The convention of contracting nearest indices need be violated only rarely throughout this book, and we shall expressly note when it happens.

A similar convention holds regarding ordering of indices when doubly indexed objects are constructed from singly indexed ones. This will occur most frequently in conjunction with derivative symbols, such as $\partial/\partial \underline{z}$ or simply $\underline{\partial}$. Our convention on ordering means that

$$(\partial \underline{v}/\partial \underline{z})_{ij} = \partial v_i/\partial z_j, \quad \text{but} \quad (\underline{\partial}\, \underline{v})_{ij} = \left(\underline{\underline{\partial v}}\right)_{ij} = \partial_i v_j = \partial v_j/\partial z_i.$$

It should not take long to become accustomed to this notation, which will prevent us from encumbering equations with unnecessary indices and transpose symbols.

Lastly, throughout the book we shall use the terms "dynamical system," "vector field," and "dynamic" interchangeably. (After all, if the subject is "dynamics," must not a "dynamic" be an instance of what the subject is about?)

PROBLEMS

1.1 A question of historical research: Did Newton actually write $F = ma$, or did he use $F = dp/dt$, and $p = m\,dx/dt$? And why would it matter?

1.2 Show that the set $\lfloor \theta, L \rceil$, defined in Eq. (1.3), is indeed an equivalence class.

1.3 Prove that the level set P_W is an equivalence class of points.

1.4 Two children are swinging. The first pushes himself by kicking the ground; the second is pushed from behind by a parent. Which method is more effective? (Examine the question by drawing pictures, not writing equations.)

1.5 That the frequency of a pendulum vanishes at $E = I\omega_o^2$ is motivated by the idea that orbits which approach the hyperbolic fixed point must take longer to complete their cycle. As stated in the text, "by continuity, orbits must move slowly in the vicinity of the hyperbolic fixed point." Why doesn't this same argument prove that the frequency in the vicinity of the elliptic fixed point is also zero.

1.6 Expand the pendulum energy about the elliptic and hyperbolic fixed points and verify that local orbits are ellipses and hyperbolae. Connect the parameters of these to those of the pendulum. Similarly expand the vector field about these points to obtain the "linearized" equations of motion in the vicinity of the fixed points.

1.7 A one-dimensional anharmonic oscillator on R is described by the second order differential equation

$$\ddot{x} + x^{2n-1} = 0,$$

where n is a positive integer. Show that the solution satisfying $x(0) = 1$, $\dot{x}(0) = 0$ is given implicitly by

$$t = \sqrt{n} \int_{x(t)}^{1} \frac{du}{\sqrt{1 - u^{2n}}}$$

and describe the connection with elliptic integrals.

1.8 Not all manifolds are global cross products, prime examples of which include the (infinite) Möbius strip and the Klein bottle. Provide an atlas for one or the other. These two are examples of nonorientable manifolds. Is it possible for such to be the phase space of a dynamical system?

1.9 Describe an abstract dynamical system for which S^1 is both the configuration manifold *and* the phase space. Sketch a vector field on the cylindrical tangent bundle of S^1, $S^1 \times R$. Can you tell from your sketch why a vector field is called a "cross section" of the tangent bundle?

1.10 Let $\Phi \equiv \phi_{t+T\,t}$ be a period advance map of a periodic dynamical system. Show that (a) the Chapman-Kolmogorov equation implies that the subset of maps $\{\Phi^n \mid n \in Z\}$ forms a group and (b) the different groups obtained by varying the base time, t, are isomorphic to each other.

1.11 Show that the maps $\phi_\tau \equiv \phi_{t+\tau\,t}$ associated with an autonomous system form a group. (Hint: Prove Eq. (1.6) from the Chapman-Kolmogorov equation.)

1.12 Consider the autonomous vector field, $v(x) = -x^2$, defined on the phase space of reals, $\mathcal{P} = R$. Is $\dot{x} = v(x)$ the equation of motion of a dynamical system? In what way was our discussion of the Chapman-Kolmogorov equation inadequate?

1.13 Let $g : \mathcal{P} \longrightarrow R$ be an observable, and let $f_k : R \longrightarrow R$ for $k = 1, 2, \ldots n$. If ϕ is a flow map, show that $\phi(f_1 \circ f_2 \circ \cdots \circ f_n \circ g) = f_1 \circ f_2 \circ \cdots \circ f_n \circ (\phi g)$.

CHAPTER 2

Basic Concepts

> If one were to tell a man about things of which he had no previous knowledge, nor ever seen depicted in any way, he would know no more about them than if one had not spoken.
>
> — St. John of the Cross

In this chapter we shall develop more completely the concepts of vector fields and maps, which model physical systems. Because of their importance, and because they are the simplest, we begin with a quick review of linear dynamics. Spectral theory provides the fundamental tools for characterizing their orbit spaces (flows). Bifurcations of linear systems foreshadow the later study of local bifurcations, popularly called "catastrophes," of nonlinear systems. Special attention will be paid to periodic linear systems, which are closely connected with linear stability analysis of circular accelerators and storage rings. We shall describe the Floquet theory for such systems of arbitrary dimension. This section will end with a discussion of resonances and their consequences for the orbit space.

Brief treatments of a few nonlinear models comprise the subject matter of the second section. Our intention here is not to be comprehensive; as was pointed out in the Preface, there are already a number of books on the market that study these models in great detail, and the bulk of this one is reserved for applications more directly linked to accelerators and beams. We want only to illustrate a few of the phenomena associated with nonlinear systems, especially the existence of attractors, chaotic or otherwise. The simple Van der Pol oscillator will be our vehicle for demonstrating the utility of "qualitative analysis" of dynamical systems as a means of gaining information without having to solve the "orbit-history" problem.

With these examples behind us, the third section covers in more detail the phase space concepts that were introduced in the last chapter. Emphasis is placed on developing the ideas that vector fields are geometrical objects and that the ordinary differential equations, which physicists call "equations of motion," are associated with the components of these fields on specific coordinate charts. A new object, the so-called Lie bracket, will make an appearance as a natural consequence of this discussion. This section closes with a treatment of phase space volume, highlighting the distinction between conservative and nonconservative dynamical systems. Specific topics to be discussed include (a) the (generalized) Vlasov equation, (b) Poincaré's recurrence theorem, and (c) chaos.

2.1 LINEAR MODELS

Real (or complex, for that matter) linear dynamical systems are those whose phase space is R^N (or C^N), for some N, and whose vector field can be expressed globally in the form

$$\dot{\underline{z}} = \underline{v}(\underline{z}, t) = \underline{\underline{A}}(t) \cdot \underline{z}, \tag{2.1}$$

on a (global) coordinate chart, with a real (complex) matrix $\underline{\underline{A}}$. These systems are not only important in their own right — having extensive application in physics and engineering — but also because much of the analysis of nonlinear systems involves a process of "linearization" in neighborhoods of critical points: fixed points, periodic orbits, resonant orbits, attractors, and the like. $\underline{\underline{A}}$ need not be autonomous — we do not require that $\partial \underline{\underline{A}}/\partial t = 0$ — but it must be constant over all phase space. Nonetheless, we shall begin with an overview of autonomous linear dynamics.

2.1.1 Autonomous

Without a doubt, the archetypal dynamical system for physicists is the one-degree-of-freedom damped harmonic oscillator. We shall use this simple model to introduce some key ideas, including bifurcations, exponential maps, and normal form coordinates.

Bifurcations of a Harmonic Oscillator Phase space of a harmonic oscillator is the Cartesian plane, R^2, with interpretation

$$\underline{z} = \begin{pmatrix} x \\ p \end{pmatrix} \in R^2.$$

With momentum defined as $p = m\dot{x}$, and including a damping term, Newton's law takes the familiar form,

$$\dot{p} = -\Gamma p - m\omega^2 x,$$

where m is the mass, Γ the damping constant, and ω the frequency of the oscillator; alternatively, $m\omega^2$ is the "spring constant." We write this now as the linear vector field,

$$\underline{v}(\underline{z}) = \dot{\underline{z}} = \begin{pmatrix} \dot{x} \\ \dot{p} \end{pmatrix} = \begin{pmatrix} 0 & 1/m \\ -m\omega^2 & -\Gamma \end{pmatrix} \underline{z} \equiv \underline{\underline{A}} \cdot \underline{z}. \tag{2.2}$$

In order to examine the flow, or orbit space, implied by Eq. (2.2) we find the eigenvalues of $\underline{\underline{A}}$.

$$\det[\underline{\underline{A}} - \lambda \underline{\underline{1}}] = \lambda^2 + \Gamma\lambda + \omega^2 = 0$$

$$\Rightarrow \lambda_\pm = \frac{1}{2}\left[-\Gamma \pm \sqrt{\Gamma^2 - 4\omega^2}\right] \tag{2.3}$$

$$\text{Eigenvectors: } \underline{e}_\pm \propto \begin{pmatrix} 1 \\ \lambda_\pm m \end{pmatrix}$$

We assume that $\omega \neq 0$, so that the origin itself is the one and only fixed point. (If $\omega = 0$, then $\lambda_+ = 0$, and the line $p = 0$ would be a continuum of fixed points.) The two variables, ω and Γ, are "control parameters" of the system rather than coordinates; that is, they are not considered dynamical variables but are imposed externally. Only one is essential, as the other can be absorbed in setting a scale for "time." The topological character of the flow thus should depend only on the ratio, ω/Γ, sometimes called the Q of the oscillator. Consider the nature of the flow for various regions in the (ω, Γ) "control space." For $\Gamma \gg |2\omega|$, we have $\lambda_+ \approx 0^-$ and $\lambda_- \approx -\Gamma$. Each eigenvector of $\underline{\underline{A}}$ spans a one-dimensional invariant manifold. Orbits on one of these, corresponding to λ_-, move rapidly (since Γ is large) toward the origin, while those on the other approach the origin more slowly. Any other orbit is, by linearity, a superposition of these two motions: it will be dominated by movement along the "fast" direction until it approaches the vicinity of the "slow" invariant manifold. The two one-dimensional invariant submanifolds thus form a separatrix, partitioning phase space into the two regions seen in Figure 2.1. The orbits in one region overshoot the origin and must reverse themselves, while those in the other move monotonically toward the origin.

If we now reduce the control parameter Γ, a bifurcation occurs when $\Gamma = |2\omega|$: $\underline{\underline{A}}$ possesses only one eigenvalue and one eigenvector. The middle region of monotonic flow toward the origin has vanished; all orbits, except the origin itself, overshoot. At the bifurcation point, $\Gamma = |2\omega|$, $\underline{\underline{A}}$ cannot be diagonalized by a similarity transformation. At best, it is brought to its Jordan normal form, with the eigenvalues on the diagonal but a 1 in the off-diagonal position,[1] via the transformation,

$$\underline{\underline{V}} \equiv \begin{pmatrix} 1 & 0 \\ -m\omega & m \end{pmatrix}$$

$$\underline{\underline{V}}^{-1}\underline{\underline{A}}\underline{\underline{V}} = \begin{pmatrix} -\Gamma/2 & 1 \\ 0 & -\Gamma/2 \end{pmatrix}.$$

If we continue to decrease Γ below 2ω, the eigenvalues and, by implication, the eigenvectors of $\underline{\underline{A}}$ will become complex. The lack of real eigenvalues means that there is no longer any one-dimensional invariant subspace, and all orbits *spiral* in toward the origin. Note that the transition to this behavior has been set up by the bifurcation.

[1] Readers unfamiliar with these terms can find the theory explained in most textbooks on matrix theory or linear algebra; Bellman's [11] is recommended as particularly readable.

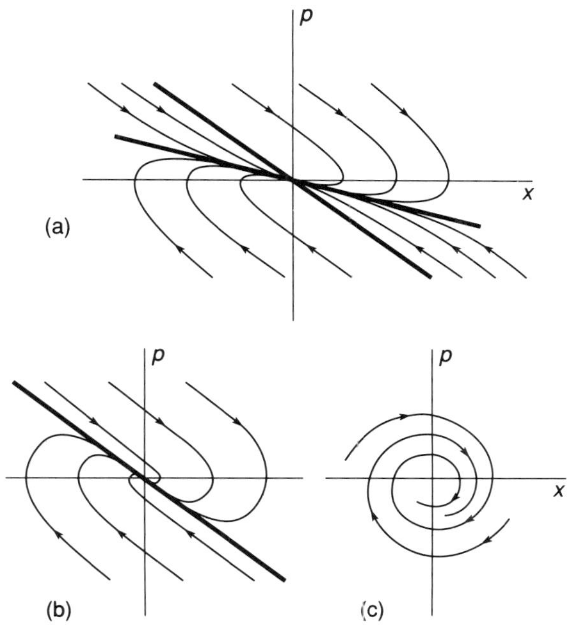

FIGURE 2.1 Phase space of a damped harmonic oscillator: (a) overdamped, (b) critically damped, (c) underdamped.

We have illustrated a "bifurcation" without explaining what the word means and shall pause here to correct that. Dynamics, or vector fields, typically depend not only on the phase space point and the orbit parameter but also on a set of "control parameters" whose values identify the particular dynamical system in a collection of like systems. Our notation has suppressed this dependence, but it is there nonetheless. To make it explicit, we might write something like $\underline{v}_{\Sigma}(\underline{z}, t)$ for the vector field components, where $\underline{\Sigma}$ stood for the set of control parameters. To say that a particular point, $\underline{\Sigma}_c$, in the control space — a particular set of values for the control parameters — is a "bifurcating point" means that any neighborhood of $\underline{\Sigma}_c$, no matter how small, contains points that produce topologically inequivalent vector fields and flows. The set of all bifurcating points is the "bifurcating set" of the dynamical system. What we have seen so far is that the control-space subset characterized by $\Gamma = 2\omega$ is a bifurcating set for the harmonic oscillator, and that it corresponds to the transition from "overdamped" to "underdamped" motion.

The spirals get tighter as $\Gamma \to 0$ until finally at $\Gamma = 0$ all of phase space splits into a collection of invariant circles, the flow of the undamped harmonic oscillator. In fact, $\Gamma = 0$ is another point of bifurcation. In principle we could continue this process and consider $\Gamma < 0$; Figure 2.2 illustrates the full set of all possible eigenvalues. Orbits would then flow away from the origin rather than toward it. (From Eq. (2.2) we see that the complete symmetry is: $\Gamma \to -\Gamma$, $t \to -t$, $p \to -p$.) Although it does not make sense to make a distinction between "local" and "global" bifurcations for linear systems, since they possess no scale, we can point out that, unlike the previous bifurcation, this one is not characterized by a degeneracy in the spectrum

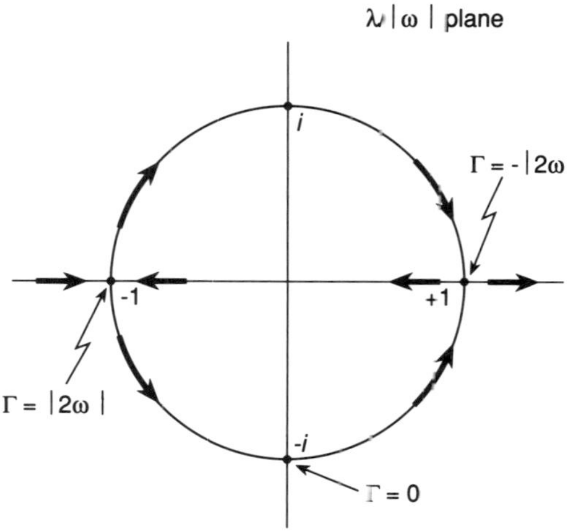

FIGURE 2.2 Trajectory of eigenvalues of the damped harmonic oscillator as Γ decreases from $+\infty$ to $-\infty$. Two bifurcations occur: (a) vanishing of the separatrix at $\Gamma = \pm|2\omega|$ and (b) a Hopf-like bifurcation, but without limit cycle, at $\Gamma = 0$.

of eigenvalues. Rather, the bifurcation takes place as the eigenvalues of $\underline{\underline{A}}$ cross the imaginary axis and the origin changes from being an attractor to being a repellor. As such, it is a variant of something called the "Hopf bifurcation." We shall see in the next section that in the presence of nonlinearities a Hopf bifurcation is accompanied by the creation of a periodic orbit.

Notice that this qualitative analysis of the flow, in which we have learned how all the orbits are arranged in phase space and how this organizational structure varies with changes in the control parameters, has taken place *without actually solving the equations of motion.* The objective of this sort of analysis is to seek out special invariant submanifolds which partition phase space and thereby organize the flow of orbits. From them we get a qualitative and semiquantitative understanding of the dynamics *as a whole,* the kind of information best suited for studying stability or developing statistics on an ensemble.

Exponential Maps Of course, obtaining formal, analytic solutions for autonomous linear systems is easy. Since the matrix $\underline{\underline{A}}$ does not depend on t, the integration of Eq. (2.2) is immediate,

$$\underline{z}(t) = e^{t\underline{\underline{A}}} \cdot \underline{z}(0) = \left(\sum_{n=0}^{\infty} \frac{t^n}{n!} \underline{\underline{A}}^n \right) \cdot \underline{z}(0), \tag{2.4}$$

and we can identify the flow map introduced in the previous chapter with the exponential $\phi_t \leftrightarrow e^{t\underline{\underline{A}}}$. In the undamped case, $\Gamma = 0$, it is easy to confirm that

$$\left(\frac{1}{\omega} \underline{\underline{A}} \right)^2 = -\underline{\underline{1}},$$

so that we can write explicitly

$$e^{t\underline{\underline{A}}} = e^{\omega t(\underline{\underline{A}}/\omega)} = \underline{\underline{1}} \cdot \cos \omega t + \left(\frac{1}{\omega}\underline{\underline{A}}\right) \cdot \sin \omega t. \tag{2.5}$$

Substituting Eq. (2.5) into Eq. (2.4) and writing out the components yield the usual introductory physics result:

$$\begin{aligned} x(t) &= x(0)\cos \omega t + \frac{p(0)}{m\omega}\sin \omega t, \\ p(t) &= p(0)\cos \omega t - m\omega x(0)\sin \omega t. \end{aligned} \tag{2.6}$$

Eq. (2.5) is easily generalized, perhaps not to great advantage. By the Cayley-Hamilton theorem, any square matrix satisfies its own characteristic equation. It follows immediately that for an $N \times N$ square matrix, $\underline{\underline{A}}$, $\exp t\underline{\underline{A}}$ is a $(N-1)^{\text{th}}$ degree polynomial in $\underline{\underline{A}}$.

$$e^{t\underline{\underline{A}}} = \sum_{k=0}^{\infty} \frac{t^k}{k!}\underline{\underline{A}}^k = \sum_{k=0}^{N-1} F_k(t)\underline{\underline{A}}^k.$$

In particular, for any 2×2 matrix $\underline{\underline{A}}$, we can write

$$e^{t\underline{\underline{A}}} = F(t)\underline{\underline{1}} + G(t)\underline{\underline{A}},$$

where, as a consequence of the Chapman-Kolmogorov equation, F and G must obey the following composition rules:

$$\begin{aligned} F(t+t') &= F(t)F(t') - \det\left[\underline{\underline{A}}\right]G(t)G(t'), \\ G(t+t') &= G(t)F(t') + F(t)G(t') + \mathrm{Tr}[\underline{\underline{A}}]G(t)G(t'). \end{aligned} \tag{2.7}$$

Normal Form Coordinates Eigencoordinates of $\underline{\underline{A}}$ take on special significance for $\Gamma = 0$. Referring to Eqs. (2.3),

$$\begin{aligned} \lambda_{\pm} &= \pm i\omega, \\ \underline{e}_{\pm} &\propto \begin{pmatrix} 1 \\ \pm i\omega m \end{pmatrix} \\ &\equiv \begin{pmatrix} 1/\sqrt{2m\omega} \\ \pm i\sqrt{m\omega/2} \end{pmatrix}, \\ \underline{\underline{A}} \cdot \underline{e}_{\pm} &= \pm i\omega \underline{e}_{\pm}. \end{aligned}$$

The reason for the particular, "arbitrary" assignment written in the middle line will be obvious in a moment. Corresponding to $\underline{e}_{\pm}$ is the dual basis, $\underline{u}_{\pm}$, satisfying $\underline{u}_a^\dagger \cdot \underline{e}_b = \delta_{ab}$.

$$\underline{u}_{\pm} = \begin{pmatrix} \sqrt{m\omega/2} \\ \pm i/\sqrt{2m\omega} \end{pmatrix}.$$

The eigencoordinates of $\underline{\underline{A}}$ are now obtained by writing $\underline{z}$ in the $\{\underline{e}_+, \underline{e}_-\}$ basis.

$$\begin{aligned} \underline{z} = \begin{pmatrix} x \\ p \end{pmatrix} &\equiv a^* \underline{e}_+ + a \underline{e}_- \\ a^* = \underline{u}_+^\dagger \cdot \underline{z} &= \sqrt{m\omega/2}x - ip/\sqrt{2m\omega} \\ a = \underline{u}_-^\dagger \cdot \underline{z} &= \sqrt{m\omega/2}x + ip/\sqrt{2m\omega} \end{aligned} \tag{2.8}$$

We recognize these complex coordinates as being those that, upon quantization, become the annihilation and creation operators of the harmonic oscillator system. They are useful in classical dynamics as well when doing resonance and perturbation theoretic calculations, as we shall see in later chapters.

Equally important is the polar (or de Moivre) representation of this system, in which we write

$$\begin{aligned} a &\equiv i\sqrt{I}e^{-i\varphi}, \\ a^* &\equiv -i\sqrt{I}e^{i\varphi}. \end{aligned} \tag{2.9}$$

The manner in which this is written will be motivated in the next chapter. For reasons that will be more apparent later, these polar coordinates, (φ, I), are called "action-angle" coordinates. They can exist only on atlases that do not contain the origin, as it is impossible to assign φ unambiguously when $I = 0$. The origin is a fixed point, however, so the entire phase space can be written as a union of the origin with this "punctured" phase space, which is itself equivalent to the two-dimensional cylinder: $R^2 - \{\underline{0}\} \simeq C$.

We note in passing that the energy of the system is

$$E = \frac{p^2}{2m} + \frac{1}{2}m\omega^2 x^2 = \omega a^* a = \omega I. \tag{2.10}$$

This expression will become very familiar in subsequent chapters.

The representation of the undamped harmonic oscillator vector field changes according to the coordinate chart used.

$$\begin{gathered} \begin{pmatrix} \dot{x} \\ \dot{p} \end{pmatrix} = \begin{pmatrix} 0 & 1/m \\ -m\omega^2 & 0 \end{pmatrix} \begin{pmatrix} x \\ p \end{pmatrix}, \\ \dot{a} = -i\omega a, \qquad \dot{a}^* = i\omega a^*, \\ \dot{I} = 0, \qquad \dot{\varphi} = \omega. \end{gathered} \tag{2.11}$$

The undamped harmonic oscillator contains no essential parameter; both parameters m and ω can be transformed away by rescaling extended phase space. Thus all (undamped) harmonic oscillators are topologically equivalent, and one can choose any of them to represent this equivalence class of vector fields. That representative is called the "normal form" of the class, and "bringing a system to its normal form" means finding a change of coordinates that transforms a particular representation of a vector field into its normal form representation. While not unique, the normal form is usually simpler than most other representations. For linear systems it is derived from the eigenvectors and eigenvalues of the system matrix. Typically, then, the complex eigencoordinates (a, a^*) are considered "normal form" coordinates for the harmonic oscillator; less frequently, the action-angle coordinates (φ, I) are sometimes called "normal form" coordinates as well. The real and imaginary parts of a have the properties: (a) both real coordinates possess the same units, and (b) imposing a Euclidean metric makes the orbits circles. Although some might insist that this be the unique set of normal form coordinates for more complicated dynamics, we shall be more flexible (and vague) in using this term. Its meaning will, it is hoped, become more clear with subsequent usage. The important thing to realize is that, although the "equations of motion" of members of the equivalence class look superficially different, they all represent the same dynamical system, the same vector field, which itself transcends each individual representation as a *geometric object*. We shall return to this point later.

Multidimensional Linear Systems Having done this in a two-dimensional phase space, we extend the picture to N dimensions. If the real matrix, $\underline{\underline{A}}$, can be diagonalized by a similarity transformation, phase space decomposes naturally into a direct sum of one- and two-dimensional invariant subspaces. As the dynamic is linear, the orbits within each of these can be viewed as the projection of the more general orbits; stated the other way around, the general orbit is built from those in the invariant subspaces by superposition.

The one-dimensional subspaces are associated with eigenvalues of $\underline{\underline{A}}$ that are on the real axis; orbits are pure expansion or contraction according to its sign, for within the subspace,

$$e^{t\underline{\underline{A}}} \leftrightarrow e^{\lambda t}.$$

(If $\lambda = 0$, we have a continuum of fixed points.)

Complex eigenvalues must come in conjugate pairs, as the matrix $\underline{\underline{A}}$ is itself real. When λ is complex, the eigenvectors are complex also, and the corresponding *real* invariant subspace is two-dimensional, as in Figure 2.1. Orbits spiral either into the origin or away from it, according to the sign of Re λ. Some terminology: the subspace spanned by eigenvectors with eigenvalues satisfying (a) *Re* $\lambda < 0$ is called the "stable" subspace, and (b) *Re* $\lambda > 0$ is called the "unstable" subspace. If it is not obvious why this is so, the reader should work out a few examples with pencil and paper until it is.

A subspace of great importance to us is that associated with purely imaginary eigenvalues, called the "center" subspace of the system. Within the space spanned by

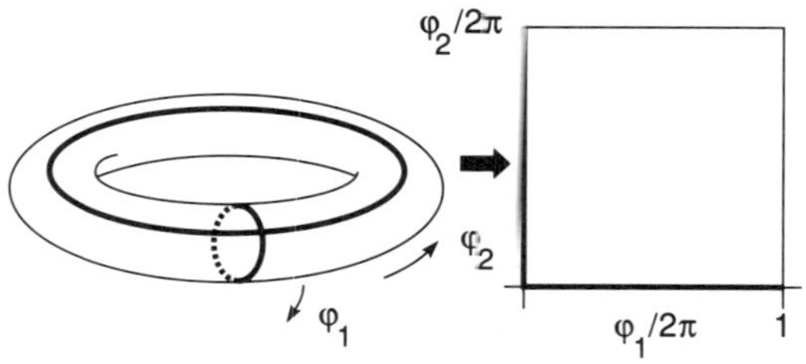

FIGURE 2.3 Angle coordinates on a torus.

eigenvectors with eigenvalues satisfying Re $\lambda = 0$, and assuming $\underline{\underline{A}}$ can be diagonalized, the origin acts like an elliptic fixed point, orbits are circular on the proper (eigen) chart of each two-dimensional invariant subspace, and the projected flow is equivalent to that of an undamped harmonic oscillator. Indeed, if there are N two-dimensional invariant subspaces with imaginary eigenvalues, then an orbit projected into each is equivalent to the circle, S^1, and projected into the entire collection of such subspaces it must be imbedded within an N-dimensional torus, $S^1 \times S^1 \times \cdots \times S^1 \equiv T^N$, and is equivalent to that of a collection of N independent (undamped) harmonic oscillators. It is more than convenient to express the orbit on this torus parametrically using the polar coordinates introduced in Eq. (2.9). The chart is punctured phase space, $R^{2N} - \{\underline{0}\}$, which is equivalent to the direct product of N cylinders. We collect the moduli of the imaginary eigenvalues into an array, $\underline{\omega} = (\omega_1, \omega_2, \ldots, \omega_N)$, and express an orbit history on this atlas as follows:

$$\underline{z}(t) = (\underline{\varphi}(t); \underline{I}(t)); \ \underline{I}(t) = \underline{I}(0); \ \underline{\varphi}(t) = \underline{\varphi}(0) + \underline{\omega}t. \tag{2.12}$$

These are parametric equations describing motion on an N-dimensional torus imbedded in a $2N$-dimensional phase space. The "action" or "amplitude" coordinates, which are constants of the motion, serve to *label the torus*, while the "angle" or "phase" coordinates locate position on the torus. (See Figure 2.3.) This angle chart is equivalent to slicing the torus along N independent directions and flattening it out to fit into a unit cube in Cartesian N-space.

You have undoubtedly recognized that this was written sloppily, as though the punctured phase space were $R^{+N} \times R^N$ rather than $R^{+N} \times T^N$. Of course, there is not one polar chart but an atlas of charts defining the polar coordinates. We can still use Eq. (2.12) as written if we recognize the angle coordinates as obeying a "mod 2π" arithmetic. That is, we set up an equivalence relationship, as was done in the previous chapter,

$$\underline{\varphi} \cong \underline{\varphi}' \quad \textbf{iff} \quad \exists \underline{n} \in Z^N : \underline{\varphi}' = \underline{\varphi} + 2\pi \underline{n}\,,$$

and think of the angle part of Eq. (2.12) as an algebraic expression in the "mod 2π" quotient space: $\underline{\varphi}/2\pi \in T^N \equiv R^N/Z^N$, topologically equivalent to an N-dimensional torus. Within this context, R^N is sometimes called the "universal covering space."

The basic "cell" representing T^N within R^N is the N-dimensional unit cube, $U^N \equiv \otimes^N[0,1) = \{\underline{x} \in R^N \mid 0 \leqslant x_k < 1, \forall k\}$.

A dynamical matrix, $\underline{\underline{A}}$, that is *not* diagonalizable generally indicates a bifurcation. We have seen one such in the transition from the overdamped to underdamped harmonic oscillator: two real eigenvalues and, more importantly, their eigenvectors coalesced as a preparatory step to going off the real axis. The problem of characterizing the flow becomes more complicated. We shall not pursue the general theory of linear classification here. Arnold [6] has written a very readable summary of the analysis of this problem using Lyapunov functions. The two major results are encapsulated within a theorem and a statement. The theorem is this:

Theorem 2.1 A necessary and sufficient condition for topological equivalence of two linear systems, all of whose eigenvalues have nonzero real parts, is that the number of eigenvalues with negative (and hence positive) real parts be the same in both systems.

The statement is this:

> "Suppose all the eigenvalues ... of the linear equation $\dot{\underline{z}} = \mathbf{A}\underline{z}$, $\underline{z} \in R^n$, $\mathbf{A} : R^n \rightarrow R^n$, are purely imaginary. Then under what conditions are two such equations topologically equivalent? The answer to this question is not known, and evidently the problem cannot be solved by presently available mathematical methods."

We have here a limitation on spectral analysis for classifying the flow.

2.1.2 Nonautonomous

If the matrix $\underline{\underline{A}}$ depends on time, we can no longer write the flow map as easily as in Eq. (2.4), the correct generalization being a time-ordered product, a rather formal object obtained from the Chapman-Kolmogorov equation by partitioning the time interval into sub-intervals and approximating the vector field to be autonomous on each sub-interval. Rather than dwell on that, we shall concentrate on the important, special case of systems for which $\underline{\underline{A}}$ is a *periodic* function, a possibly multidimensional version of Hill's equation [64, 46]. Our treatment of this problem will anticipate its application to the physics of beams in storage rings and synchrotrons. Accordingly, we shall interpret the orbit parameter not as "time," t, but as an "angle," θ, and the period will be set at 2π. Although formally it makes no difference, these conventions will ease our transition to the notation of later chapters.

Floquet's Theorem Analysis of the original, one-dimensional Hill's equation was accomplished by Floquet and is consequently called "Floquet theory." His ideas are important for understanding first order stability in circular accelerators. Indeed, the entire development of modern synchrotrons was built on the "strong-focusing" principle, explained in the landmark paper of Courant and Snyder [25], the key idea of which has Floquet theory as its foundation. A variation on this theme leads also to

Bloch's theorem, which determines a form for the solutions of Schrödinger's equation in a periodic crystal lattice.[2]

We shall work with a matrix system of arbitrary dimensionality,

$$\dot{\underline{\underline{Z}}}(\theta) = \underline{\underline{A}}(\theta) \cdot \underline{\underline{Z}}(\theta), \tag{2.13}$$

where $\underline{\underline{A}}$ is periodic,

$$\underline{\underline{A}}(\theta + 2\pi) = \underline{\underline{A}}(\theta).$$

Each column of $\underline{\underline{Z}}$ obeys an equation of the form Eq. (2.1). The advantage of going through the analysis in this manner is that we can work with a complete set of linearly independent solutions simultaneously, these being loaded into the columns of $\underline{\underline{Z}}$. Our objective is to demonstrate the following theorem.

Theorem 2.2 (Floquet, generalized) Given the matrix system of Equation (2.13) there is a solution, $\underline{\underline{Z}}(\theta)$, which can be written in the form

$$\underline{\underline{Z}}(\theta) = \underline{\underline{Q}}(\theta) \cdot \exp[\theta \underline{\underline{C}}], \tag{2.14}$$

where C is a constant matrix, the matrix $\underline{\underline{Q}}(\theta)$ is periodic,

$$\underline{\underline{Q}}(\theta + 2\pi) = \underline{\underline{Q}}(\theta),$$

and the columns of $\underline{\underline{Z}}$ are linearly independent.

Our demonstration of this theorem is at least partially constructive, so at its conclusion we shall possess a technique for actually finding such a solution. With minor variations, it follows a line of reasoning recorded by Bellman [11] and will be expedited with the use of three lemmas, which we shall state first.

Lemma 2.1 Equation (2.1) possesses a unique solution satisfying any given boundary condition, $\underline{z}(0)$.

Proofs that solutions exist can be found in standard texts on linear differential equations. That a solution is unique follows on the observation that $\underline{0}$ is a fixed point of Eq. (2.1): Let $\underline{z}^{(1)}$ and $\underline{z}^{(2)}$ be two solutions satisfying the same boundary condition. By linearity, their difference, $\Delta\underline{z} \equiv \underline{z}^{(2)} - \underline{z}^{(1)}$, is a solution that satisfies the boundary condition, $\Delta\underline{z}(0) = \underline{0}$. Therefore, by the fixed point property, $\Delta\underline{z}(t) = \underline{0}$, for all t.

[2]Despite its long history and widespread use, Floquet theory tends to be reinvented from time to time; during the 1980's its discovery was announced in *Physical Review Letters*. It is especially not well appreciated that Floquet theory is valid in more than one dimension. A prominent nonlinear dynamicist recently submitted a proposal to the U.S. Department of Energy to solve this very problem and, indeed, may have been funded to do so.

Lemma 2.2 On the phase space R^N, Equation (2.1) possesses a complete set of N linearly independent solutions.

"Complete set" means that any solution can be written as a linear superposition of the those in the set. Given the previous lemma, proving this is trivial.

Lemma 2.3 Any nonsingular square matrix possesses a logarithm. That is, given a square, nonsingular matrix, $\underline{\underline{A}}$, there exists a square matrix, $\underline{\underline{B}}$, such that $\underline{\underline{A}} = \exp \underline{\underline{B}}$.

Proof (outline): There are two cases to consider, according to whether $\underline{\underline{A}}$ is or is not diagonalizable.[3]

(Case I) If $\underline{\underline{A}}$ can be diagonalized via a similarity transformation: for some $\underline{\underline{V}}$, $\underline{\underline{A}} = \underline{\underline{VDV}}^{-1}$, where $(\underline{\underline{D}})_{ij} = \lambda_i \delta_{ij}$. Since $\underline{\underline{A}}$ is not singular, by hypothesis, no eigenvalue vanishes, and we can take $\underline{\underline{B}} = \underline{\underline{V}}^{-1} \cdot \ln \underline{\underline{D}} \cdot \underline{\underline{V}}$. (Of course, $(\ln \underline{\underline{D}})_{ij} = \ln \lambda_i \delta_{ij}$.)

(Case II) If $\underline{\underline{A}}$ is not diagonalizable, then it is brought into Jordan form by a similarity transformation: $\underline{\underline{A}} = \underline{\underline{V}}(\bigoplus_m \underline{\underline{J_m}})\underline{\underline{V}}^{-1}$, where $\underline{\underline{J_m}} = \lambda_m \underline{\underline{\mathbf{1}}} + \underline{\underline{T}}$, and $(\underline{\underline{T}})_{ij} = \delta_{i+1j}$. Each $\underline{\underline{J_m}}$ may have its own dimension; we have suppressed explicitly denoting these. What is important is that only those elements of $\underline{\underline{T}}$ just above the diagonal are nonzero. Now, each Jordan block of dimension K possesses a logarithm, by construction:

$$
\begin{aligned}
\ln \underline{\underline{J}} &= \ln\left[\lambda\left(\underline{\underline{\mathbf{1}}} + \frac{1}{\lambda}\underline{\underline{T}}\right)\right] \\
&= \ln \lambda \underline{\underline{\mathbf{1}}} + \ln\left[\underline{\underline{\mathbf{1}}} + \frac{1}{\lambda}\underline{\underline{T}}\right] \\
&= \ln \lambda \underline{\underline{\mathbf{1}}} - \sum_{n=1}^{\infty} \frac{1}{n}(-1/\lambda)^n (\underline{\underline{T}})^n \\
&= \ln \lambda \underline{\underline{\mathbf{1}}} - \sum_{n=1}^{K-1} \frac{1}{n}(-1/\lambda)^n (\underline{\underline{T}})^n.
\end{aligned}
$$

The sum is truncated, as $(\underline{\underline{T}})^K = \underline{\underline{\mathbf{0}}}$. Because of this, convergence is not an issue, and we are assured the logarithm exists. $\underline{\underline{B}}$ is then given by the following:

$$
\underline{\underline{B}} = \underline{\underline{V}}^{-1} \cdot \left(\bigoplus_m \ln \underline{\underline{J_m}}\right) \cdot \underline{\underline{V}}. \qquad \square
$$

Notice that $\underline{\underline{B}}$ may be a complex matrix. You are invited to flesh out this proof to whatever level of detail is necessary to convince yourself that this is more than mere formal manipulation of symbols.

[3]We note in passing the unlikelihood that followers of the constructive school of mathematics would accept this application of the excluded middle.

Finally, we prove the Floquet theorem itself, in the form in which we have expressed it.

Proof (Floquet theorem): To begin, let $\underline{\underline{Z}}_s$, called the "standard" solution, be the solution of Eq. (2.13) that satisfies the boundary condition,

$$\underline{\underline{Z}}_s(0) = \underline{\underline{1}}. \tag{2.15}$$

Lemma 2.1 assures us that one and only one such solution exists. We leave open the question of how to find it; in general, constructing the solution, to a specified accuracy, will require a numerical procedure. We shall show that *any* solution, $\underline{\underline{Z}}(\theta)$, of Eq. (2.13) must satisfy the following identity:

$$\underline{\underline{Z}}(\theta + 2\pi) \cdot \underline{\underline{Z}}^{-1}(2\pi) = \underline{\underline{Z}}_s(\theta). \tag{2.16}$$

To this end, consider that if $\underline{\underline{Z}}(\theta)$ is a solution, then for any constant matrix $\underline{\underline{B}}$, $\underline{\underline{Z}}^+(\theta) \equiv \underline{\underline{Z}}(\theta + 2\pi) \cdot \underline{\underline{B}}$ must be a solution as well, for

$$\begin{aligned}\dot{\underline{\underline{Z}}}^+(\theta) = \dot{\underline{\underline{Z}}}(\theta + 2\pi) \cdot \underline{\underline{B}} &= \underline{\underline{A}}(\theta + 2\pi) \cdot \underline{\underline{Z}}(\theta + 2\pi) \cdot \underline{\underline{B}} \\ &= \underline{\underline{A}}(\theta) \cdot \underline{\underline{Z}}(\theta + 2\pi) \cdot \underline{\underline{B}} \\ &= \underline{\underline{A}}(\theta) \cdot \underline{\underline{Z}}^+(\theta).\end{aligned}$$

Now, setting $\underline{\underline{B}} = \underline{\underline{Z}}^{-1}(2\pi)$, since

$$\underline{\underline{Z}}(\theta + 2\pi) \cdot \underline{\underline{Z}}^{-1}(2\pi)\Big|_{\theta=0} = \underline{\underline{1}}, \tag{2.17}$$

the desired result, Eq. (2.16), is established, as $\underline{\underline{Z}}_s(\theta)$ is the *only* solution satisfying this boundary condition. In particular, Eq. (2.16) must hold for $\underline{\underline{Z}} = \underline{\underline{Z}}_s$ itself.

Lemma 2.3 allows us to define the constant matrix $\underline{\underline{C}}$ and the matrix function, $\underline{\underline{Q}}(\theta)$, according to

$$\underline{\underline{Z}}_s(2\pi) \equiv \exp[2\pi\underline{\underline{C}}] \tag{2.18}$$

$$\underline{\underline{C}} \equiv \frac{1}{2\pi} \ln \underline{\underline{Z}}_s(2\pi)$$

$$\underline{\underline{Q}}(\theta) \equiv \underline{\underline{Z}}_s(\theta) \cdot \exp[-\theta\underline{\underline{C}}]$$

That this can be done depends on the fact that $\underline{\underline{Z}}_s(2\pi)$ is nonsingular, but this must be so, since the linearity of Hill's equation provides that

$$\underline{\underline{Z}}_s(2\pi) \cdot \underline{b} = \underline{0} \Rightarrow \underline{\underline{Z}}_s(0) \cdot \underline{b} = \underline{0} \Rightarrow \underline{b} = \underline{0}.$$

(The careful reader will have noticed already that we slipped past the requirement of nonsingularity without commenting on it when writing Eq. (2.16).) Accepting the above definitions, the Floquet theorem is proved by demonstrating that $\underline{\underline{Q}}(\theta)$ is periodic, as $\underline{\underline{Z_s}}(\theta)$ can then be identified as the desired solution. This is immediate, using Eq. (2.16).

$$\begin{aligned}\underline{\underline{Q}}(\theta + 2\pi) &= \underline{\underline{Z_s}}(\theta + 2\pi) \cdot \exp[-(\theta + 2\pi)\underline{\underline{C}}] \\ &= \underline{\underline{Z_s}}(\theta + 2\pi) \cdot \exp[-2\pi\underline{\underline{C}}] \cdot \exp[-\theta\underline{\underline{C}}] \\ &= \underline{\underline{Z_s}}(\theta + 2\pi) \cdot \underline{\underline{Z_s}}^{-1}(2\pi) \cdot \exp[-\theta\underline{\underline{C}}] \\ &= \underline{\underline{Z_s}}(\theta) \cdot \exp[-\theta\underline{\underline{C}}] \\ &= \underline{\underline{Q}}(\theta)\end{aligned}$$

Apart from the confirmation left to the reader, this completes the proof. □

In retrospect it seems almost obvious that the Floquet theorem must be correct; because the vector field is linear, the period-advance map, $\phi_{\theta+2\pi,\theta}$, must also be linear. Notice that neither $\underline{\underline{C}}$ nor $\underline{\underline{Q}}$ are, as yet, uniquely determined: for example, we could add $in\underline{\underline{1}}$ to $\underline{\underline{C}}$, for any integer n. More generally, we could add any matrix $\underline{\underline{\mathcal{N}}}$ satisfying (a) $[\underline{\underline{C}}, \underline{\underline{\mathcal{N}}}] = \underline{\underline{0}}$ and (b) $\exp(2\pi\underline{\underline{\mathcal{N}}}) = \underline{\underline{1}}$.

Auxiliary Relations Because $\underline{\underline{Z_s}}$ is nonsingular, any solution of the periodic, linear dynamical system,

$$\underline{v}(\underline{z}, \theta) = \underline{\underline{A}}(\theta) \cdot \underline{z},$$

can be written as a linear superposition of the columns of $\underline{\underline{Z_s}}$,

$$\underline{z}(\theta) = \underline{\underline{Z_s}}(\theta) \cdot \underline{z}(0), \tag{2.19}$$

where we have used the boundary condition, Eq. (2.15). The importance of Floquet's theorem is the ease with which the long-term stability of $\underline{z}$ can be determined from the eigenvalues of $\underline{\underline{C}}$. Evaluating successive period-advance maps, we have

$$\begin{aligned}\phi_{2\pi}^{n} : \underline{z}(\theta) \longmapsto \underline{z}(\theta + 2\pi n) &= \underline{\underline{Q}}(\theta + 2\pi n) \cdot \exp[(\theta + 2\pi n)\underline{\underline{C}}] \cdot \underline{z}(0) \\ &= \left(\underline{\underline{Q}}(\theta) \cdot \exp[\theta\underline{\underline{C}}]\right) \cdot \exp[2\pi n\underline{\underline{C}}] \cdot \underline{z}(0) \\ &= \underline{\underline{Z_s}}(\theta) \cdot \left(\exp[2\pi\underline{\underline{C}}]\right)^{n} \cdot \underline{z}(0).\end{aligned}$$

Setting $\theta = 0$ in the above, we get

$$\underline{z}(2\pi n) = \left(\exp[2\pi n\underline{\underline{C}}]\right) \cdot \underline{z}(0) = \left(\exp[2\pi\underline{\underline{C}}]\right)^{n} \cdot \underline{z}(0) = \left(\underline{\underline{Z_s}}(2\pi)\right)^{n} \cdot \underline{z}(0). \tag{2.20}$$

All the n dependence is absorbed into a power of the single exponential factor, $\exp[2\pi\underline{\underline{C}}]$. Whether orbits diverge or remain bounded depends only on the eigenvalues of $\underline{\underline{C}}$, called the "Floquet exponents" of the system. The stability analysis of periodic linear systems reduces to constructing $\underline{\underline{C}}$ and finding these numbers. If they are all pure imaginary, then the origin is an elliptic fixed point, and the entire phase space is stable in the sense that all orbits will remain bounded. If some of the Floquet exponents have nonzero real parts, then the origin is hyperbolic and, if a real part is positive, the corresponding subspaces will contain unstable orbits, that is, orbits that are not confined to a finite region of phase space.

If the phase space is $\mathcal{P} = R^2$, orbit stability is determined by a simple, celebrated criterion, which follows upon the observation that the characteristic equation for the eigenvalues of any 2×2 matrix, $\underline{\underline{Z_s}}(2\pi)$, is

$$\lambda^2 - \lambda \operatorname{Tr} \underline{\underline{Z_s}}(2\pi) + \det \underline{\underline{Z_s}}(2\pi) = 0.$$

The solutions are off the real axis provided that (a) $\det \underline{\underline{Z_s}}(2\pi) > 0$ and (b) $|\operatorname{Tr} \underline{\underline{Z_s}}(2\pi)| < 2\sqrt{\det \underline{\underline{Z_s}}(2\pi)}$, so that the discriminant is negative. We shall see in Chapter 3 that for Hamiltonian systems, $\det \underline{\underline{Z_s}}(2\pi) = 1$, so that eigenvalues are on the unit circle when

$$|\operatorname{Tr} \underline{\underline{Z_s}}(2\pi)| = |\lambda_1 + \lambda_2| < 2. \tag{2.21}$$

The period-advance map written in Eq. (2.18) is identical to a 2π "time"-advance map of an effective, *autonomous* linear system with matrix $\underline{\underline{A}}^{\text{eff}} \equiv \underline{\underline{C}}$, and the collection of points defined in Eq. (2.20) would be obtained by sampling a continuous orbit generated by $\underline{\underline{A}}^{\text{eff}}$. By our previous observations in connection with Eq. (2.12), if all orbits are bounded *and* $\underline{\underline{C}}$ is diagonalizable with no vanishing eigenvalues: (a) eigenvalues of $\underline{\underline{C}}$ are pure imaginary, (b) eigenvalues of $\underline{\underline{Z_s}}(2\pi)$ lie on the unit circle, (c) the dimension of phase space must be even, say, $2N$, and (d) the collection of Eq. (2.20) must lie on an N-dimensional torus. These few observations are critically important. Take the time to convince yourself that they are correct!

A time-evolution matrix that "propagates" a solution from one arbitrary value of θ to another is written easily in terms of $\underline{\underline{Z_s}}(\theta)$ by using Eq. (2.19) twice.

$$\begin{aligned}
\underline{z}(\theta_2) &= \underline{\underline{Z_s}}(\theta_2) \cdot \underline{z}(0) \\
&= \left(\underline{\underline{Z_s}}(\theta_2) \cdot \underline{\underline{Z_s}}^{-1}(\theta_1)\right) \cdot \left(\underline{\underline{Z_s}}(\theta_1) \cdot \underline{z}(0)\right) \\
&= \underline{\underline{M}}(\theta_2, \theta_1) \cdot \underline{z}(\theta_1), \\
\underline{\underline{M}}(\theta_2, \theta_1) &\equiv \underline{\underline{Z_s}}(\theta_2) \cdot \underline{\underline{Z_s}}^{-1}(\theta_1).
\end{aligned} \tag{2.22}$$

Notice the obvious identity: $\underline{\underline{M}}(\theta_1, \theta_2) = \underline{\underline{M}}^{-1}(\theta_2, \theta_1)$. We can push this one step further and generalize $\underline{\underline{Z_s}}(2\pi)$ to a family of equivalent, period-advance matrices.

$$\begin{aligned}
\underline{\underline{M_{2\pi}}}(\theta) \equiv \underline{\underline{M}}(\theta + 2\pi, \theta) &= \underline{\underline{M}}(\theta, 0) \cdot \underline{\underline{Z_s}}(2\pi) \cdot \underline{\underline{M}}(0, \theta) \\
&= \underline{\underline{Z_s}}(\theta) \cdot \underline{\underline{Z_s}}(2\pi) \cdot \underline{\underline{Z_s}}^{-1}(\theta)
\end{aligned}$$

This can be written more suggestively as follows:

$$\underline{\underline{M}}_{2\pi}(\theta) \cdot \underline{\underline{Z}}_s(\theta) = \underline{\underline{Z}}_s(\theta) \cdot \underline{\underline{Z}}_s(2\pi). \tag{2.23}$$

If this matrix equation is written in the (possibly complex) representation in which $\underline{\underline{Z}}_s(2\pi)$ is diagonal, then on the right-hand side, each column of $\underline{\underline{Z}}_s(\theta)$ is simply multiplied by the appropriate eigenvalue. That is, in this representation, the columns of $\underline{\underline{Z}}_s(\theta)$ are eigenvectors of the period-advance map, $\underline{\underline{M}}_{2\pi}(\theta)$:

$$\underline{\underline{M}}_{2\pi}(\theta) \cdot \left(\underline{\underline{Z}}_s(\theta)\right)_{\cdot i} = e^{2\pi\lambda_i} \left(\underline{\underline{Z}}_s(\theta)\right)_{\cdot i}, \tag{2.24}$$

where each λ_i is an eigenvalue of $\underline{\underline{C}}$.

Thin Lens FODO Lattice Familiarity breeds contempt: it is possible to be insufficiently impressed by the conclusion of Floquet's theorem. To illustrate its truly remarkable nature we consider a model near and dear to the heart of all accelerator physicists: the thin lens FODO lattice. One can think of it as an optics model, an infinite, periodic lattice of alternating focusing and defocusing lenses, as sketched in Figure 2.4(a). A single "FODO cell" of this array comprises a focusing lens (F), a "drift" section of length l (O), a defocusing lens (D), and another drift (O).

To place this within the language of phase space let s, the longitudinal position along the lattice, be the orbit parameter, and let the phase space coordinates be $\underline{z} = (x, x')$, the transverse position and slope ($x' = dx/ds$) of the ray. Orbits are described by a one-parameter family of "time"-evolution maps of the type given in Eq. (1.6), $\phi_s : R^2 \longrightarrow R^2, \underline{z}(0) \longmapsto \underline{z}(s)$. It is constructed by catenating two types of elementary, linear maps. The first models the bending of a ray as it passes through a lens. In the sudden, impulse approximation of a "thin" lens, this is expressed as a "kick" applied to the slope coordinate. If the magnitude of the kick is simply proportional to the transverse "position" coordinate, the map is linear and written as a matrix equation:

$$\begin{aligned}\phi_F : \begin{pmatrix} x \\ x' \end{pmatrix} &\longmapsto \begin{pmatrix} 1 & 0 \\ -1/f & 1 \end{pmatrix} \cdot \begin{pmatrix} x \\ x' \end{pmatrix} \equiv \underline{\underline{M}}_F \cdot \underline{z} \\ \phi_D : \begin{pmatrix} x \\ x' \end{pmatrix} &\longmapsto \begin{pmatrix} 1 & 0 \\ 1/f & 1 \end{pmatrix} \cdot \begin{pmatrix} x \\ x' \end{pmatrix} \equiv \underline{\underline{M}}_D \cdot \underline{z}. \end{aligned} \tag{2.25}$$

Here, f is the focal length of the lens; the subscripts F and D refer to passage through focusing and defocusing elements, respectively. The second elementary map corresponds to traversal of an open "drift" space between lenses. A distance s downstream from a lens, the map is written as follows:

$$\phi_s : \begin{pmatrix} x \\ x' \end{pmatrix} \longmapsto \begin{pmatrix} 1 & s \\ 0 & 1 \end{pmatrix} \cdot \begin{pmatrix} x \\ x' \end{pmatrix} \equiv \underline{\underline{M}}_O(s) \cdot \underline{z}.$$

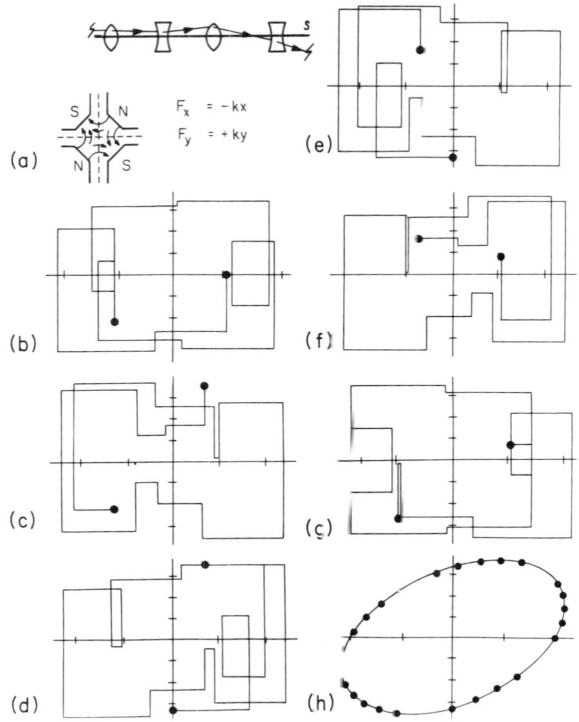

FIGURE 2.4 The thin lens FODO lattice comprises an infinite sequence of alternating focusing and defoccusing lenses. An orbit's structure is revealed not by following it continuously but by sampling it periodically.

By catenating this with the maps ϕ_F and ϕ_D we can follow an orbit as a function of the parameter s, as illustrated in Figures 2.4(b)–(g). Each frame shows the orbit as it passes through eight FODO cells, the endpoint of one frame becoming the starting point of the next.

In this model, the Floquet matrix, $\underline{\underline{Z}}_s(2\pi)$, is the map describing traversal through a single FODO cell.

$$\underline{\underline{Z}}_s(2\pi) = \underline{\underline{M}}_C = \underline{\underline{M}}_O(l) \cdot \underline{\underline{M}}_D \cdot \underline{\underline{M}}_O(l) \cdot \underline{\underline{M}}_F = \begin{pmatrix} 1 - l/f - l^2/f^2 & l(2 + l/f) \\ -l/f^2 & 1 + l/f \end{pmatrix}$$

Notice that $\det(\underline{\underline{M}}_C) = 1$, so that this linear map is area-preserving.[4] We can therefore apply the stability criterion of Eq. (2.21) to determine that the central periodic orbit is elliptic provided that $l < 2f$; otherwise, it is hyperbolic. In the former case, Floquet's theorem tells us that periodic orbit samples must lie on an ellipse (one-dimensional torus). And indeed if the endpoints of the orbit fragments in Figures 2.4(b)–(g) are plotted together they are seen to lie on this ellipse, as shown in Figure 2.4(h).

[4] In two dimensions, this is equivalent to saying that it is "symplectic," but we shall go into that in the next chapter.

Furthermore, if orbit samples are taken periodically starting from *any* point s along the array, those samples also will lie on any invariant torus (ellipse). The eccentricity and orientation of these ellipses will depend on s, and the s-dependent parameters that determine these are called "lattice functions."

Stop to appreciate how the orbit's structure is revealed by using the period-advance map (Figure 2.4(h)), *not* by following the orbit continuously (Figures 2.4(b)–(g)). This is a theme that will be repeated: generally speaking, maps expose structure better than flows. Indeed, the proof of Floquet's theorem itself relied on constructing a period-advance map. Were it not for the map, that such structure could arise from the orbit seen in Figures 2.4(b)–(g) would seem almost unbelievable.

Accelerator Lattice Functions As an accelerator model, one that we shall study in more detail later, the FODO lattice corresponds to a sequence of quadrupole magnets, the "rays" to the orbits of charged particles. The magnetic field of such magnets is a linear function of the transverse offset from their center. It was the application of Floquet theory to such models that gave rise to the "alternating gradient," or "strong focusing," principle that made much of high energy physics possible. Following Courant and Snyder [25], a special notation was employed, giving rise to what are called "accelerator lattice functions." Familiarity with that notation is essential as a point of contact with almost all of the accelerator literature. Here we shall concern ourselves only with periodic, linear dynamics in one degree of freedom, that is, over a two-dimensional phase space; in the next chapter we shall extend this to a four-dimensional space.

Most beam physicists first see Hill's equation in its more traditional form of a second order differential equation:

$$\frac{d^2x}{ds^2} + K(s)x = 0. \tag{2.26}$$

The change in orbit parameter, from θ to s, may here be viewed as no more than symbolic; in fact, it represents a change in scale, according to $ds = R\,d\theta$, where $2\pi R$ is the circumference of a circular accelerator. Apart from this trivial change, this second order equation is equivalent to Eq. (2.13), one possible way of writing it being

$$\frac{d}{ds}\begin{pmatrix} x \\ p \end{pmatrix} = \begin{pmatrix} 0 & 1 \\ -K(s) & 0 \end{pmatrix}\begin{pmatrix} x \\ p \end{pmatrix}. \tag{2.27}$$

We shall leave to Chapter 3 the task of describing the physics behind the real function, $K(s)$ in terms of magnetic fields. Our principal concern here is that, as a consequence of this equation, the matrix of the period-advance map, $\underline{\underline{M_{2\pi}}}(0) = \underline{\underline{Z_s}}(2\pi)$, is unimodular.[5] To see this easily, recall that $\underline{\underline{Z_s}}(2\pi)$ was constructed from solutions

[5]This result is actually a special case of the more general "symplectic condition," for Hamiltonian systems, a topic we shall address in the next chapter. See also Eq. (2.84).

to the Hill's equation. Let us denote its two columns as $(x_1 p_1)^T$ and $(x_2 p_2)^T$. Then,

$$\det \underline{\underline{Z_s}}(\theta) = x_1 p_2 - x_2 p_1 = (x_1 p_1) \cdot \begin{pmatrix} 0 & 1 \\ -1 & 0 \end{pmatrix} \cdot \begin{pmatrix} x_2 \\ p_2 \end{pmatrix}$$

$$\frac{1}{R}\frac{d}{d\theta} \det \underline{\underline{Z_s}}(\theta) = (x_1 p_1) \cdot \begin{pmatrix} 0 & -K \\ 1 & 0 \end{pmatrix} \cdot \begin{pmatrix} 0 & 1 \\ -1 & 0 \end{pmatrix} \cdot \begin{pmatrix} x_2 \\ p_2 \end{pmatrix}$$
$$+ (x_1 p_1) \cdot \begin{pmatrix} 0 & 1 \\ -1 & 0 \end{pmatrix} \cdot \begin{pmatrix} 0 & 1 \\ -K & 0 \end{pmatrix} \cdot \begin{pmatrix} x_2 \\ p_2 \end{pmatrix}$$
$$= 0.$$

Therefore, $\det \underline{\underline{Z_s}}(2\pi) = \det \underline{\underline{Z_s}}(0) = 1$. Combining this with Eq. (2.23), we see that, as one would expect, this property is shared by the entire family of period-advance maps: $\det \underline{\underline{M_{2\pi}}}^{[\theta]} = 1$, for all θ.[6] This means that the matrices $\underline{\underline{C}}^{[\theta]} = \ln(\underline{\underline{M_{2\pi}}}^{[\theta]})/2\pi$ have zero trace. (See Problem 2.41.) Denoting the eigenvalues of $\underline{\underline{C}}$ as $\pm i\nu$, we can write it as follows:[7]

$$\underline{\underline{C}}^{[\theta]} \equiv \nu \underline{\underline{J}}^{[\theta]} = \nu \begin{pmatrix} \alpha^{[\theta]} & \beta^{[\theta]} \\ -\gamma^{[\theta]} & -\alpha^{[\theta]} \end{pmatrix}. \tag{2.28}$$

The eigenvalue is factored out so that $\underline{\underline{J}}$ will have eigenvalues $\pm i$, will be unimodular ($\det \underline{\underline{J}} = 1$), and will satisfy $\underline{\underline{J}}^2 = -\underline{\underline{1}}$ (by the Cayley-Hamilton theorem).[8] This allows us to express the period-advance maps in the form

$$\underline{\underline{M_{2\pi}}}^{[\theta]} = \exp\left(2\pi \underline{\underline{C}}^{[\theta]}\right) = \exp\left(2\pi\nu \underline{\underline{J}}^{[\theta]}\right) = \cos 2\pi\nu \underline{\underline{1}} + \sin 2\pi\nu \underline{\underline{J}}^{[\theta]}.$$

We also note that $2\cos 2\pi\nu = \mathrm{Tr}\, \underline{\underline{M_{2\pi}}}^{[\theta]}$, for any θ, as expected from Eq. (2.21).

The eigenvectors of $\underline{\underline{M_{2\pi}}}^{[\theta]}$ are the eigenvectors of $\underline{\underline{J}}^{[\theta]}$; the one with eigenvalue $+i$ is

$$\begin{pmatrix} \beta \\ i - \alpha \end{pmatrix} \propto \begin{pmatrix} i + \alpha \\ -\gamma \end{pmatrix}. \tag{2.29}$$

Recall from Eqs.(2.23) and (2.24) that the solutions to Hill's equations incorporated in $\underline{\underline{Z_s}}(\theta)$ themselves become eigenvectors of $\underline{\underline{C}}$ upon diagonalization. That is, let $\underline{\underline{U}}$

[6]For the remainder of this discussion I shall denote the parametric dependence of the period-advance map and of its matrix elements on θ with square brackets. This is done to avoid confusion with using θ as the orbit parameter for solutions to the Hill's equation.

[7]If the motion is linearly unstable, then $i\nu$ is on the real, not imaginary, axis.

[8]Notice that this particular trick will not work in higher dimensions. The notation was established in the classic paper of Courant and Snyder [25].

be a matrix that diagonalizes $\underline{\underline{C}} = \underline{\underline{C}}^{[0]}$:

$$\underline{\underline{U}}^{-1} \cdot \underline{\underline{C}} \cdot \underline{\underline{U}} = \begin{pmatrix} i\nu & 0 \\ 0 & -i\nu \end{pmatrix}.$$

Then, by the development above, Eq. (2.14), and Eq. (2.24),

$$\begin{aligned} \underline{\underline{Z}}_s(\theta) \cdot \underline{\underline{U}} &= \underline{\underline{Q}}(\theta) \cdot \exp[\theta \underline{\underline{C}}] \cdot \underline{\underline{U}} \\ &= \underline{\underline{Q}}(\theta) \cdot \underline{\underline{U}} \cdot \underline{\underline{U}}^{-1} \cdot \exp[\theta \underline{\underline{C}}] \cdot \underline{\underline{U}} \\ &= \left(\underline{q}(\theta) e^{i\nu\theta} \ \underline{q}^*(\theta) e^{-i\nu\theta} \right), \\ \underline{q}(\theta) &= \begin{pmatrix} \xi(\theta)\beta^{[\theta]} \\ \xi(\theta)(i - \alpha^{[\theta]}) \end{pmatrix} e^{i\tilde{\psi}(\theta)}, \end{aligned} \tag{2.30}$$

where ξ and $\tilde{\psi}$ are functions to be determined; of course, we make ξ real and positive. The periodicity of $\underline{\underline{Q}}$ requires that both ξ and $\tilde{\psi}$ be periodic functions. The two columns of $\underline{\underline{Z}}_s(\theta) \cdot \underline{\underline{U}}$ are called the "Floquet solutions" to Hill's equation. Normally, the phases are combined and a nonperiodic, or multi-valued, phase $\psi \equiv \tilde{\psi} + \nu\theta$ defined. The ambiguity in defining ν is resolved by requiring that the map $\theta \longmapsto \exp(i\tilde{\psi}(\theta))$ have winding number zero.

We now substitute $x = \xi\beta \exp i\psi$ and $p = \xi(i - \alpha) \exp i\psi$ into Eq. (2.27). In what follows, primes will indicate d/ds. The condition $x' = p$ immediately yields

$$\begin{aligned} \beta\psi' &= 1, \text{ and} \\ \xi'\beta + \xi\beta' &= -\alpha\xi. \end{aligned} \tag{2.31}$$

From $p' = -Kx$ we see that $\text{Im}(p'/x) = \text{Im}(p' \exp^{(-i\psi)}) = 0$, which provides

$$\xi' = \alpha\xi\psi',$$

which, combined with Eq. (2.31), yields the result

$$\beta' = -2\alpha.$$

Finally, substitute this back into Eq. (2.31) to get

$$\xi'\beta + \frac{1}{2}\xi\beta' = 0 \Rightarrow \xi = \sqrt{k/\beta},$$

where k is an arbitrary constant. This brings the Floquet solution to its traditional form in accelerator theory:

$$\begin{aligned} x &= \sqrt{\beta k}\, e^{i\psi}, \\ \alpha x + \beta p &= \sqrt{\beta k}\, i e^{i\psi}. \end{aligned} \tag{2.32}$$

Of course, taking the real and imaginary parts provides linearly independent, real Floquet solutions. The lattice function β is sometimes called the "envelope function," reflecting its role in bounding the possible values of x. That x is more directly observable than p accounts for the popularity of this representation. We leave as an exercise to the reader the task of repeating the above analysis starting from the second representation in Eq. (2.29).

2.1.3 Maps, Tori, and Resonance

A fundamental building block of instability is resonance. In this section we introduce this crucial concept within the context of linear systems and examine its phase space structure.

Resonant Orbits We have mentioned that periodically sampled points from a bounded orbit of a periodic linear dynamic lie on a torus: for autonomous dynamics— which are periodic with no smallest period—the entire orbit is imbedded in a torus; for nonautonomous dynamics this need not be the case. For future reference, we shall refer to this infinite set of points as the "data set," with apologies to everyone who has undergone the ordeal of obtaining *real* data. Consider the following question: Does the data set pass arbitrarily closely to every point on its torus? Put another way, is the closure of the data set the complete torus or only a subset? In the former case, mathematicians speak of the data set being a "dense" subset of the torus.

One obvious case for which the data set cannot "fill" the torus densely occurs when the orbit is itself periodic, say, with (smallest) period T. Applying Eq. (2.12) to the universal covering space, R^N, periodicity means that $\underline{\varphi}(T) - \underline{\varphi}(0) = 2\pi\underline{\nu}$, where $\underline{\nu}$ is an array of integers. The frequencies, $\underline{\omega}$, must therefore satisfy the "periodicity condition,"

$$\underline{\omega} = (2\pi/T)\underline{\nu}.$$

The integers $\underline{\nu}$ are the "winding numbers" of the orbit around the torus; they count the numbers of times the orbit winds around each dimension of the torus in one period. (Accelerator physicists call these the "tunes.")

"Resonance" is a weaker condition that also will prevent an orbit from filling its torus. Staying with autonomous dynamics for a moment, suppose that the frequencies satisfy

$$\underline{m} \cdot \underline{\omega} = 0 \tag{2.33}$$

for some array of integers, $\underline{m}$. The orbit is not necessarily periodic but, according to Eq. (2.12), it is constrained to lie within the invariant subset, $\underline{m} \cdot \underline{\varphi} = \text{constant}$. Thinking of this as a hyperplane in the R^N, the smallest $\underline{\varphi}_0 = 2\pi\underline{\nu}$ satisfying $\underline{m} \cdot \underline{\varphi}_0 = 0$ can serve as the diagonal of a modular supercell. ($\underline{\varphi}_0$ need not lie on the orbit passing through the origin. If it does, then we recover the case of periodicity discussed above.) If one translates the supercell, the entire hyperplane can be covered by the segment within it. It is therefore sufficient to translate the hyperplanar segments falling within the supercell back to the base unit cell, U^N, to have a complete representation of the resonance hyperplane; a few examples are sketched in Figure 2.5. Since there are only a finite number of these, they cannot fill the base cell densely.

Eq. (2.33) is still stronger than we need, especially in the context of the nonautonomous, periodic systems discussed in the last section. We relax it once more and consider a resonance condition of the form

$$\underline{m} \cdot \underline{\omega} + \frac{2\pi}{T} n = 0, \quad \text{or} \quad \underline{m} \cdot \underline{\nu} + n = 0, \tag{2.34}$$

where n is an integer and, as always, $\underline{m} = (m_1, m_2, \ldots, m_N)$ is a multiplet of integers. In a sense, this is equivalent to Eq. (2.33) in a compact extended phase space if we recognize that $n = n \cdot 1$, extend $\underline{\nu}$ to $(\underline{\nu}, 1)$, and interpret $1 = d\theta/dt$. In order to visualize the implications of this condition in phase space, we sample the orbit at the regular intervals $t_k = kT$ using Eq. (2.12) one last time.

$$\begin{aligned} \underline{\varphi}\big|_{t=kT} &= \underline{\varphi}\big|_{t=0} + k\underline{\omega}T = \underline{\varphi}\big|_{t=0} + 2\pi k\underline{\nu}, \\ \frac{1}{2\pi}\left(\underline{\varphi}\big|_{t=kT} - \underline{\varphi}\big|_{t=0}\right) &= k\underline{\nu} \simeq \lfloor k\underline{\nu} \rfloor. \end{aligned} \tag{2.35}$$

Here, $\lfloor x \rfloor$ represents the fractional part of x, and for any N-tuple, $\underline{\alpha}$,

$$\begin{aligned} \lfloor \underline{\alpha} \rfloor &\equiv (\lfloor \alpha_1 \rfloor, \lfloor \alpha_2 \rfloor, \ldots, \lfloor \alpha_N \rfloor) \\ &= \underline{\alpha} \bmod U^N, \end{aligned}$$

where $U^N \equiv \bigotimes^N [0, 1)$ is the N-dimensional unit cube.

If the $\underline{\omega}$ of an invariant torus obeys a resonance condition, as in Eq.(2.34), then from Eq. (2.35) we get

$$\frac{1}{2\pi}\left(\underline{m} \cdot \underline{\varphi}\big|_{t=kT} - \underline{m} \cdot \underline{\varphi}\big|_{t=0}\right) \simeq \lfloor k\underline{m} \cdot \underline{\nu} \rfloor = \lfloor -kn \rfloor = 0. \tag{2.36}$$

This means that the data set once again lies in the $(N - 1)$-dimensional subtorus $\underline{m} \cdot \underline{\varphi} = \text{constant}$, as was illustrated in Figure 2.5. If $\underline{\omega}$ satisfies two inequivalent resonance conditions, then they are contained within the intersection of two $(N - 1)$-dimensional subtori, the connected parts of which are $(N - 2)$-dimensional subtori. In general, if k independent resonance conditions are satisfied, orbit samples lie on a family of $(N - k)$-dimensional tori; if $k = N$, the orbit is periodic. (There is no

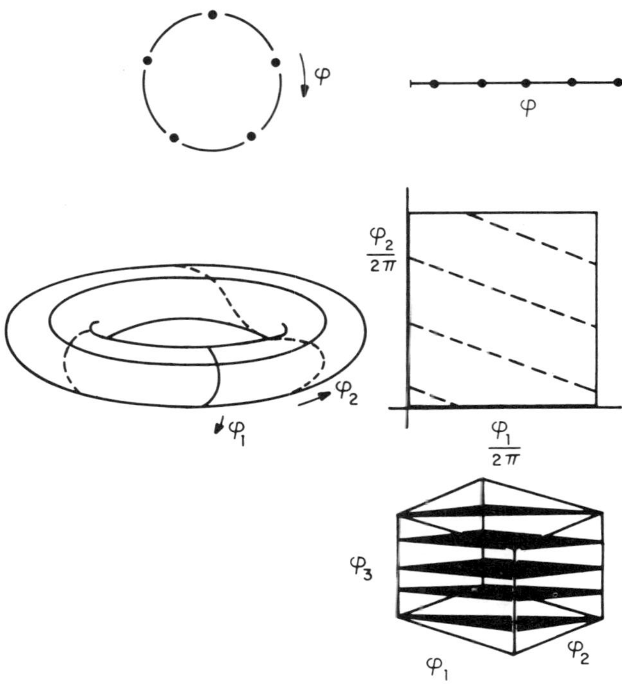

FIGURE 2.5 Regularly sampled points on a resonant orbit lie on an $(N-1)$-dimensional submanifold of the N-dimensional torus: (a) an $\omega_1 + 3\omega_2$ resonance, (b) $-\omega_1 + \omega_2 + 4\omega_3$, a six-dimensional phase space resonance.

substitute for sitting down with a pencil and paper and sketching a few of these for yourself.)

By iterating Eq. (2.24) we see that if $\underline{\lambda} = i\underline{\nu} = i\underline{\omega}T/2\pi$ the data sets associated with autonomous and nonautonomous dynamics are identical. There is no way of distinguishing between the two systems by using the discrete orbit samples alone.

Linear Ergodic Theorem On the other hand, as we shall now show, if $\underline{\omega}$ obeys no resonance condition of the type in Eq. (2.34) then the sampled points will densely *and uniformly* fill the N-dimensional invariant torus; the condition is as weak as we need to make it. We express this very nontrivial result formally as the fundamental theorem:

Theorem 2.3 (Linear ergodic) If $\underline{\nu}$ is nonresonant, then the set of points

$$S \equiv \{\lfloor k\underline{\nu} \rfloor \mid k \in Z\}$$

is uniformly distributed over U^N.

A matter of terminology: interestingly enough, if an orbit is *not* resonant, then it is said to be "quasiperiodic," a definition that seems completely backward. Assuredly, the *resonant* orbits are the ones that should be awarded this appelation, but history and mathematicians have deemed it otherwise.

In general, there are more calculations than proofs in this text. However, this particular theorem is so basic and so important that it is worthwhile to sketch at least the outline of a proof. To understand it, we must first describe a little more carefully what is meant by a "uniform distribution" of points in phase space. This requires that we first specify a measure over phase space, relative to which densities can be gauged. The Liouville volume element associated with a single harmonic oscillator is $dxdp$, and we have seen already that $dx\,dp = dI\,d\varphi$. Since all $dI_k = 0$ on the torus, if we use the measure

$$d\mu\big|_{\text{torus}} = d\varphi_1 d\varphi_2 \cdots d\varphi_N,$$

then uniform density relative to $d\mu$ on the torus means uniform density relative to the Liouville measure, $\prod_{k=1}^{N} dx_k\, dp_k$. Furthermore, apart from $(2\pi)^N$ normalization, this is the usual Lebesque measure on U^N. *A set of points on an invariant torus in phase space will be uniformly distributed relative to the Liouville measure if and only if the corresponding points on the basic cell, U^N, are uniformly distributed relative to ordinary Lebesque measure.* Proving Theorem 2.3 is therefore indeed sufficient to demonstrate the uniform distribution of sampled points on the invariant torus. To be definite, we shall assume a normalization: $\mu(U^N) = 1$, which means that we associate a point in the unit cube, U^N, with the coordinates $\underline{\varphi}/2\pi$.

The use of Liouville measure is not arbitrary: it is the measure that is invariant under the system's dynamics, a proposition that we shall examine in more detail later. This makes it an acceptable candidate for studying phase-space-density related questions.

To pin down the meaning of "uniformly distributed," consider finite subsets of the set S,

$$S_l \equiv \{\lfloor k\underline{\omega} \rfloor \mid k = 1 \ldots l\} \subset S,$$

and let $B \subset U^N$ be any μ-measurable subset of U^N. We denote the number of elements in a set W as $\mathbf{card}(W)$. Then the statement "S is uniformly distributed" means that

$$\lim_{l \to \infty} l^{-1} \mathbf{card}(S_l \cap B) = \mu(B),$$

for all μ-measurable B. Counting the number of elements in a set, B, can be turned into an analytic problem by introducing the characteristic functions of B.

$$K_B(\underline{x}) \equiv \begin{cases} 0, & \underline{x} \notin W \\ 1, & \underline{x} \in W \end{cases}.$$

The exact condition defining uniform distribution is then written

$$\lim_{l\to\infty} l^{-1} \sum_{k=1}^{l} K_B(\lfloor k\underline{\nu}\rfloor) = \mu(B), \quad \text{for all } B.$$

In words, the fraction of points falling within a subset is equal to its size.[9]

Skeleton of a Proof (Linear Ergodic Theorem) The characteristic function of any measurable subset of U^N can be expanded in Fourier series.

$$K_B(\underline{x}) = \sum_{\underline{m}} c_{\underline{m}} e^{2\pi i \underline{m}\cdot\underline{x}} \quad \text{almost everywhere over } U^N.$$

We then evaluate

$$\begin{aligned}
\lim_{l\to\infty} \frac{1}{l} \sum_{k=1}^{l} K_B(\lfloor k\underline{\nu}\rfloor) &\overset{a.e.}{=} \lim_{l\to\infty} \frac{1}{l} \sum_{k=1}^{l} \sum_{\underline{m}} c_{\underline{m}} e^{2\pi i \underline{m}\cdot\lfloor k\underline{\nu}\rfloor} \\
&= \sum_{\underline{m}} c_{\underline{m}} \left[\lim_{l\to\infty} \frac{1}{l} \sum_{k=1}^{l} e^{2\pi i k \underline{m}\cdot\underline{\nu}} \right].
\end{aligned}$$

Since $\underline{\nu}$ is nonresonant, by hypothesis, for all $\underline{m} \neq \underline{0}$ we have

$$\begin{aligned}
\left| \frac{1}{l} \sum_{k=1}^{l} \left(e^{2\pi i \underline{m}\cdot\underline{\nu}} \right)^k \right| &= \frac{1}{l} \left| \frac{e^{2\pi i (l+1)\underline{m}\cdot\underline{\nu}} - e^{2\pi i \underline{m}\cdot\underline{\nu}}}{e^{2\pi i \underline{m}\cdot\underline{\nu}} - 1} \right| \\
&\leqslant \frac{1}{l} \frac{2}{\left| e^{2\pi i \underline{m}\cdot\underline{\nu}} - 1 \right|} \\
&= \frac{1}{l} \frac{1}{|\sin \pi \underline{m}\cdot\underline{\nu}|} \\
&\to 0, \quad \text{as } l \to \infty.
\end{aligned}$$

Therefore the only term that survives is $c_{\underline{0}}$. But

$$c_{\underline{0}} = \int_{U^N} d^N x K_B(\underline{x}) = \mu(B)$$

So we get the desired result:

$$\lim_{l\to\infty} \frac{1}{l} \sum_{k=1}^{l} K_B(\lfloor k\underline{\nu}\rfloor) = \mu(B). \qquad \square$$

[9]This can be extended to noncompact spaces by requiring that the fraction of points falling within a subset be *proportional* to its size.

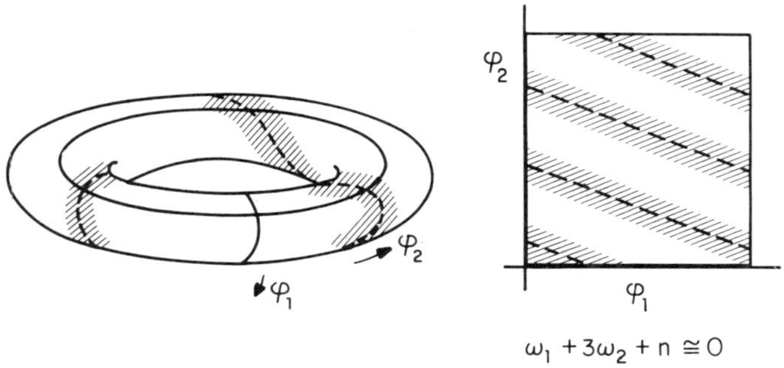

FIGURE 2.6 Filling a torus with an orbit near resonance.

Combining this with our previous discussion of resonant orbits, we have a *purely geometric description of resonance* based on the dimensionality of the closure of a set of orbit samples. Resonance corresponds to dimensional collapse, if you will: resonant orbits lie on lower dimensional tori, nonresonant orbits fill invariant N-tori. This distinction is somewhat unsettling, as the resonant orbits form a dense subset of the phase space, and it is constructively impossible to tell exactly if a set of frequencies is either "on" or "off" resonance. The physical effect can be observed by watching a set of orbit samples develop for a $\underline{\nu}$ that was "close to" but not exactly "on" resonance; then you would see a speckled band form in the vicinity of the resonance line and slowly expand until it filled the torus. (See Figure 2.6.) For a small number of samples, the influence of the nearby resonance would be evident, but it would eventually wash out as the number increased without bound. Samples from an orbit that is "far from" resonance, or at least from a low order resonance, seem to populate the torus uniformly from the beginning: no discernible microstructure ever develops.

Let us stress again that, both for autonomous and nonautonomous systems, this structure becomes evident *only when we focus on the data set of points periodically sampled from the orbit,* that is, on the period-advance map rather than the continuous flow. That this is so more than justifies considering maps in their own right as a fundamental tool for understanding dynamical systems.

First Digits There is a delightful application of the linear ergodic theorem to the "first digits problem," which falls within a subrealm of number theory sometimes called the "geometry of numbers." Let M and b be fixed integers with M not a rational power of b and $b \geqslant 3$. Let d_k represent the first digit of M^k in base b, and consider the sequence

$$\mathcal{D} = \{d_0, d_1, d_2, d_3, \ldots\}.$$

What fraction of the elements of $\mathcal{D}$ have a given value, d, where $1 \leqslant d < b$? (To express this more exactly, we must consider partial sets,

$$\mathcal{D}_N = \{d_0, d_1, d_2, d_3, \ldots, d_N\} \subset \mathcal{D}.$$

Let $n(d;N)$ be the number of elements in $\mathcal{D}_N$ having the value d. The question is: What, if anything, is the value of $p(d) \equiv \lim_{N\to\infty} n(d;N)/N$?) The answer is $p(d) = \log_b(1 + 1/d)$. If the first digit of M^k in base b is d, then

$$\exists p \geqslant 0 : d \times b^p \leqslant M^k < (d+1) \times b^p.$$

Taking the logarithm yields the inequalities

$$p \leqslant k \cdot \log_b M - \log_b d < p + \log_b(1 + 1/d),$$

which in turn means that

$$\lfloor k \cdot \log_b M - \log_b d \rfloor < \log_b(1 + 1/d).$$

Provided M is not a rational power of b, so that we can apply the ergodic theorem, the fraction of numbers $M^k, k = 1, 2, \ldots,$ whose base b expansion begins with the digit d is therefore $\log_b(1 + 1/d)$. Notice that these numbers sum to unity, as they should.

$$\begin{aligned}\sum_{d=1}^{b-1} \log_b(1 + 1/d) &= \log_b \left(\prod_{d=1}^{b-1} \frac{d+1}{d} \right) \\ &= \log_b b \\ &= 1.\end{aligned}$$

2.2 NONLINEAR MODELS

Linear dynamics do not possess microstructure. The scaling law that they satisfy,

$$\forall \lambda \in R : \underline{v}(\lambda \underline{z}, t) = \lambda \underline{v}(\underline{z}, t),$$

assures that all of phase space looks the same; to understand a linear dynamic, one need only examine a small neighborhood of the origin. Because nonlinear systems are freed from this scaling constraint, they can and invariably do possess much more interesting structures; the behavior of orbits in separate regions of phase space can be very different. In the next few sections we shall examine a few simple, nonlinear dynamics to illustrate the techniques of "qualitative analysis," a methodology for studying the topological structure of the orbit space—the space of orbit paths—without actually solving the equations of motion. This will be done using both vector fields and maps. In the latter case, we work from the map itself; we need not construct a presumed, underlying vector field nor integrate it to obtain what was already given.

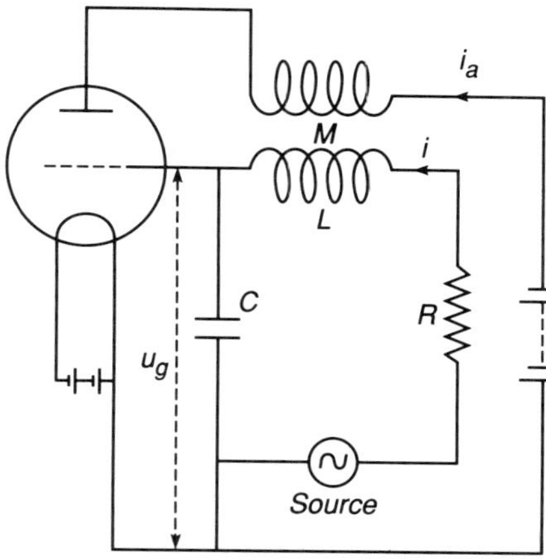

FIGURE 2.7 Electrical circuit motivating Van der Pol's differential equation. (Taken from Davis [27].)

2.2.1 Van der Pol Oscillator

Van der Pol's oscillator was one of the earlier *simple* nonlinear dynamics whose study hinted at the variety of behavior one could expect from such systems. It was originally intended to model the behavior of an electric circuit containing a triode, as sketched in Figure 2.7. For certain values of the parameters, the circuit was known to spontaneously go into self-excited oscillations due to the feedback to the triode. Using a suitably scaled coordinate, x, a second order differential equation describing the charge on the capacitor can be written as follows:

$$\ddot{x} + \gamma f(x)\dot{x} + x = E_0 \sin \omega t. \tag{2.37}$$

We shall consider only the homogeneous system, $E_0 = 0$, in which case the system looks much like that of a damped harmonic oscillator except that the damping coefficient now depends on the coordinate, x, through the nonlinear feedback function $f(x)$. We recast this as a vector field on the phase space, $\mathcal{P} = R^2$, by writing an equivalent set of coupled first order equations. This can be done in (at least) two ways:

$$\text{(a)}\quad \begin{aligned} \dot{x} &= p \text{ and} \\ \dot{p} &= -x - \gamma f(x)p \end{aligned} \qquad \text{or} \qquad \text{(b)}\quad \begin{aligned} \dot{x} &= p - \gamma F(x) \text{ and} \\ \dot{p} &= -x, \end{aligned} \tag{2.38}$$

where F is an integral of f. Decomposition (a) seems more natural, but we shall use (b).[10] We require that f be an even function, which makes F odd. The lowest order nontrivial possibility can be written

$$F(x) = -\mu x + x^3,$$
$$f(x) = -\mu + 3x^2.$$

[10]There is a minor analogy to electrodynamics here, where the momentum is defined in terms of both velocity and a gauge field (see Section 3.1.5), one that even extends to a nonunique definition of F.

We shall adopt this in what follows; appropriately scaled, it conforms to a phase space representation of the original Van der Pol equation.

Hopf Bifurcation: Small γ Analysis We shall assume that $\gamma > 0$, so that contraction (damping) occurs in those regions for which $f(x) > 0$, and expansion (antidamping) in those for which $f(x) < 0$; in particular, the regions $|x| > \sqrt{\mu/3}$ are contracting. The only fixed point is the origin, and its linearized dynamic is

$$\frac{d}{dt}\begin{pmatrix} x \\ p \end{pmatrix} = \begin{pmatrix} \gamma\mu & 1 \\ -1 & 0 \end{pmatrix}\begin{pmatrix} x \\ p \end{pmatrix} \equiv \underline{\underline{A}} \cdot \begin{pmatrix} x \\ p \end{pmatrix}.$$

This is equivalent to Eq. (2.2), as can be established by setting $m = \omega = 1, \Gamma = -\gamma\mu$, and rotating the coordinates, so that $x \longrightarrow p$ and $p \longrightarrow -x$.

Making these substitutions in Eq. (2.3) yields the eigenvalues of our present matrix, $\underline{\underline{A}}$.

$$\lambda_{\pm} = \tfrac{1}{2}\left(\gamma\mu \pm \sqrt{(\gamma\mu)^2 - 4}\right).$$

Spectral analysis of the origin's linearized flow proceeds as with the damped harmonic oscillator. If we think of fixing γ at some positive value and allowing μ to vary, then the Hopf bifurcation occurs at $\mu = 0$, when the eigenvalues cross the imaginary axis and the origin switches between being attracting and repelling. Unlike the damped harmonic oscillator, however, this is a nonlinear system, and the linearized flow does not scale to all regions of phase space.

To understand what does happen, we first express the Van der Pol vector field using a polar coordinate chart, (r, φ), where

$$x = r\sin\varphi \quad \text{and} \quad p = r\cos\varphi, \tag{2.39}$$

essentially the chart of Eq. (2.9). Substituting these into Equation (2.38b) provides the field's components using these coordinates.

$$\begin{aligned} \dot{r}/r &= \gamma\mu\sin^2\varphi(1 - (r^2/\mu)\sin^2\varphi), \\ \dot{\varphi} &= 1 + \gamma\mu\sin\varphi\cos\varphi(1 - (r^2/\mu)\sin^2\varphi). \end{aligned} \tag{2.40}$$

Clearly, as we already knew, $\dot{r} < 0$ everywhere when $\mu < 0$, and the origin is an attractor whose basin of attraction is the entire phase space. For $\mu > 0$, consider the direction of the vector field on a small circle, C_0, surrounding the origin at radius $r_0 \lll \sqrt{\mu}$. (Please refer to Figure 2.8 throughout this treatment.) On $C_0, \dot{r} > 0$ and near transition, where μ is positive but sufficiently small, $\dot{\varphi} \approx 1 > 0$. The field is therefore directed outward and "clockwise." If we try the same argument on a much larger circle, C_1, with radius $r_1 \ggg \sqrt{\mu}$, we see that the field is directed inward and "clockwise" *except* where it intersects the strip $-\sqrt{\mu} < x < \sqrt{\mu}$. We adjust the curve with two small bumps so that the vector field does not point outward anywhere on (the modified) C_1. This can be done, for example, by leaving the circle at point A and following the orbit through A until it turns around and returns to the circle at point B. By construction, the segment from A to B is everywhere parallel to the vector field. For r_1 suffi-

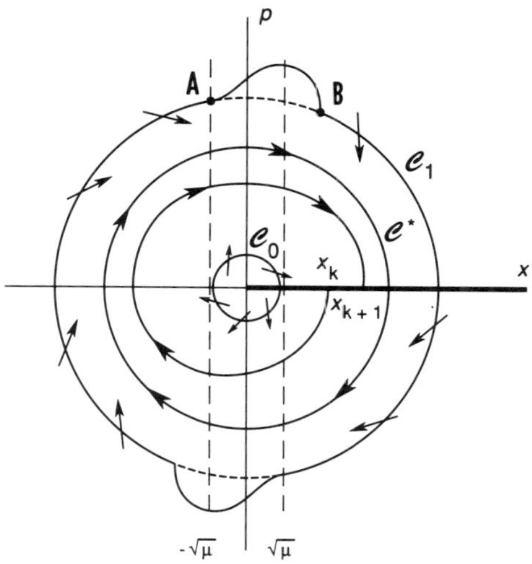

FIGURE 2.8 The limit cycle of the Van der Pol oscillator is inferred by an argument whose generalization leads to the Poincaré-Bendixson theorem.

ciently large, this orbit will surely return to the circle before returning to the line $x = \sqrt{\mu}$. A similar construction is used for the second bump on the opposite side of C_1.

By this construction, C_0 and C_1 bound an annular region in phase space such that the vector field does not point outward anywhere on its boundary. No orbit can leave the annulus once it is inside. If, in addition, we assume that γ is sufficiently small that $\dot{\varphi} > 0$ everywhere within the annulus, then it is intuitively reasonable that there must exist *at least one* periodic orbit, C^*, which is an attractor for *at least some* of the annulus. Those who are satisfied with this can skip to the next paragraph. For the others, we can sketch a proof of its existence using a surface of section. Of course, we have great freedom in choosing the surface, but to be definite, let it be the semiaxis $L_> = \{(x,p) \mid x > 0 \text{ and } p = 0\}$. Every orbit intersects $L_>$ an infinite number of times, since $\dot{\varphi} > 0$. In this way the surface of section produces a map, $\phi : [r_0, r_1] \rightarrow [r_0, r_1]$, obtained by following an orbit around the annulus. That is, for $x \in [r_0, r_1]$, $\phi(x) = y$ if and only if the orbit intersecting $L_>$ at $(x, 0)$ intersects it again for the first time at $(y, 0)$. By following the orbit through $(r_0, 0)$ we generate an infinite sequence, $x_0, x_1, x_2, \ldots$, where $x_0 = r_0$ and $x_{k+1} = \phi(x_k)$. Because (a) the vector field is continuous and continuously differentiable (e.g., no cusp at $(r_0, 0)$), (b) there are no fixed points (since $\dot{\varphi} \neq 0$ anywhere in the annulus), and (c) the orbit cannot cross itself (for then $\underline{v}$ would be multivalued at a point), it follows that $x_k < x_{k+1}$ for all k. Since the sequence has r_1 as an upper bound, there must be a least upper bound, say, x^*, and $\lim_{k\to\infty} x_k = x^*$. We can then argue that $\phi(x^*) = x^*$, since if this were not the case: (a) $\phi(x^*) > x^*$, and (b) by continuity for some x_n sufficiently close to x^* we would have $\phi(x_n) = x_{n+1} \approx \phi(x^*) > x^*$,[11] which violates

[11]This is, of course, not mathematically pure, but I hope it gets the idea across. The reader can express this to whatever level of rigor desired. Another detail to be considered is that we have not established the continuity property of orbits.

the definition of x^* as an upper bound. Thus the orbit through x^* is periodic, and we shall designate its closed path (a Jordan curve) as C^*. Furthermore, a few moments' thought are sufficient to establish that the sequence of Poincaré iterates of any orbit intersecting C_0 must have x^* as its limit, so that C^* is an attractor with the region between C_0 and C^* as at least part of its basin of attraction. Such an orbit is called an "ω-limit cycle"; if the periodic orbit were a repellor — that is, if it became an attractor under time reversal — it would be an "α-limit cycle."

Clearly a similar argument holds starting from C_1, leading to a limit cycle, C^{**}, with basin of attraction bounded by C_1. We shall, however, cut short the argument at this point and leave it to the reader to establish that $C^{**} = C^*$. The principal point we wish to make here is the fruitfulness of using a map to reveal orbit structure, much as was the case with the FODO model of Section 2.1.2.

If the nonlinearity is weak, that is, for γ "sufficiently small," the limit cycle is approximately circular, $r^2 \equiv x^2 + p^2 = \text{constant}$, and its radius can be estimated roughly via a simple integral identity.[12] First divide $\dot{p}$ by $\dot{x}$ to get

$$\frac{dp}{dx} = \frac{-x}{p - \gamma F(x)},$$

which upon a little rearrangement can be written as an equation among differentials (or, more properly, differential forms):

$$p\,dp + x\,dx = d\left(\frac{1}{2}r^2\right) = \gamma F(x)\,dp.$$

Since the limit cycle is an orbit, this is satisfied everywhere on it, and we can integrate around the cycle, C^*, to obtain

$$\oint_{C^*} F(x)\,dp = \frac{1}{\gamma}\oint_{C^*} d\left(\frac{1}{2}r^2\right) = 0.$$

Temporarily accepting that the limit cycle is approximately circular, say, $r = r_c$, we substitute the polar coordinates of Eq. (2.39) to obtain our estimate.

$$\begin{aligned}
0 &= \oint_{C^*} dp(-\mu x + x^3) \\
&\approx \oint_{C^*} (-r_c \sin\varphi\, d\varphi)(-\mu r_c \sin\varphi + r_c^3 \sin^3\varphi) \\
&= -\pi r_c^2\left(\frac{3}{4}r_c^2 - \mu\right) \\
r_c &\approx \sqrt{4\mu/3}. \qquad (2.41)
\end{aligned}$$

[12]Much of what follows repeats a line of reasoning found in Gilmore [40].

The period of the limit cycle locks onto the value assumed at its birth,

$$\begin{aligned} T = \oint_{C^*} dt &= -\oint_{C^*} \frac{dp}{x} \\ &\approx -\oint_{C^*} \frac{d(r_c \cos\varphi)}{r_c \sin\varphi} \\ &= \oint_{C^*} d\varphi = 2\pi, \end{aligned}$$

consistent with the observation that $\omega = 1$ near the origin at transition.

A self-consistency condition necessary to validate the approximation is obtained using Eq. (2.40). In order for $\dot{r}/r \ll \dot{\varphi}$ when $r^2 \approx |\mu|$ it is necessary that $\gamma \ll 1/|\mu|$. This would be satisfied even for fairly large γ near the point of bifurcation, $\mu \approx 0$, with restriction to smaller values as μ increases in magnitude.

The creation of a limit cycle that grows out of a fixed point as μ passes from 0^- to 0^+ is more characteristic of the Hopf bifurcation than the example of the damped harmonic oscillator considered earlier. The geometric argument we used to demonstrate its existence can be generalized to prove the Poincaré-Bendixson theorem, which we write in the form given by Guckenheimer and Holmes [43].

Theorem 2.4 (Poincaré-Bendixson) A nonempty, compact ω- or α-limit set of a planar flow, which contains no fixed points, is a closed orbit.

This theorem exists in more general forms, such as that recorded in Abraham and Marsden [1], expressed in terms of vector fields on two-dimensional manifolds.

Limit Cycle: Large γ Analysis In the opposite extreme, when the nonlinearity is strong ($\gamma \longrightarrow \infty$), the limit cycle loses its circular shape and begins to behave more like a hysteresis loop in phase space. For large γ the momentum range of the limit cycle scales with γ, so it is convenient to anticipate this by defining a scaled coordinate, z, according to $p \equiv \gamma z$. The vector field components on this chart are

$$\begin{aligned} \dot{x} &= \gamma(z - F(x)), \\ \dot{z} &= -\frac{1}{\gamma}x. \end{aligned} \tag{2.42}$$

With a little rearrangement we see that the orbit slope function obeys the equation

$$(z - F(x))\frac{dz}{dx} = -\frac{1}{\gamma^2}x$$

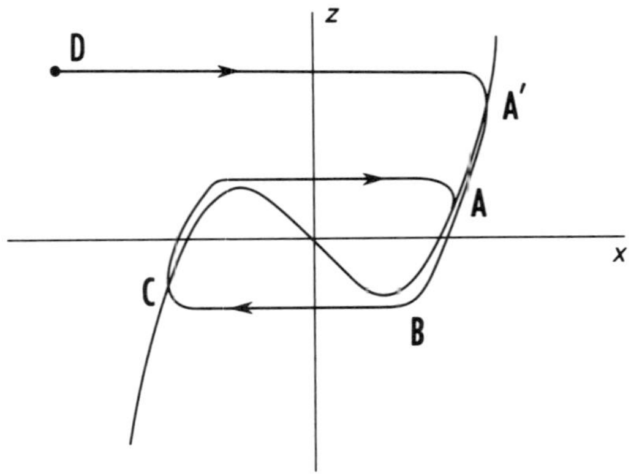

FIGURE 2.9 The limit cycle behaves like a hysteresis in phase space in the limit $\gamma \rightarrow \infty$.

As $\gamma \rightarrow \infty$, the right-hand side can be made vanishingly small for regions bounded in x. The consequences for the left-hand side are that either

$$z \approx F(x) \quad \text{or} \tag{2.43}$$

$$dz/dx \approx 0. \tag{2.44}$$

It is useful to consider the obvservable which is the difference between z and $F(x)$.

$$\begin{aligned} u &\equiv z - F(x), \\ \dot{x} &= \gamma u, \\ \dot{u} &= \dot{z} - f(x)\dot{x} \\ &= -\frac{1}{\gamma}x - \gamma f(x)u \\ &\approx -\gamma f(x)u, \quad \text{provided } f(x) \neq 0, \\ d(u^2)/dt &= -2\gamma f(x)u^2. \end{aligned} \tag{2.45}$$

In regions of substantial slope (of F), u^2 is decreasing if the slope is positive and increasing if negative.

The picture that emerges is sketched in Figure 2.9. If one starts at an arbitrary point, shown as point D, removed from the curve $z = F(x)$, the system moves exceedingly rapidly almost parallel to the x-axis, according to Eq. (2.45) and Eq. (2.44), so as to close the gap. Arriving approximately at A′ the system overshoots the curve slightly but remains close (see Eq. (2.43)) as it moves now exceedingly slowly (see Eq. (2.42)) from A′ to B, the point of zero slope. Upon passing point B the system detaches itself from the curve, and as it is now in a region of negative $f(x)$ it begins to move rapidly to point C. The remaining piece of the orbit, from C back to the vicinity of A, is essentially a rotated version of what has gone before. The transient portion of the orbit, from D to A, proceeds so rapidly that the system seems to lock itself into its limit cycle instantaneously.

Finally, we can estimate the period of the limit cycle in this limit as well. Referring to Figure 2.9, and ignoring the negligible transit times for the horizontal portions of the path, we have

$$T \approx 2\int_A^B dt = -2\gamma\int_A^B \frac{dz}{x} = -2\gamma\int_A^B \frac{dF(x)}{x}$$
$$= -2\gamma\int_{x_A}^{x_B} dx\left(\frac{-\mu}{x^2} + 3x\right) = 2\gamma\left(\frac{3}{2}x^2 - \mu\ln x\right)\Big|_{x_B}^{x_A}.$$

From the condition $f(x_B) = 0$ we obtain $x_B = \sqrt{\mu/3}$, and from $F(x_A) = F(-x_B)$ we obtain $x_A = 2\sqrt{\mu/3}$, so that

$$T \approx (3 - 2\ln 2)\gamma\mu \approx 1.6\gamma\mu.$$

What we have described, of course, is a square-wave oscillator. That a system could exhibit such behavior, independent of its initial conditions and over a wide range of its parameters, came as something of a revelation. Nicolis and Prigogine [76] used the term "self-organization" to describe this tendency of *damped, driven* systems to exhibit structured behavior spontaneously. Within the phase space models that we are considering, it arises from the existence of an attractor, in this case, a limit cycle.

2.2.2 Lorenz Equation

We have seen two types of attractors so far: the point attractor, such as the origin in the damped harmonic oscillator, and the limit cycle, as in the Van der Pol oscillator. Theorems like the Poincaré-Bendixson theorem indicate that no other types of attractors are possible for flows (not maps) on two-dimensional manifolds. However, in three or more dimensions it is easy to devise vector fields with attractors that are neither points nor limit cycles. These have been called "strange" attractors, probably for the same reason that strange particles got their name — their existence came as something of a surprise — but also because they generally exhibit "chaotic" behavior. One of the simplest, and most famous, dynamics to possess a strange attractor is represented by the Lorenz equations, a three-parameter, three-dimensional vector field with quadratic nonlinearities.

$$\begin{pmatrix}\dot{u}\\ \dot{v}\\ \dot{w}\end{pmatrix} = \begin{pmatrix}-\sigma & \sigma & 0\\ \rho & -1 & 0\\ 0 & 0 & -\beta\end{pmatrix}\cdot\begin{pmatrix}u\\ v\\ w\end{pmatrix} + \begin{pmatrix}0\\ -uw\\ uv\end{pmatrix}.$$

The equation was originally written in a simplified study of the behavior of amplitudes of Fourier modes in a fluid dynamics problem; whether it was successful in that context need not concern us.

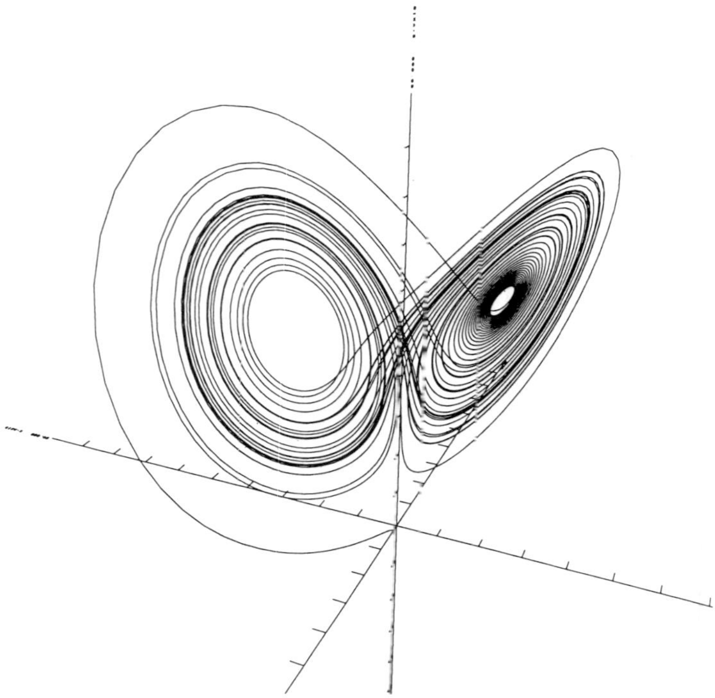

FIGURE 2.10 Strange attractor of the Lorenz equations.

The system possesses a π-rotation symmetry: $x \rightarrow -x$, $y \rightarrow -y$, $z \rightarrow z$, and it is contracting when $\sigma + \beta > -1$ (see Problem 2.40). In what follows, we shall assume all parameters are positive. One obvious fixed point, and therefore attractor, is the origin; it is not difficult to verify that the other nontrivial fixed points, if they exist, have coordinates $z_o = \rho - 1$ and $x_o = y_o = \pm\sqrt{\beta z_o}$. Therefore, for $0 < \rho < 1$, the origin is the only point attractor. The two fixed points spawned by the (pitchfork; see Fig. 5.1) bifurcation at $\rho = 1$ are point attractors provided that

$$1 < \rho < \rho_b = \frac{\sigma(\sigma + \beta + 3)}{\sigma - \beta - 1}.$$

This is confirmed by first order expansion about the fixed points and is left as an exercise (Problem 2.19). More importantly, for $\rho > \rho_b$ a strange attractor exists, a piece of which is shown in Figure 2.10. It is neither a point nor a limit cycle, but all orbits converge on it as $t \rightarrow \infty$.[13] An orbit near the attractor winds around either lobe, moving toward a fixed point until suddenly it is repelled out of the "plane" of the lobe and falls onto the other one. This process repeats back and forth in a seemingly random, though deterministic, fashion. This "randomness" is a manifestation of the chaotic nature of the orbit, an attribute we shall examine more closely in another

[13]This statement can be made more precise, but there is no advantage here in doing so.

section. The two points to be noted here are (a) how even trivial nonlinearities can give rise to extraordinary dynamical behavior, and (b) how confining attention to two-dimensional dynamics can constrain one's intuition about how dynamical systems behave.

2.2.3 Maps

We have seen and shall continue to see how Poincaré and period-advance maps of a dynamical system are indispensable for revealing the topological structure of its orbit space. Much insight has been gained by studying the orbits of maps themselves, without reference to an underlying continuous vector field. Many of these — such as logistic maps or linear automorphisms of a torus — while interesting in their own right have no connection with accelerator physics. We shall consider two which do: one, which we shall call the "bucket map" and which is very similar to the "standard map" that has received so much attention during the last decade, is a basic model for longitudinal motion in an accelerator; the other is a Hénon map, which we shall later connect with the introduction of sextupolar magnetic fields into a storage ring.

Bucket Map The principal function of an accelerator is to increase the energy of a beam of charged particles. This is done by passing them through a cavity in which a radio-frequency (RF) electromagnetic field is excited. The particle is made to cross the cavity when the phase of the electric vector is such that its energy increase has the desired value. For our purposes at present, the rest of the synchrotron — the "ring": dipoles, quadrupoles, drift sections, correction elements, and so forth — can be thought of as a black box device for returning particles to the cavity.

We shall model this process in a simple but fundamental manner. Assume that the RF cavity imparts to the proton an instantaneous energy kick of size

$$\delta E = eV \sin \varphi,$$

where φ is the phase of the electric field at the time of passage, and $|V|$ is the maximum voltage. (We shall adopt the convention that $eV > 0$.) To remain reasonable, we assume that the energy gain is designed to be the same at every crossing. In its turn, this means that the phase should have the same value, say, φ^s, at each crossing; φ^s is called the "synchronous" phase.

Naively, we might expect $\varphi^s = \pi/2$ to be a desireable value, as this maximizes the energy kick per turn. That this is not normally done is due to considerations of stability. The time required for a particle to traverse the ring depends on its energy.[14] However, a beam is not purely monochromatic: the energies of its constituent particles fall into a distribution having some nonzero width. It is, accordingly, impossible for the phase to be exactly synchronous with all of them simultaneously. The stability question is this: If the cavity and an orbiting particle start out close to synchrony, will they remain so throughout the acceleration period? Put another way, will the energy

[14]It also depends on its transverse motion; we are neglecting that here.

of all the particles in an almost but not quite monochromatic beam stay close to the ideal, "synchronous energy" or drift away from it.

We shall label the state of the particle with (1) the phase of the field at its crossing (abusing the language slightly, we shall speak of the "phase of the particle") and (2) the energy of the proton immediately *after* its crossing. The energy increment from one cycle to the next is easily written:

$$E_k = E_{k-1} + eV \sin \varphi_k.$$

where the subscript counts cycles. We are interested not in the energy itself but in its deviation from the currently synchronous energy, the one that grows arithmetically according to the above recursion with $\varphi_k = \varphi^s$, for all k. Let $y = E - E^s$ be this difference. By subtraction, the y sequence is generated by the recursion[15]

$$y_k = y_{k-1} + eV(\sin \varphi_k - \sin \varphi^s). \qquad (2.46)$$

Let the cycling time, at cycle k, of a particle with energy E be represented by $2\pi/\omega_k(E)$.[16] This is written in a form that emphasizes the current "cycling frequency," $\omega_k(E)$. In a low energy beam, a particle of higher energy traverses the ring more quickly, by virtue of its greater velocity; at high energies, all particles in the beam travel virtually at the speed of light, so that the variations in cycle time are dominated by differences in the particles' path lengths. The paths of higher momentum particles will have a larger radius of curvature passing through the dipoles. Unless designed specifically to avoid it, this generally means that their total path length is larger, so that the cycle time *increases* with energy. The energy at which one passes from one regime to the other is called the "transition energy" of the ring. Whenever it is convenient to be specific, we shall assume that the synchronous energy is above transition in what follows; the reader can modify the results for energies below transition.

The RF cavity oscillates at some multiple of the synchronous frequency, say, $\omega_k^{\text{rf}} = h\omega_k(E_k^s) = h\omega_k^s$, where h is an integer.

$$\begin{aligned}\varphi_k &= \varphi_{k-1} + \frac{2\pi}{\omega_{k-1}(E_{k-1})} h\omega_{k-1}^s \\ &= \varphi_{k-1} + 2\pi h \left(\frac{\omega_{k-1}(E_{k-1}^s)}{\omega_{k-1}(E_{k-1}^s + y_{k-1})} - 1 \right) \quad (\text{mod}\, 2\pi).\end{aligned} \qquad (2.47)$$

The combination of Eq. (2.46) and Eq. (2.47) defines a mapping in the two-dimensional cylindrical phase space, with local coordinates $\underline{z} = (\varphi, y)$. The "orbit history" task is to iterate this map and generate the sequence, $\{(\varphi_k, y_k) \mid k \in Z^+\}$. Some sample orbits, using physically realistic parameters, are shown in Figure 2.11. We shall study their behavior in more detail later. For now, we content ourselves with demonstrating two or three simple properties

[15] Or is it "iteration?"

[16] See earlier footnote.

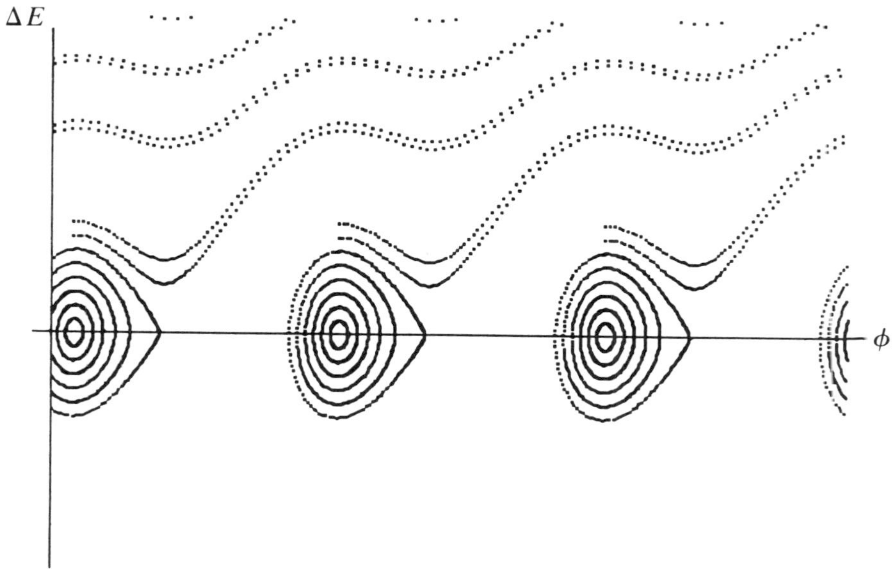

FIGURE 2.11 Orbits of the "bucket map" that models the longitudinal phase space of an synchrotron.

Consider first the location and classification of fixed points. From Eq. (2.45), $y_k = y_{k-1}$ either when $\varphi_k = \varphi^s$ or when $\varphi_k = \pi - \varphi^s$; from Eq. (2.47), $\varphi_k = \varphi_{k-1}, \bmod 2\pi$, provided that $y_{k-1} = 0$. The two fixed points are therefore located at $(\varphi, y) \simeq (\varphi^s, 0)$ or $(\pi - \varphi^s, 0)$. Whether they be elliptic or hyperbolic is determined by linearizing the map about the fixed points.

$$\underline{\phi}(\underline{z}) = \underline{\phi}(\underline{z}_o + (\underline{z} - \underline{z}_o)) \approx \underline{z}_o + (\underline{z} - \underline{z}_o) \cdot \underline{\underline{\partial\phi}}(\underline{z}_o). \tag{2.48}$$

The Jacobian matrix is obtained by writing the differential forms of Eqs.(2.46) and (2.47):[17]

$$\begin{aligned} d\varphi_k &= d\varphi_{k-1} - 2\pi h \frac{\omega_{k-1}(E^s_{k-1})\omega'_{k-1}(E^s_{k-1} + y_{k-1})}{\omega^2_{k-1}(E^s_{k-1} + y_{k-1})}\, dy_{k-1} \\ &\equiv d\varphi_{k-1} - \alpha_{k-1}(y_{k-1})\, dy_{k-1}, \\ dy_k &= dy_{k-1} + eV \cos\varphi_k\, d\varphi_k \\ &= (1 - \alpha_{k-1}(y_{k-1})eV\cos\varphi_k)\, dy_{k-1} + eV\cos\varphi_k\, d\varphi_{k-1}, \end{aligned}$$

$$\underline{\underline{\partial\phi}} = \frac{\partial(\varphi_k, y_k)}{\partial(\varphi_{k-1}, y_{k-1})} = \begin{pmatrix} 1 & -\alpha_{k-1}(y_{k-1}) \\ eV\cos\varphi_k & 1 - \alpha_{k-1}(y_{k-1})eV\cos\varphi_k \end{pmatrix}.$$

Notice that *the map is area-preserving everywhere:* $\det(\underline{\underline{\partial\phi}}) = 1$, identically. It is not a dissipative map; it possesses no attractors. We therefore can use the criterion given in Eq. (2.21) to determine stability of the fixed points, the condition for an elliptic

[17]The quantities α_k that appear here have nothing to do with the α introduced in Section 2.1.2.

fixed point at $\underline{z}_o = (\varphi^s, 0)$ being

$$|\,\text{Tr}(\underline{\underline{\partial\phi}}(\underline{z}_o))| = |2 - \alpha_{k-1}(0)eV\cos\varphi^s| < 2. \tag{2.49}$$

Provided the term is sufficiently small, this will be satisfied if $\omega'(E^s)\cos\varphi^s > 0$. Accordingly, below transition we choose $\varphi^s \in (-\pi/2, \pi/2)$, and we make the opposite choice above transition. By substituting $\varphi^s \to \pi - \varphi^s$ we see that the other fixed point must then be hyperbolic.

Before leaving this topic we make two observations. The first is that one normally sees this map in the accelerator literature with $\omega(E)$ expanded about the synchronous energy.

$$\begin{aligned}\omega(E) + O(y^2) &= \omega^s + \left(\frac{d\omega}{dE}\right)^s y \\ &= \omega^s + \left(\frac{d\omega}{dp}\right)^s \left(\frac{dp}{dE}\right)^s y \\ &= \omega^s + \left(\frac{\omega E}{p^2c^2}\eta\right)^s y.\end{aligned}$$

The dimensionless parameter $\eta = (p/\omega)(d\omega/dp)$ is the "frequency slip factor" of the ring. If the final form is considered exact, then it effectively defines the function $\eta(y)$. Using this, Eq. (2.47) is frequently replaced with the following approximation:

$$\varphi_k \cong \varphi_{k-1} - 2\pi h\left(\frac{E\eta}{p^2c^2}\right)^s_{k-1} y_{k-1}. \tag{2.50}$$

Second, in this form the mapping is closely related to one of the most extensively studied maps of all time: the "standard map,"

$$\begin{aligned}y_k &= y_{k-1} + K\sin\varphi_{k-1}, \\ \varphi_k &= \varphi_{k-1} + y_k.\end{aligned}$$

Apart from the extra term, $\sin\varphi^s$, present in Eq. (2.46) and a trivial renormalization, they are virtually identical. (The indices could be brought into compliance by defining y_k to be the energy defect just before *entering* the RF cavity.) However, many discussions of the standard map include one extra modification that makes it different: y, as well φ, is treated as a modular coordinate, so that the mapping is interpreted as existing on a torus. In addition, those who have analyzed this map have generally concentrated on setting K to values large enough so that much of the phase space is occupied with chaotic orbits, as seen in Figure 2.12. Because of these two choices, most of the vast literature on the standard map is not directly applicable to synchrotron dynamics.

Hénon Maps The logistic map is a one-dimensional quadratic map over R. Its two-dimensional generalization on the phase space R^2 is of the form

$$\underline{\phi}(\underline{z}) = (P_1(\underline{z}), P_2(\underline{z})),$$

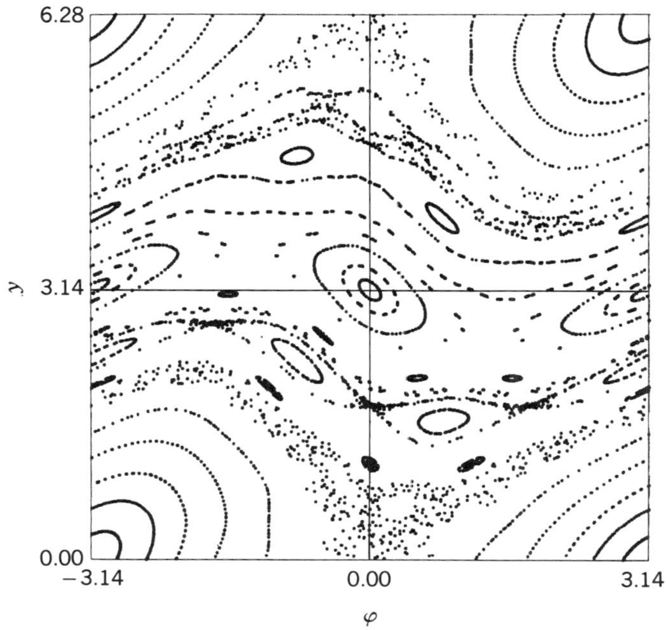

FIGURE 2.12 Typically displayed orbits generated by the standard map; the parameter K is set to 0.97.

where P_1 and P_2 are polynomials of degree 2 in the components of $\underline{z} = (z_1, z_2)$. If the Jacobian, $\det(\partial\underline{\phi}/\partial\underline{z})$, is constant over R^2, then the map is called a "Hénon" map, named for the one who systematically studied their properties. In particular, Hénon showed that any such map is topologically equivalent to the following two-parameter family of maps:

$$\underline{\phi}^{(a,b)} : \begin{pmatrix} z_1 \\ z_1 \end{pmatrix} \longmapsto \begin{pmatrix} z_2 + 1 - az_1^2 \\ bz_1 \end{pmatrix}, a, b \in R. \tag{2.51}$$

We may think of this as a normal form representation for Hénon maps. It is trivial to show that its constant Jacobian is $\det(\partial\underline{\phi}/\partial\underline{z}) = -b$. In general, for $b > 0$ the map is dissipative, and indeed the map possesses an attractor. Whether it be regular or "strange" depends in a nontrivial fashion on the parameters of the map.

Our interest in such quadratic maps concerns one of some importance to accelerator physics. We shall see in later chapters that the dynamics obtained by inserting a (thin) sextupole magnet into the storage ring can be modeled by a quadratic Poincaré map of the form

$$\underline{\phi} : \begin{pmatrix} z_1 \\ z_2 \end{pmatrix} \longmapsto \begin{pmatrix} \cos 2\pi\nu & \sin 2\pi\nu \\ -\sin 2\pi\nu & \cos 2\pi\nu \end{pmatrix} \begin{pmatrix} z_1 \\ z_2 - \lambda z_1^2 \end{pmatrix}. \tag{2.52}$$

The sextupole produces the quadratic kick, $-\lambda z_1^2$, and the rest of the storage ring provides the rotation matrix. λ is actually an inessential parameter; it can be transformed away via the rescaling $\underline{z} \longrightarrow \lambda\underline{z}$. We include it here only for future reference. The

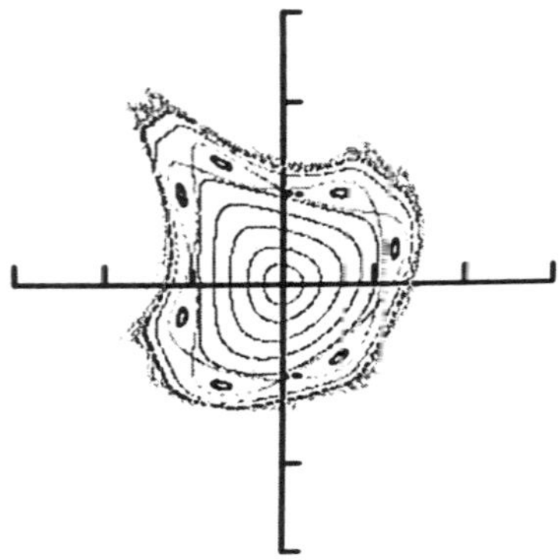

FIGURE 2.13 Orbits of an area-preserving Hénon map, which models the effect of inserting a thin sextupole into a storage ring.

topological structure of orbits depends only on the "tune" parameter, ν. A few representative samples are shown in Figure 2.13. The structure that appears has become almost a paradigm for understanding the behavior of maps in accelerator physics and, more generally, a large family of area-preserving (or symplectic) maps. Near the (fixed point) origin the orbits lie on tori — circles in two dimensions — much like those of a harmonic oscillator, except that, like the pendulum, the frequency of winding around the torus is not the same for all orbits: it depends on amplitude, that is, on the particular torus that the orbit is on. At these small amplitudes, this is the principal effect of the quadratic term. At larger amplitudes, tori with resonant frequencies break, and one sees the formation of separatrices, the stable and unstable manifolds of hyperbolic resonant orbits. Generically, these separatrices are chaotic, although that is not visually apparent on this scale for the ones at smaller amplitude. At larger amplitudes, however, their chaotic behavior is more readily discerned. Eventually one reaches an outer stability boundary, sometimes called the "dynamic aperture"[18] or "chaotic sea," which, while not sharply defined, borders the phase space region occupied by orbits that diverge quickly, due to the quadratic kick. Remarkably, even within the chaotic sea there can exist islands of stability, isolated outposts of serenity surrounded by turbulent chaos. This picture is not universal, of course: notice the behavior near $\nu = 1/3$ in Figure 5.3, in which the transition to the chaotic sea occurs very suddenly.

We shall return to study the physics behind these orbit structures in Chapters 3 and 4.

2.3 DYNAMICS

Having been exposed to a number of simple examples, we are ready to develop the subject more abstractly. This is done in preparation for applying the concepts

[18]This concept and especially the use of this term are still controversial, however.

and methods to more complicated, and more realistic, problems in later chapters. We shall go through this quickly: as in Chapter 1, our principal goal is conceptual understanding, not rigor.

2.3.1 Phase Space

We have confronted already the fact that one cannot speak of arithmetic on orbits in a geometric sense without introducing the concept of tangent spaces and tangent bundles. We need to explore a little more carefully the ways that the components of tangent vectors transform when transitions are made between different charts of an atlas, that is, when one changes coordinates.

Transformations We want to develop a little more the ideas that a vector field is a geometric object, and that equations of motion are merely its component representations on a particular chart (coordinate system). Changing charts transforms these components, much as rotating the basis of a vector space alters the representations of vectors. The most trivial change that can be made is the orbit parameter itself, say, $t \to t'$. The transformation of the vector field is obtained by the chain rule.

$$\underline{v}' = \frac{d\underline{z}}{dt'} = \left(\frac{dt}{dt'}\right)\frac{d\underline{z}}{dt} = \left(\frac{dt}{dt'}\right)\underline{v}. \tag{2.53}$$

A constant rescaling, for example, would be accomplished by multiplying the vector field by the constant. An instance of this is "time" reversal: $t \to -t$, $\underline{v} \to -\underline{v}$. It is not necessary that t' be a linear function of t, but it is necessary that it be monotonic. It would not do to have a point for which dt/dt' either vanishes or becomes infinite.

More generally, consider a change of coordinates, $\underline{z} \mapsto \underline{z}'$, on phase space, or possibly extended phase space. Requiring that the new components be interpretable as transformed derivatives and applying the chain rule to this situation produce the new components,

$$\underline{v}' = \dot{\underline{z}}' = (\partial\underline{z}'/\partial\underline{z}) \cdot \dot{\underline{z}} = (\partial\underline{z}'/\partial\underline{z}) \cdot \underline{v}. \tag{2.54}$$

But this is precisely equivalent to identifying them as belonging to a *geometric object,* a "real" vector of the type we all learned about in high school. This can be seen by observing that the "natural" basis vectors connected with coordinates $\underline{z}$ transform like the operator $\partial/\partial\underline{z}$. That is, the coordinate transformation induces a transformation of the natural basis vectors, $\{\vec{e}_k \mid k = 1 \ldots N\} \mapsto \{\vec{e}'_k \mid k = 1 \ldots N\}$, according to,

$$\vec{e}'_k \leftrightarrow \partial/\partial z'_k = \sum_j (\partial z_j/\partial z'_k)\partial/\partial z_j \leftrightarrow \sum_j (\partial z_j/\partial z'_k)\vec{e}_j. \tag{2.55}$$

This rule is, in fact, nothing more than what we all learned as sophomores under another guise — something that probably was not stressed as much as it should have been. As an example, consider the "egg carton transformation" illustrated in

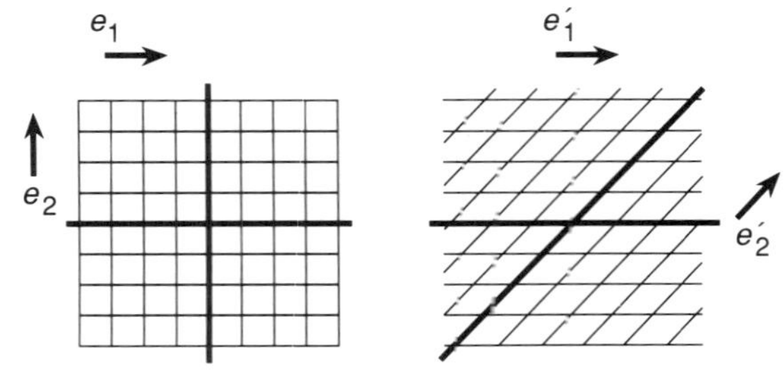

(a) (b)

FIGURE 2.14 Egg carton transformation. Changing coordinates induces a change in basis vectors.

Figure 2.14. The equations describing this transformation are

$$\begin{aligned} z_1 &\equiv z'_1 + \cos\theta\, z'_2, \\ z_2 &\equiv \sin\theta\, z'_2. \end{aligned}$$

From the Jacobian matrix,

$$\partial \underline{z}/\partial \underline{z}' = \begin{pmatrix} 1 & \cos\theta \\ 0 & \sin\theta \end{pmatrix},$$

and Eq. (2.55) we get the correspondence

$$\begin{aligned} \vec{e}'_1 &= \vec{e}_1, \\ \vec{e}'_2 &= \cos\theta \vec{e}_1 + \sin\theta \vec{e}_2. \end{aligned}$$

A more familiar transformation arises in the change from Cartesian to spherical coordinates. Reverting to the most common notation, we have

$$\begin{aligned} x &= r\sin\theta\cos\varphi, \\ y &= r\sin\theta\sin\varphi, \\ z &= r\cos\theta, \end{aligned}$$

from which we get

$$\frac{\partial(x,y,z)}{\partial(r,\theta,\varphi)} = \begin{pmatrix} x/r & r\cos\theta\cos\varphi & -r\sin\theta\sin\varphi \\ y/r & r\cos\theta\sin\varphi & r\sin\theta\cos\varphi \\ z/r & -r\sin\theta & 0 \end{pmatrix}.$$

Combined with Eq. (2.56), this yields the usual result.

$$\vec{e}_r = \frac{1}{r}(x\hat{i} + y\hat{j} + z\hat{k}) = \hat{r},$$
$$\vec{e}_\theta = r(\cos\theta\cos\varphi\hat{i} + \cos\theta\sin\varphi\hat{j} - \sin\theta\hat{k}) = r\hat{\theta},$$
$$\vec{e}_\varphi = r(-\sin\theta\sin\varphi\hat{i} + \sin\theta\cos\varphi\hat{j}) = r\hat{\varphi}.$$

Had we used $(r, r\theta, r\varphi)$ as the new coordinates, $\vec{e}_{r\theta}$ and $\vec{e}_{r\varphi}$ would have been dimensionless as well.

The *components* of a vector transform contravariantly to the basis vectors. If we identify $\vec{v} = \vec{v}'$ as a geometric object, then we immediately have the following:

$$\vec{v} = \sum_j v_j\vec{e}_j = \sum_k v'_k\vec{e}'_k = \sum_{j,k} v'_k(\partial z_j/\partial z'_k)\vec{e}_j,$$
$$v_j = \sum_k(\partial z_j/\partial z'_k)v'_k, \quad \text{or in matrix form,}$$
$$\underline{v} = (\partial\underline{z}/\partial\underline{z}')\cdot\underline{v}'.$$

Comparing this with Eq. (2.54) demonstrates that these geometric transformation laws are indeed the same as those required to assure that vector fields be interpreted as time derivatives, and therefore that dynamical systems are, in fact, geometric objects. The two expressions,

$$\vec{v} = \sum_k v_k\vec{e}_k \Longleftrightarrow \mathbf{V} = \underline{v}\cdot\underline{\partial}, \tag{2.56}$$

can be thought of as two ways of writing the same thing.

There is no need to restrict Eq. (2.56) or Eq. (2.54) to real coordinates; they would work just as well with complex coordinates, such as those introduced in Eq. (2.8). Consider, for example, the harmonic oscillator vector field from Eq. (2.2), with $\Gamma = 0$:

$$\underline{v}(\underline{z}) = \begin{pmatrix} p/m \\ -m\omega^2 x \end{pmatrix}, \quad \mathbf{V} = p/m\frac{\partial}{\partial x} - m\omega^2 x\frac{\partial}{\partial p}.$$

Transforming to the complex coordinates of Eq. (2.8), we use Eq. (2.54) directly.

$$\begin{aligned}\underline{v}' = (\partial\underline{z}'/\partial\underline{z})\cdot\underline{v} &= (\partial(a, a^*)/\partial(x, p))\cdot\underline{v} \\ &= \begin{pmatrix} \sqrt{m\omega/2} & i/\sqrt{2m\omega} \\ \sqrt{m\omega/2} & -i/\sqrt{2m\omega} \end{pmatrix}\cdot\begin{pmatrix} p/m \\ -m\omega^2 x \end{pmatrix} \\ &= i\omega\begin{pmatrix} -a \\ a^* \end{pmatrix}.\end{aligned}$$

This provides the vector field identity,

$$\mathbf{V} = p/m\frac{\partial}{\partial x} - m\omega^2\frac{\partial}{\partial p} = -i\omega a\frac{\partial}{\partial a} + i\omega a^*\frac{\partial}{\partial a^*}. \tag{2.57}$$

2.3.2 Vector Fields and Brackets

We now shall interpret Eq. (2.56) not as a symbolic device but as the definition of a geometric operator acting on the space of observables (real-valued, smooth (?) functions over phase space). Properties of this operator will introduce us to Lie algebras essential for understanding the perturbation theories to be discussed in Chapter 4.

Lie Derivative Given any observable, g, and local representation, $\underline{v}$, of the vector field, we define the linear operator, $\mathbf{V}$, as (see Eq. (1.9) and above)

$$\mathbf{V}g = h, \quad \text{where} \quad h(\underline{z}) = \underline{v}(\underline{z}) \cdot \frac{\partial g(\underline{z})}{\partial \underline{z}}. \tag{2.58}$$

$\mathbf{V}g$ is called the "Lie derivative" of g. Notice that (a) $h = \mathbf{V}g$ is again an observable, at least locally, so that $\mathbf{V}$ indeed acts like a linear operator on the space of observables (treated as a vector space), and (b) by Eq. (2.56), $\mathbf{V}$ is a geometric object: it is chart independent, as a vector should be, since

$$\underline{v}(\underline{z}) \cdot \frac{\partial}{\partial \underline{z}} = \underline{v}'(\underline{z}') \cdot \frac{\partial}{\partial \underline{z}'}.$$

The Lie derivative has a natural interpretation as either a directional derivative or a "time" derivative, through the application of the chain rule.

$$\dot{g} = \dot{\underline{z}} \cdot \frac{\partial g}{\partial \underline{z}} = \underline{v} \cdot \underline{\partial} g = \mathbf{V}g. \tag{2.59}$$

This equation provides the simplest argument that $\mathbf{V}$ *must* be a geometric, coordinate-independent entity, for $\dot{g}$ cannot possibly depend on the coordinates used to evaluate it. In fact, Eq. (2.59) is the starting point for more exact, formal treatments of the theory of tangent spaces and tangent bundles. From it is *defined* what one literally means by the tangent space attached to a manifold at a point.

Equation (2.59) works for nonautonomous systems as well if we interpret $\underline{z}$ as belonging to extended phase space, say, $\underline{z} = (\underline{u}, t)$. If we let τ be the orbit parameter, then

$$(\mathbf{V}g)(\underline{z}) = \dot{\underline{z}} \cdot \frac{\partial g}{\partial \underline{z}} = \dot{\underline{u}} \cdot \frac{\partial g}{\partial \underline{u}} + \frac{dt}{d\tau}\frac{\partial g}{\partial t}.$$

Identifying the dynamical variable, t, with the orbit parameter, τ, via $dt/d\tau = 1$, recovers the usual formula for the "total," or sometimes "convective," derivative of the function. Especially worth noting are the coordinates themselves, treated as observables. Let $\underline{I} : \underline{z} \longmapsto \underline{z}$ be the identity function, with components, $I_k(\underline{z}) = z_k$.[19] We set $g = I_k$ above to obtain

$$\dot{I}_k(\underline{z}) = (\mathbf{V} I_k)_{(\underline{z})} = (\underline{v} \cdot \underline{\partial} I_k)_{(\underline{z})} = v_k(\underline{z}), \tag{2.60}$$

as expected. This is a way to express the equations of motion "covariantly."

A little care must be taken in interpreting this time derivative. One naively thinks of it as the derivative as viewed by an observer traveling with the system. It is better to think of it in terms of a hypothetical omniscient, outside observer logging the orbit history. At each value of the orbit parameter, τ, this observer records the value of g at the system point, thereby obtaining the graph $\{(\tau, g(\underline{z}(\tau))) \mid \tau \in R\} \subset R^2$. It is the derived function of that graph that is meant by $\dot{g}$. More properly, it is the total collection of all such derivatives, one for each possible orbit of the vector field. $\dot{g}$ is itself an observable that can be evaluated at any point on phase space, its value being associated with the one and only orbit passing through that point.

If $\dot{g} = 0$ everywhere on a particular orbit, then g is called an "orbit invariant"; if $\dot{g}$ vanishes everywhere on phase space, then g is a "dynamical invariant," or "constant of the motion." Notice that a constant of the motion need not be constant: it can depend explicitly both on phase space and on the orbit parameter.

Exponential Maps If $\mathbf{V}$ is autonomous — or we are working in extended phase space, so that it possesses no explicit dependence on the orbit parameter (time) — and g is itself a time-independent observable, then we can view Eq. (2.59) as an operator equation on the linear vector space of observables with an immediate formal solution.

$$g_t = e^{t\mathbf{V}} g_0. \tag{2.61}$$

We write g_t with a subscript to emphasize that this represents a one-parameter family of observables, $\{g_t \mid t \in R\}$.

This is nothing more than an extension of Taylor's formula to orbits. For example, consider that the one-dimensional Taylor's formula states that

$$g(z + s) = \sum_{k=0}^{\infty} \frac{g^{[n]}(z)}{n!} s^n = \left(e^{s d/dz} g\right)_{(z)},$$

and that the multidimensional version of this is

$$g(\underline{z} + \underline{s}) = \left(e^{\underline{s} \cdot \underline{\partial}} g\right)_{(\underline{z})}.$$

[19]Properly speaking, each component of $\underline{I}$ is a projector onto a component of $\underline{z}$.

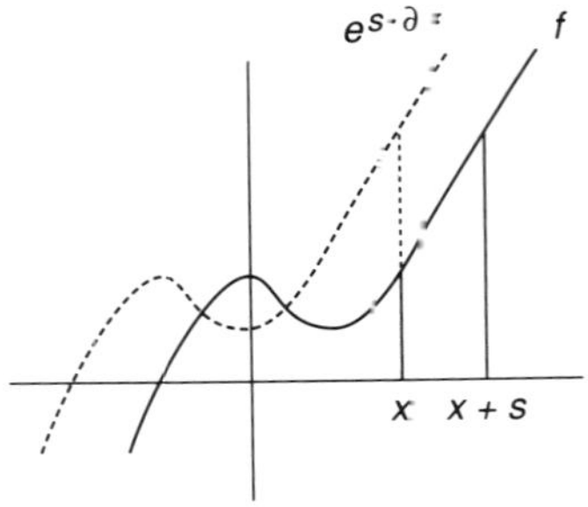

$f(x+s) = (e^{s\cdot\partial} f)_{(x)}$

FIGURE 2.15 Schrödinger and Heisenberg pictures of classical mechanics.

These are precisely the correspondences we expect by applying Eq. (2.61) in the case of a constant vector field, for then $\underline{s} = t\underline{v}$, and indeed, identifying $g_0 = g$,

$$\begin{aligned} g_t(\underline{z}) &= g(\underline{z} + t\underline{v}) \\ &= \left(e^{t\underline{v}\cdot\underline{\partial}} g\right)_{(\underline{z})} \\ &= \left(e^{t\mathbf{V}} g_0\right)_{(\underline{z})} . \end{aligned}$$

The situation is illustrated in Figure 2.15. Whether one interprets the vector field as moving the system point in phase space or as performing transformations among the observables is a matter of preference: the results are the same. This is precisely the sort of duality we have in quantum theory between the Schrödinger and Heisenberg "pictures," depending on whether we think of state vectors or operators as being the dynamical objects.

For nonconstant, autonomous vector fields, the generalized Taylor's expansion *along an orbit* is, according to Eq. (2.61),

$$g(\underline{z}(t)) = g_t(\underline{z}(0)) = \left(e^{t\mathbf{V}} g\right)_{(\underline{z}(0))} . \tag{2.62}$$

In particular, this holds for the observables that are the coordinate functions themselves. Let $\underline{I} : \underline{z} \longmapsto \underline{z}$ be the identity function, with components $I_k(\underline{z}) = z_k$.[20] We understand that operators act component-wise: for example, $(e^{\mathbf{W}}\underline{I})_k = e^{\mathbf{W}} I_k$. Then we can write

$$\underline{z}(t) = \underline{I}(\underline{z}(t)) = \underline{I}_t(\underline{z}(0)) = \left(e^{t\mathbf{V}} \underline{I}\right)_{(\underline{z}(0))} . \tag{2.63}$$

This provides us with an identification of the flow map, $\phi_t \leftrightarrow \exp(t\mathbf{V})\underline{I}$; this little identity is the formal construction that expresses the orbit as a *mapping* with the

[20] See previous footnote.

"initial conditions" as its argument. It enables us to rewrite Eq. (2.62) as follows:

$$\left(e^{t\mathbf{V}} g\right)_{(\underline{z}(0))} = g(\underline{z}(t)) = g\left(\left(e^{t\mathbf{V}} \underline{I}\right)_{(\underline{z}(0))}\right).$$

And now, since the argument is the same on either side, we express this formally in terms of composition of functions,

$$e^{t\mathbf{V}} g = g \circ \left(e^{t\mathbf{V}} \underline{I}\right). \tag{2.64}$$

Although this appears a bit formal it will turn out to be useful later. (In addition, it is good exercise to start thinking in terms of the classical "Heisenberg picture.")

Repeating Eq. (2.64) leads to a surprising result.

$$\begin{aligned} e^{t\mathbf{V}} e^{t\mathbf{W}} g &= e^{t\mathbf{V}} \left(g \circ \left(e^{t\mathbf{W}} \underline{I}\right)\right) \\ &= \left(g \circ \left(e^{t\mathbf{W}} \underline{I}\right)\right) \circ \left(e^{t\mathbf{V}} \underline{I}\right) \\ &= g \circ \left(e^{t\mathbf{W}} \underline{I}\right) \circ \left(e^{t\mathbf{V}} \underline{I}\right). \end{aligned} \tag{2.65}$$

Composition must be applied in *reversed order* to that of the exponential operators. (Compare this with Eq. (1.8).)

An example or two will serve to make these ideas more palatable. Suppose that we take the coordinates, $\underline{z}$, as the observables to be plugged into Eq. (2.61). We shall consider two simple examples on $\mathcal{P} = R^2$, where

$$\underline{z} = \begin{pmatrix} x \\ p \end{pmatrix}.$$

Consider the dynamic expressing a constant force field,

$$\underline{v}(\underline{z}) = \begin{pmatrix} p/m \\ -ma \end{pmatrix}.$$

We shall take x as our observable and apply $\mathbf{V} = \underline{v} \cdot \underline{\partial}$ to it.

$$\mathbf{V}x = \left(p/m\frac{\partial}{\partial x} - ma\frac{\partial}{\partial p}\right)x = p/m,$$

$$\mathbf{V}^2 x = \mathbf{V}(p/m) = -a,$$

and, of course, for all $n > 2$ we have $\mathbf{V}^n x = 0$. Evaluating the exponential,

$$\begin{aligned} x_t = \exp(t\mathbf{V})x_0 &= (\mathbf{1} + t\mathbf{V} + \tfrac{1}{2}t^2\mathbf{V}^2)x_0 \\ &= x_0 + (p_0/m)t - \tfrac{1}{2}at^2. \end{aligned}$$

For the harmonic oscillator we have, from Eq. (2.2),

$$\underline{v}(\underline{z}) = \begin{pmatrix} 0 & 1/m \\ -m\omega^2 & 0 \end{pmatrix} \underline{z} = \begin{pmatrix} p/m \\ -m\omega^2 x \end{pmatrix}.$$

Again we apply $\mathbf{V}$ to the observable x.

$$\mathbf{V}x = \left(p/m \frac{\partial}{\partial x} - m\omega^2 x \frac{\partial}{\partial p} \right) x = p/m,$$

$$\mathbf{V}^2 x = \mathbf{V}(p/m) = -\omega^2 x.$$

By iterating these two equations, we find

$$(t\mathbf{V})^{2n} x = (-1)^n (\omega t)^{2n} x,$$

$$(t\mathbf{V})^{2n+1} x = (-1)^n (\omega t)^{2n+1} p/m\omega.$$

We can now write the full solution.

$$\begin{aligned} x_t &= \exp(t\mathbf{V}) x_0 = \sum_{n=0}^{\infty} \frac{1}{n!} (t\mathbf{V})^n x_0 \\ &= \sum_{n=0}^{\infty} \frac{1}{2n!} (t\mathbf{V})^{2n} x_0 + \sum_{n=0}^{\infty} \frac{1}{(2n+1)!} (t\mathbf{V})^{2n+1} x_0 \\ &= \left(\sum_{n=0}^{\infty} (-1)^n \frac{(\omega t)^{2n}}{2n!} \right) x_0 + \left(\sum_{n=0}^{\infty} (-1)^n \frac{(\omega t)^{2n+1}}{(2n+1)!} \right) p_0/m\omega \\ x_t &= x_0 \cos \omega t + \frac{p_0}{m\omega} \sin \omega t. \end{aligned} \tag{2.66}$$

This result is in complete agreement with and appears formally identical to Eq. (2.6). However, the interpretation is very different. In Eq. (2.6), the symbols $x(0)$ and $p(0)$ represent real numbers; in Eq. (2.66) they represent observables, real-valued functions over $\mathcal{P}$. (Again, the difference between these two points of view is analogous to the difference between the Schrödinger and Heisenberg pictures in quantum mechanics.) This explains how we can start from $\exp(t\mathbf{V})x(0)$ and end with an answer that contains both $x(0)$ and $p(0)$.

If $\mathbf{V}$ is *not* autonomous, we cannot write Eq. (2.61) as given, because we cannot assume that $[\mathbf{V}(t_1), \mathbf{V}(t_2)] = 0$ for all t_1 and t_2. Again in analogy to quantum theory (or vice versa) we may employ a formal construction called a "time-ordered product." Like Riemann integration, one partitions the time interval into small subintervals and approximates $\mathbf{V}(t) \approx$ constant within each subinterval. Invoking the Chapman-Kolmogorov equation, Eq. (1.4) or Eq. (1.5), the time-evolution map is written as a catenation of exponential maps of the form Eq. (2.61), ordered from earlier to later times. In the limit of infinitesimally small subintervals the resultant map should approach the system's exact time-evolution map.

Commutators, Brackets, and Lie Algebras Having gone so far as to associate a vector field with an operator, it is natural to examine the commutator of two such operators. Interestingly enough, it also is a vector field. Given two vector fields, **U** and **V**, we proceed to evaluate the action of their commutator on an observable, g.

$$\begin{aligned}[\mathbf{U},\mathbf{V}]g = \mathbf{U}(\mathbf{V}g) - \mathbf{V}(\mathbf{U}g) &= \underline{u}\cdot\underline{\partial}(\underline{v}\cdot\underline{\partial}g) - \underline{v}\cdot\underline{\partial}(\underline{u}\cdot\underline{\partial}g)\\ &= \underline{u}\cdot\underline{\underline{\partial v}}\cdot\underline{\partial}g - \underline{v}\cdot\underline{\underline{\partial u}}\cdot\underline{\partial}g\\ &\quad + \mathrm{Tr}\left[(\underline{uv} - \underline{vu})\cdot\underline{\underline{\partial\partial g}}\right]\\ &= (\underline{u}\cdot\underline{\underline{\partial v}} - \underline{v}\cdot\underline{\underline{\partial u}})\cdot\underline{\partial}g\end{aligned}$$

We can write this in a purely operator form:

$$\begin{aligned}[\mathbf{U},\mathbf{V}] &= [\underline{u}\cdot\underline{\partial}, \underline{v}\cdot\underline{\partial}] = (\underline{u}\cdot\underline{\underline{\partial v}} - \underline{v}\cdot\underline{\underline{\partial u}})\cdot\underline{\partial}\\ &\equiv \mathbf{W} \equiv \underline{w}\cdot\underline{\partial}.\end{aligned} \tag{2.67}$$

W is simply another vector field whose representative components are obtainable from those of **U** and **V**. We overload the bracket notation of the commutator by extending it to these local components.

$$\underline{w} \equiv [\underline{u}, \underline{v}] \quad \textbf{means} \quad \underline{w} = \underline{u}\cdot\underline{\underline{\partial v}} - \underline{v}\cdot\underline{\underline{\partial u}}. \tag{2.68}$$

We shall call this the "Lie bracket" of $\underline{u}$ with $\underline{v}$. In a one-dimensional phase space it is identical to the Wronskian between two functions, u and v; in this sense, it is a multidimensional extension of the Wronskian to the world of vector fields.

It is not difficult to show that the Lie bracket is a binary operation on the space of vector fields: it is required only to show that the right-hand side of Eq. (2.67) is again a geometric object. Its key algebraic properties are:

Bilinearity: $[\underline{u}, \underline{v} + c\underline{w}] = [\underline{u}, \underline{v}] + c[\underline{u}, \underline{w}]$ and

$[\underline{u} + c\underline{v}, \underline{w}] = [\underline{u}, \underline{w}] + c[\underline{v}, \underline{w}]$.

Antisymmetry: $[\underline{u}, \underline{v}] = -[\underline{v}, \underline{u}]$.

Jacobi identity: $[\underline{u}, [\underline{v}, \underline{w}]] + [\underline{v}, [\underline{w}, \underline{u}]] + [\underline{w}, [\underline{u}, \underline{v}]] = \underline{0}$.

Any binary operation on a vector space that possesses these properties belongs, by definition, to a "Lie algebra." The most obvious example of this is the commutator algebra defined on the linear operators acting on any vector space; another is the cross product defined on the space of three-dimensional vectors. That the Lie bracket is bilinear and antisymmetric is obvious. The easiest way of establishing that it obeys the Jacobi identity is by way of inheritance from the commutator of vector fields, using

$$[\mathbf{U}, [\mathbf{V}, \mathbf{W}]] = ([\underline{u}, [\underline{v}, \underline{w}]])\cdot\underline{\partial},$$

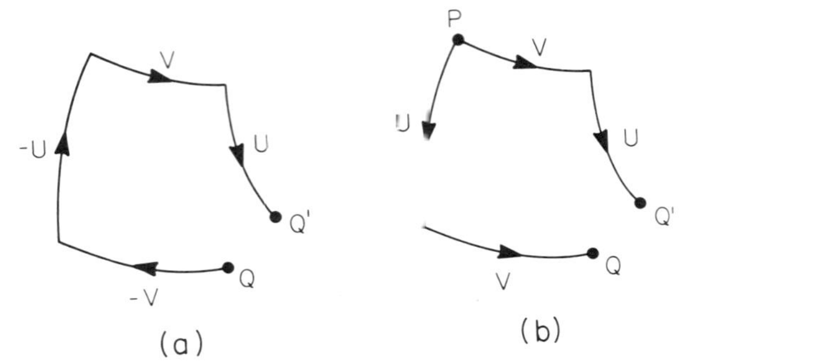

FIGURE 2.16 The orbit defect associated in following a small loop generated by two vector fields.

and the fact that $\underline{a} \cdot \underline{\partial} = 0$ implies that $\underline{a} = \underline{0}$. Indeed, it makes sense to overload the bracket notation only because it, like the commutators, *is* a Lie algebra. Equation (2.67) then becomes a statement of morphism between two algebraic structures.

Loop Defects The commutator of two vector fields has a simple geometric interpretation, which is illustrated in Figure 2.16. Given two vector fields, **U** and **V**, imagine that we could switch them on and off as desired. Consider following a path through phase space, as in Figure 2.16(a), traced out in four steps: (a) first by following an orbit of $-\mathbf{V}$ for a *time* τ,[21] (b) subsequently following the orbit of $-\mathbf{U}$ for the same *time* τ, (c) then the orbit of $\mathbf{V}$, and (d) $\mathbf{U}$ through the same *time* interval. An orbit starting at point Q, say, could return to a point $Q' \neq Q$. We want to evaluate the defect to lowest order.

The calculation is done more easily by considering two paths, beginning from point P, as shown in Figure 2.16(b). Following the figure, we define the orbit segments, 1, 2, 1′, and 2′, as follows:

$$
\begin{aligned}
\dot{\underline{z}}_1 &= \underline{u}(\underline{z}_1), & \underline{z}_1(0) &= \underline{P}; \\
\dot{\underline{z}}_2 &= \underline{v}(\underline{z}_2), & \underline{z}_2(\tau) &= \underline{z}_1(\tau); \\
\dot{\underline{z}}_{1'} &= \underline{v}(\underline{z}_{1'}), & \underline{z}_{1'}(0) &= \underline{P}; \\
\dot{\underline{z}}_{2'} &= \underline{u}(\underline{z}_{2'}), & \underline{z}_{2'}(\tau) &= \underline{z}_{1'}(\tau).
\end{aligned}
$$

Our goal is evaluation of the difference[22] $\underline{Q} - \underline{Q}'$, that is, $\underline{z}_2(2\tau) - \underline{z}_{2'}(2\tau)$. First note that (a) the chain rule, or the Lie derivative, provides: if $\dot{\underline{z}} = \underline{u}(\underline{z})$ then $\ddot{\underline{z}} = \underline{u}(\underline{z}) \cdot \underline{\underline{\partial u}}(\underline{z})$; and (b) we shall employ the notation $\underline{v}(\underline{z}) \cdot \underline{\underline{\partial u}}(\underline{z}) \equiv (\underline{v} \cdot \underline{\underline{\partial u}})_{(\underline{z})}$.

With these observations, we begin by writing the orbit segments in Figure 2.16(b) explicitly to second order in τ.

$$
\begin{aligned}
\underline{z}_1(\tau) &= \underline{P} + \tau \underline{u}(\underline{P}) + \tfrac{1}{2}\tau^2 (\underline{u} \cdot \underline{\underline{\partial u}})_{(\underline{P})} + O(\tau^3), \\
\underline{z}_2(2\tau) &= \underline{z}_1(\tau) + \tau \underline{v}(\underline{z}_1(\tau)) + \tfrac{1}{2}\tau^2 (\underline{v} \cdot \underline{\underline{\partial v}})_{(\underline{z}_1(\tau))} + O(\tau^3).
\end{aligned}
$$

[21] Equivalently, one can think of following an orbit of **V** for a *time* $-\tau$, that is, going backward into its previous history.

[22] Of course, to do this arithmetic we work within a representation, that is, we are working within an open subset of some R^N.

We must identify more carefully the first and second order terms in the latter equation.

$$\begin{aligned}
\underline{v}(\underline{z}_1(\tau)) &= \underline{v}(\underline{P} + \tau\underline{u}(\underline{P}) + O(\tau^2)) \\
&= \underline{v}(\underline{P}) + \tau(\underline{u} \cdot \underline{\underline{\partial v}})_{(\underline{P})} + O(\tau^2), \\
(\underline{v} \cdot \underline{\underline{\partial v}})(\underline{z}_1(\tau)) &= (\underline{v} \cdot \underline{\underline{\partial v}})_{(\underline{P})} + O(\tau).
\end{aligned}$$

Putting the pieces together, we obtain

$$\begin{aligned}
\underline{z}_2(2\tau) &= \underline{P} + \tau\underline{u}(\underline{P}) + \tfrac{1}{2}\tau^2(\underline{u} \cdot \underline{\underline{\partial u}})_{(\underline{P})} \\
&\quad + \tau\underline{v}(\underline{P}) + \tau^2(\underline{u} \cdot \underline{\underline{\partial v}})_{(\underline{P})} \\
&\quad + \tfrac{1}{2}\tau^2(\underline{v} \cdot \underline{\underline{\partial v}})_{(\underline{P})} + O(\tau^3) \\
&= \underline{P} + \tau(\underline{u}(\underline{P}) + \underline{v}(\underline{P})) \\
&\quad + \tau^2(\tfrac{1}{2}(\underline{u} \cdot \underline{\underline{\partial u}})_{(\underline{P})} + \tfrac{1}{2}(\underline{v} \cdot \underline{\underline{\partial v}})_{(\underline{P})} + (\underline{u} \cdot \underline{\underline{\partial v}})_{(\underline{P})}) + O(\tau^3).
\end{aligned}$$

We can obtain the corresponding calculation for $\underline{z}_{2'}(2\tau)$ by simply interchanging $\underline{u}$ and $\underline{v}$ in the above. When we take the difference, all terms that are symmetric under this interchange are canceled, which leaves

$$\underline{Q} - \underline{Q}' = \underline{z}_2(2\tau) - \underline{z}_{2'}(2\tau) = \tau^2((\underline{u} \cdot \underline{\underline{\partial v}})_{(\underline{P})} - (\underline{v} \cdot \underline{\underline{\partial u}})_{(\underline{P})}) + O(\tau^3).$$

This provides our geometric interpretation of the Lie bracket as the leading coefficient in the loop defect.

$$[\underline{u}, \underline{v}] = \lim_{\tau \to 0} \frac{1}{\tau^2}(\underline{Q} - \underline{Q}')$$

Having done this calculation the hard way, we shall demonstrate a more elegant approach for autonomous systems. According to Eq. (2.63) the evolution of coordinates as we traverse the loop in Figure 2.16(a) is governed by the exponential map. Keeping in mind that coordinates are observables and therefore, according to Eq. (2.65), that we must evaluate the operations in reverse order, we can write

$$\begin{aligned}
\underline{Q} &= e^{\tau\mathbf{U}}e^{\tau\mathbf{V}}\underline{P} \\
&= (1 + \tau\mathbf{U} + \tfrac{1}{2}\tau^2\mathbf{U}^2)(1 + \tau\mathbf{V} + \tfrac{1}{2}\tau^2\mathbf{V}^2)\underline{P} \\
\underline{Q}' &= e^{\tau\mathbf{V}}e^{\tau\mathbf{U}}\underline{P} \\
&= (1 + \tau\mathbf{V} + \tfrac{1}{2}\tau^2\mathbf{V}^2)(1 + \tau\mathbf{U} + \tfrac{1}{2}\tau^2\mathbf{U}^2)\underline{P}
\end{aligned}$$

Expanding the operator products and collecting terms,

$$\underline{Q} - \underline{Q}' = \tau^2[\mathbf{U}, \mathbf{V}]\underline{P} + O(\tau^3),$$

which is equivalent to the previous result.

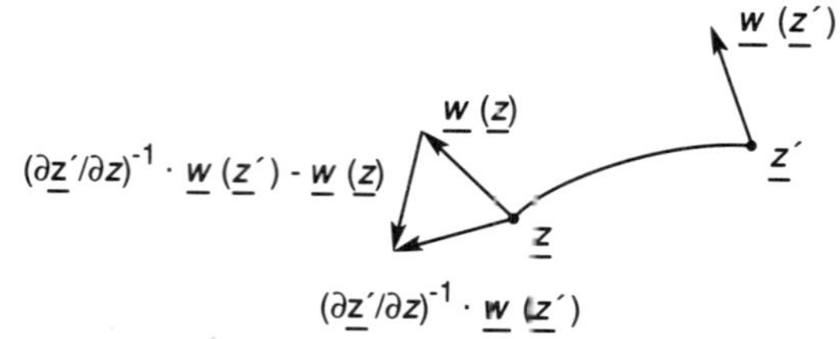

FIGURE 2.17 Illustrating the Lie derivative of a vector field.

Lie Derivative The Lie bracket has another geometric interpretation: it is the Lie derivative of a vector field. We have already introduced the Lie derivative of a scalar observable, g, in Eq. (2.58) and Eq. (2.59) and connected it with rate of change of g as calculated by an observer logging an orbit of the dynamical system. Expressed as a limit,

$$(\mathbf{V}g)_{\underline{z}(t)} = \lim_{\tau \to 0}(g(\underline{z}(t+\tau)) - g(\underline{z}(t))).$$

In extending this to vector fields, we must be careful to account for the different sets of basis vectors used to evaluate components at different locations in phase space. Let an orbit evolve according to a dynamic, $\mathbf{V} = \underline{v} \cdot \underline{\partial}$, and let $\underline{z} \longmapsto \underline{z}'$ under the time-evolution map, $\phi_{t+\tau,t}$. In logging the variation in the "observed" value of a second vector field, $\mathbf{W} = \underline{w} \cdot \underline{\partial}$, it is not correct to merely take the difference,

$$\underline{w}(\underline{z}') - \underline{w}(\underline{z}).$$

Rather, we must take the change in basis vectors into account and use (see Figure 2.17 and Eq. (2.54))

$$(\partial \underline{z}'/\partial \underline{z})^{-1} \cdot \underline{w}(\underline{z}') - \underline{w}(\underline{z}).$$

The "Lie derivative of $\underline{w}$ along $\underline{v}$," $L_{\underline{v}}\underline{w}$, is then defined by taking the limit.

$$(L_{\underline{v}}\underline{w})_{\underline{z}} \equiv \lim_{\tau \to 0} \frac{1}{\tau}\{(\partial \underline{z}'/\partial \underline{z})^{-1} \cdot \underline{w}(\underline{z}') - \underline{w}(\underline{z})\}. \tag{2.69}$$

The calculation proceeds by expanding everything to first order in τ.

$$\begin{aligned} \underline{z}' &= \underline{z} + \tau\underline{v}(\underline{z}) + O(\tau^2), \\ (\partial \underline{z}'/\partial \underline{z})^{-1} &= \underline{\mathbf{1}} - \tau \partial \underline{v}/\partial \underline{z} + O(\tau^2), \\ \underline{w}(\underline{z}') &= \underline{w}(\underline{z} + \tau\underline{v}(\underline{z})) + O(\tau^2) \\ &= \underline{w}(\underline{z}) + \tau\underline{v}(\underline{z}) \cdot \underline{\underline{\partial w}} + O(\tau^2). \end{aligned}$$

Substituting these pieces back into Eq. (2.69) and expanding provide the following result:

$$(L_{\underline{v}} \underline{w}) = \underline{v} \cdot \underline{\underline{\partial w}} - \underline{w} \cdot \underline{\underline{\partial v}} = [\underline{v}, \underline{w}]. \tag{2.70}$$

The Lie bracket is equivalent to the Lie derivative extended to the components of a vector field.

2.3.3 Three Theorems

By and large, the emphasis in this text is on calculations, not theorems and proofs. Most of those that do appear, such as the linear ergodic theorem, have been included because of their (subjectively perceived) importance to understanding the behavior of orbits. In this section we shall discuss a few others of which the reader should be aware, even though we shall make no further direct use of them.

We are not going to present these theorems with the precision they deserve; students with a greater mathematical background may be offended by such things as our blurring the distinctions between homeomorphisms and C^1 diffeomorphisms under the umbrella category of "charts" or between a phase space and its local chart. These and other instances of sloppiness can be at most partially justified only on the grounds that, as in Chapter 1, our objective is to provide a quick, broad-brushed rendering[23] without becoming enmeshed in details. Those who wish to explore details can find them laid out in many texts and papers.

Existence We have been blithely assuming that it makes sense to interpret arbitrary vector fields as derivatives of orbits. Given a vector field, $\underline{v}(\underline{z}, t)$, over an extended phase space, $(\underline{z}, t) \in \mathcal{P}^e$, are we assured there actually exists an orbit function, say, $\underline{\psi} : R \longrightarrow \mathcal{P},\ t \longmapsto \underline{\psi}(t) \equiv \underline{z}(t)$, which satisfies $\dot{\underline{z}}(t) = \underline{v}(\underline{z}(t), t)$? It turns out that a fruitful way of attacking this question is to consider its integral form: if $\underline{z}(t)$ exists and satisfies $\underline{z}(t_0) = \underline{z}_0$, then

$$\underline{z}(t) = \underline{z}_0 + \int_{t_0}^{t} d\tau\ \underline{v}(\underline{z}(\tau), \tau). \tag{2.71}$$

Think of this as a recursive equation that $\underline{z}(t)$ must satisfy. Starting from $\underline{z}_0(t) \equiv \underline{z}_0$, we use the right-hand side recursively to generate a sequence of "approximate" orbits, called "Picard approximants."

$$\underline{z}_{n+1}(t) = \underline{z}_0 + \int_{t_0}^{t} d\tau\ \underline{v}(\underline{z}_n(\tau), \tau). \tag{2.72}$$

[23]Think of it as the William Alexander approach to mathematics.

For example, consider the autonomous vector field $\underline{v}(\underline{z}, t) = \underline{z}$. Letting $t_0 = 0$, the sequence of Picard approximants are

$$\begin{aligned}
\underline{z}_1(t) &= \underline{z}_0 \cdot (1 + t), \\
\underline{z}_2(t) &= \underline{z}_0 \cdot (1 + t + \tfrac{1}{2}t^2), \\
&\vdots \\
\underline{z}_n(t) &= \underline{z}_0 \cdot \sum_{k=0}^{n} t^k/k!
\end{aligned}$$

In the limit, $\lim_{n\to\infty} \underline{z}_n(t) = \underline{z}_0 e^t$, which is indeed the orbit passing through $\underline{z}_0$ at $t = 0$.

By introducing a metric topology on the space of orbits it can be shown that the sequence of Picard approximants is a Cauchy sequence and therefore converges to a solution of Eq. (2.71) *provided that* $\underline{v}$ is at least continuously differentiable. The key idea, borrowed from functional analysis, is to establish that the recursion of Eq. (2.72) defines a contraction mapping on the space of orbits. Our statement of the existence theorem, which might also be termed the "fundamental theorem" of ordinary differential equations, is copied almost verbatim from Arnold [6].

Theorem 2.5 (Existence) Suppose that $\underline{v}(\underline{z}, t)$ is continuously differentiable in a neighborhood of a point $(\underline{z}_p, t_0)$ of extended phase space. Then, given any $\underline{z}_0$ "sufficiently close" to $\underline{z}_p$, there is a neighborhood of t_0 in which a solution of Equation (2.71) exists. Moreover, this solution depends continuously on $\underline{z}_0$, the "initial point."

There is great care taken in the wording of this theorem to emphasize its local nature. In general it may not be possible to extend an orbit globally; it need not exist for all $t \in R$. (See, for example, Problem 1.12.)

Interestingly enough, although we are frequently assured in physics that solutions to problems "exist," it is not guaranteed that they are everywhere computable functions of their initial conditions [78]. In such cases, the meaning of "existence" is problematic.

Hartman-Grobman An autonomous dynamical system possesses a fixed point at $\underline{z}_o$ if the vector field vanishes there: $\underline{v}(\underline{z}_c) = \underline{0}$. In such cases we naively study the flow's local behavior by expanding the field about the fixed point and retaining only the linear term.

$$\underline{v}(\underline{z}) + O((\underline{z} - \underline{z}_o)^2) = \underline{v}(\underline{z}_o) + (\underline{z} - \underline{z}_o) \cdot \underline{\underline{\partial v}}(\underline{z}_o) = \underline{\underline{\partial v}}^T(\underline{z}_o) \cdot (\underline{z} - \underline{z}_o).$$

The Hartman-Grobman theorem makes this activity legitimate *in certain cases* and removes it from the level of approximation to an exact equality by establishing the existence of a chart on which the dynamic looks linear. For convenience, we assume that the origin of this chart represents the fixed point.

Theorem 2.6 (Hartman-Grobman; paraphrased) If the autonomous vector field, **V**, vanishes at a *hyperbolic* fixed point, $\mathbf{p}_o$ of phase space, then there is a local chart covering $\mathbf{p}_o$ such that (a) $\underline{0}$ is the coordinate of $\mathbf{p}_o$, and (b) **V** is represented in a linear form:

$$\underline{v}(\underline{z}) = \underline{\underline{A}} \cdot \underline{z}, \quad \textbf{where} \quad \underline{\underline{A}} = (\underline{\underline{\partial v}})^T_{(0)}.$$

Recall that a hyperbolic fixed point is one for which no eigenvalue of $\underline{\underline{\partial v}}$ is on the imaginary axis. The Hartman-Grobman theorem possesses an extension to Poincaré or time-evolution maps in the vicinity of a periodic orbit. By iteration, we first construct the map, ϕ, for which the periodic orbit becomes a fixed point. The theorem asserts the existence of a chart on which the fixed point is at the origin and the map is linear,

$$\phi(\underline{z}) = \underline{\underline{A}} \cdot \underline{z}, \quad \textbf{where} \quad \underline{\underline{A}} = (\underline{\underline{\partial \phi}})^T_{(0)},$$

provided that the fixed point is hyperbolic; since we are now considering maps, this means that no eigenvalue of $\underline{\underline{A}}$ can lie on the unit circle.

The term "normal form" is frequently used to indicate particularly simple representations of a vector field. The linear representation of the Hartman-Grobman theorem may be called a "normal form" of the vector field, although the term is usually applied in a more restricted manner. Many perturbation theories are basically attempts to find the normal-form charts of the Hartman-Grobman theorem. We shall study some of these theories in a later chapter.

If the fixed point is neither hyperbolic nor elliptic — that is, if some eigenvalues, λ, of the flow (map) are on the imaginary axis (unit circle) and some are not — then, as was done earlier, one can separate the R^N phase space of the normal form, linearized flow into subspaces corresponding to the different types of eigenvalues: (a) the stable subspace, spanned by eigenvectors with eigenvalues satisfying $\text{Re}\,\lambda < 0$ ($|\lambda| < 1$); (b) the unstable subspace, spanned by eigenvectors with $\text{Re}\,\lambda > 0$ ($|\lambda| > 1$); and (c) the center subspace, spanned by eigenvectors with $\text{Re}\,\lambda = 0$ ($|\lambda| = 1$). The local stable and unstable manifolds of the flow (map) at $\mathbf{p}_o$ are tangent to these subspaces, as one would expect; what is left is the center manifold, about which the Hartman-Grobman theorem says little.

It is disturbing that the theorem does not pin down the center manifold as completely as the stable and unstable manifolds. It is easy to understand what happens when $\underline{\underline{\partial v}}$ possesses a zero eigenvalue. For example, consider the two trivial vector fields over the reals, $v(x) = x^2$ and $u(x) = x^3$. The origin is a nonhyperbolic fixed point for both of them, as $\partial v(0) = \partial u(0) = 0$. The linearized vector field vanishes identically, but the flow cannot be characterized, even locally, by this linear part: both vector fields have the same linear behavior yet the origin is a repellor for u and neither an attractor nor a repellor for v. Knowledge about the nonlinear part of the field is necessary to classify the flow. But what is the difficulty when $\underline{\underline{\partial v}}$ ($\underline{\underline{\partial \phi}}$) is not singular but merely possesses eigenvalues on the imaginary axis (unit circle)? The requirement that the fixed point (or periodic orbit) be hyperbolic is unsettling.

Physicists have come to rely on elliptic behavior: we are incurably committed to dynamical systems that behave like a collection of independent harmonic oscillators near the bottom of a potential well, and it has become an implicit article of faith that linearizing dynamics about these fixed points is a correct way of studying their local behavior. The Hartman-Grobman theorem states that to justify this strategem we really need to know something about the higher order terms before proceeding. It is possible for the nonlinear part of the vector field, no matter how small, to be such that the local flow cannot be represented linearly. To know whether this is the case or not requires knowledge of more than the linearized dynamic.

Straightening Theorem Given a vector field over an N-dimensional phase space, the existence theorem provides for local orbits of the field. We can imagine a local chart one of whose coordinates lines up along the orbits; that is, the orbits' paths are the intersections of $N - 1$ coordinate leaves.[24] On such a chart $N - 1$ components of the vector field would vanish identically. We can imagine further shifting the leaves of the remaining coordinate so as to make the final coordinate constant, independent of the point of evaluation. That such things can be done is the content of the straightening theorem (aka straightening out theorem, rectification theorem).

Theorem 2.7 Let $\underline{v}(\underline{z})$ be an autonomous vector field over an N-dimensional phase space, let $\underline{z}_0$ be a nonsingular point, and let $k \in \{1, 2, \ldots, N\}$ of $\underline{v}$ (i.e., $\underline{v}(\underline{z}_0) \neq 0$). There exists a neighborhood (open set) of $\underline{z}_0$ and a chart covering this neighborhood such that $\underline{v}(\underline{z}) = v\partial/\partial z_k$, where $v \in R$ is just a number.

Of course, the coordinate determining the direction of $\underline{v}$ can be chosen arbitrarily. By applying this to extended phase space we can enlarge the scope of the theorem to nonautonomous dynamics. It is then simplest to let the "time" coordinate carry the field, in which case all its nonextended phase space components would vanish.

The theorem is deeper than we've made it sound. Its proof requires a detailed study of orbits' differentiability and gets to the heart of what is meant by a "coordinate."

2.3.4 Chaos

At irregular intervals scientists expanded their domain of inquiry by quantifying a selection of qualitative experiences: clocks aided in replacing the impressions of "fast" and "slow" with more exact notions of "time," "velocity," and "acceleration"; "heavy" and "light" were superseded by "force" and "mass"; "hot" and "cold" gave way to very subtle concepts of "temperature," "energy," and "entropy"; and "colors" became frequencies and wavelengths in a spectrum, at least to physicists if not to artists. Corresponding to each new idea was a procedural way of evaluating it in terms of numbers.

[24] A "coordinate leaf" is the $(n - 1)$-dimensional manifold obtained by fixing the value of one coordinate. Put another way, it is the "level set" of that coordinate. The term "leaf" goes along with the idea that coordinates "foliate" the space.

The observation of "chaotic" behavior belongs to this category of quantified experience. The subject of chaos achieved an almost cult popularity during the 1980s, much as catastrophe theory did in the 1970s.[25] We have already encountered chaotic orbits in the preceding sections. Indeed, dynamical systems that do not possess chaotic orbits are rare exceptions: in the metric space of vector fields they occupy a set of measure zero. We shall not attempt a comprehensive overview of chaos; there is already an abundant supply of books on the subject — ranging in scope from very popular, graphics-dominated treatments to those less popular and more complete — and we are not going to focus on chaotic behavior in subsequent chapters. However, it is important to understand what the word means. Exactly what objective, quantifiable property of dynamical systems is "chaos" supposed to describe?

Answering this is not a trivial matter.[26] The fundamental property of a classical dynamical system, after all, is its predictability. The past determines the future. The quintessential statement of this was expressed by Laplace, who proclaimed, with the hubris we have since come to expect from scientists, that given the initial positions and velocities of all the particles in the universe, he could calculate the future, thus giving rise to the phrase "Laplacian determinism." The advent of quantum mechanics pulled us out of this trap: Nature was suddenly seen to be nondeterministic, although this was more a matter of interpretation than of mathematical formalism. The theory predicted deterministic evolution of wave functions, but the interpretation of those wave functions led to a probabilistic model of events in space-time. Laplacian determinism had died, at least among mainstream physicists.

However, it was not necessary to supersede classical mechanics in order to kill determinism. That our understanding has changed dramatically since the time of Laplace is exemplified by the following statement, in which Chirikov [21] speculates on the deeper consequences of chaotic systems.

> It is worth noting that such a motion has been searching for since long ago with the purpose of foundation of the statistical mechanics.... Could it be that in Nature there are no such "genuine" random processes as we fancy them? In my opinion ... it is not excluded that no "more random" processes than ... the motion of a K-system do really exist.

Classical mechanics has changed so drastically that Chirikov, among others, suggests that classical chaos can account even for quantum phenomena. Whether one believes this or not[27] it is important to look at classical dynamics in this new light.

The material in this section is not critical to the rest of the book. To those who want to bypass it we offer an informal, three-sentence description of "chaos." Suppose that one wants to predict whether or not at time t_f an orbit of a dynamical system is enclosed in some region of phase space, say, $\mathcal{R} \subset \mathcal{P}$. How accurately does one need to know the state at $t = t_i$ in order to do this? If there are regions, $\mathcal{R}$, such that the

[25]There is some danger in this, and I am tempted to deliver an amateur essay on the sociology of science but shall resist.

[26]Nor is the answer unique. There are several independent definitions of the word "chaotic."

[27]And I do not.

number of significant figures required of the initial coordinates grows exponentially with $t_f - t_i$, then the dynamic is "chaotic."

The discussion that follows is based on lines laid down by several authors, but particularly as recorded in Billingsley [18], Sinai [85], and Arnold and Avez [7].

Partitions Period-advance maps and Poincaré maps are instances of discrete dynamical systems (DDS), which formally consist of four pieces,

$$\text{DDS} \equiv \langle \mathcal{P}, \phi, \gamma, \mu \rangle.$$

Here, $\mathcal{P}$ is the phase space, $\phi : \mathcal{P} \to \mathcal{P}$ is a mapping, γ is a σ-algebra of measurable subsets of $\mathcal{P}$, and μ is a measure defined on γ. For practical purposes, what this means is that γ is a collection of subsets of $\mathcal{P}$ closed under intersection and union and every member of γ has a "volume," which is its measure. If $\mathcal{P}$ has finite measure, then we normalize so that $\mu(\mathcal{P}) = 1$. For Hamiltonian systems, a topic we shall get to in the next chapter, it is natural to assume that μ is the invariant Liouville measure.

Ergodic theory deals with the properties of discrete dynamical systems when ϕ preserves the measure, as is the case with Hamiltonian systems.

$$\forall a \in \gamma : \phi^{-1} a \in \gamma \quad \text{and} \quad \mu(a) = \mu(\phi^{-1} a).$$

By writing the condition in this way, we do not require ϕ to be invertible: ϕ^{-1} is defined on the power set of $\mathcal{P}$, and the symbol $\phi^{-1}a$ represents the subset of points that map into points in a under ϕ.

We now come to the heart of the matter: chaos and unpredictability can exist in deterministic systems simply because it is impossible for us to measure quantities with infinite accuracy, and any observations that we make must be finitely expressible. Thus the model appropriate for our observations of a classical system is not $\mathcal{P}$ but a *partition* of $\mathcal{P}$.

DEF: PARTITION. "α is a measurable partition of $\mathcal{P}$" **means:** (a) α is a set of measurable subsets of $\mathcal{P}$, $\alpha \subseteq \gamma$; (b) these subsets are disjoint: $\forall a, a' \in \alpha$: **either** $a = a'$ **or** $a \cap a' = \emptyset$; and (c) the union of all the sets in the partition is phase space itself: $\mathcal{P} = \bigcup_{a \in \alpha} a$.

Constructing a partition amounts to breaking phase space up into "cells," each of which is a measurable, recordable "macrostate" of the system.

Partitions naturally form a partially ordered set by reason of inclusion.

$$\alpha \geqslant \beta \quad \textbf{means} \quad \forall a \in \alpha \exists b \in \beta : a \subseteq b.$$

This is read, "the partition α is a refinement of the partition β." Thus $\alpha \geqslant \beta$ if α is obtained by chopping the cells of β into smaller pieces. The most and least refined partitions are:

$$\text{trivial partition: } \varepsilon \equiv \{\{x\} \mid x \in \mathcal{M}\},$$

$$\text{the universe: } U \equiv \{\mathcal{M}\}.$$

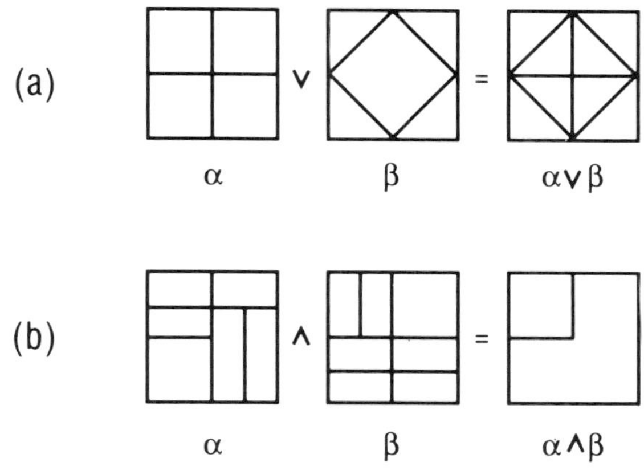

FIGURE 2.18 The algebra of partitions.

In a sense, it is unfair to include ε as a possible partition, since it is nondenumerable, but it serves as an upper bound for all partitions. Clearly, any other partition α of $\mathcal{P}$ must satisfy $\varepsilon \geqslant \alpha \geqslant U$. Therefore any two partitions possess both an upper bound and a lower bound, and it is not surprising that they will also have a least upper bound (lub) and a greatest lower bound (glb). The binary operations corresponding to finding the lub and glb impose an algebraic structure on the set of partitions, transforming it into what algebraists call a "lattice,"[28] whose operations are called "meet" and "join."

$$\begin{aligned}
\textbf{join: } \alpha \vee \beta &\equiv \textbf{lub}(\alpha, \beta) \\
&= \{a \cap b \mid a \in \alpha \quad \text{and} \quad b \in \beta\} \\
\textbf{meet: } \alpha \wedge \beta &\equiv \textbf{glb}(\alpha, \beta).
\end{aligned}$$

These operations are illustrated in Figure 2.18. Unlike the join, there seems to be no binary operator, **O**, acting on sets, such that $\alpha \wedge \beta = \{a \,\mathbf{O}\, b \mid a \in \alpha \text{ and } b \in \beta\}$.

Tracking the evolution of a DDS means observing the state of the system after each "time step," after each iteration of the mapping. Let x represent the initial state of the system. After k sampling intervals the state will be $\phi^k x$. Postulate an apparatus for observing the system whose mathematical model is a partition, α. After each iteration we make an observation and record which cell in α the system occupies. The data from such a sequence of measurements can be encoded into a string of symbols, $a_0\, a_1\, a_2\, a_3 \cdots a_n$, which is interpreted

$$x \in a_0 \in \alpha \textbf{ and } \phi x \in a_1 \in \alpha \textbf{ and } \phi^2 x \in a_2 \in \alpha$$
$$\ldots \textbf{ and } \phi^n x \in a_n \in \alpha,$$

[28]Which is very different from what solid state (condensed matter) physicists call a lattice; which is different from what accelerator physicists call a lattice.

where each a_k is in the partition α. This is equivalent to

$$x \in a_0 \in \alpha \text{ and } x \in \phi^{-1}a_1 \in \phi^{-1}\alpha \text{ and } x \in \phi^{-2}a_2 \in \phi^{-2}\alpha$$
$$\ldots \text{ and } x \in \phi^{-n}a_n \in \phi^{-n}\alpha, \tag{2.73}$$

which, in its turn, is equivalent to

$$x \in \bigcap_{k=0}^{n} \phi^{-k}a_k \in \bigvee_{k=0}^{n} \phi^{-k}\alpha.$$

Our sequence of observations is therefore equivalent to making *a single observation of the initial state* of the system using a partition that is more highly refined than the one provided by our apparatus, α. (Of course, we are assuming classical mechanics throughout: our observations do not disturb the system in any way.) Provided that we design α intelligently, the larger the number of iterations the more refined is the superpartition, $\bigvee_{k=0}^{n} \phi^{-k}\alpha$, and the greater the precision with which we know x.

The critical question is this: What is the expected (or average) *information* contained in such a sequence of observations?

Quantum Partitions Before attacking this question, we shall indulge in an aside on the topic of quantum mechanics. The partitions and lattices that we have been considering can be thought of as a "proposition system" over phase space. A partition models an apparatus for making finitely realizable observations, and the "proposition" associated with a cell is simply the observation that the state of the system must lie within the cell. The lattice algebraic structure of "meet" and "join" defined above was explicitly constructed from set-theoretic operations on these subsets. One consequence of this is that "meet" is distributive over "join,"

$$x \wedge (y \vee z) = (x \wedge y) \vee (x \wedge z), \tag{2.74}$$

just as set-theoretic intersection is distributive over set-theoretic union. However, one can develop a theory of lattices abstractly as algebraic structures on partially ordered sets, with "meet" and "join" defined as least upper bounds and greatest lower bounds, without introducing an underlying set-theoretic model. Not all abstract lattices are distributive; the simplest example of one that is not is diagrammed in Figure 2.19. The five elements of this lattice are labeled ε, x, y, z, and U, and the partial ordering is indicated by arcs, with the higher elements being larger than the lower ones. By inspection, $x \wedge (y \vee z) = x \wedge \varepsilon = x$ while $(x \wedge y) \vee (x \wedge z) = U \vee U = U$, so, as claimed, the lattice is not distributive.

However, it can be proved that all abstract lattices satisfy the following rule, which serves as a generalization of the distributive property:

$$\text{If} \quad x \leqslant z, \quad \text{then} \quad x \vee (y \wedge z) \leqslant (x \vee y) \wedge z.$$

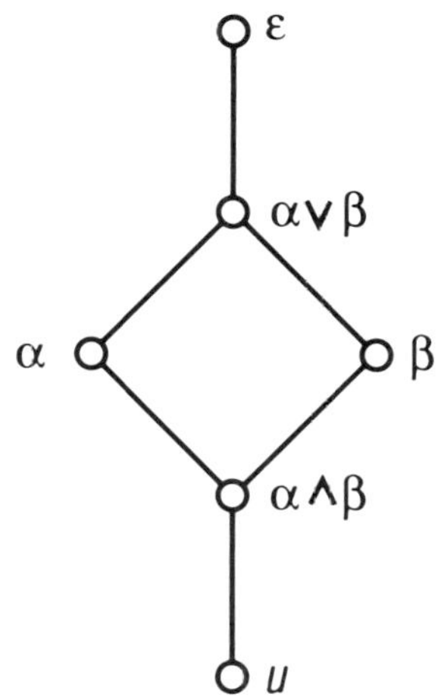

FIGURE 2.19 The simplest nondistributive lattice.

Lattices for which the inequality in the conclusion of this property is tightened to an equality are called modular lattices.

The point is this: distributive lattices model proposition systems of classical mechanics, while nondistributive, modular lattices model proposition systems of quantum mechanics. We do not have the space to consider this fully, but to get a flavor of what is involved — and how it is related to nondistributive logic, with its rejection of the law of the distributed middle — consider the two-slit experiment of quantum mechanics: photons pass one at a time through two slits and accumulate to form a double-slit interference pattern on a screen. Let y be the proposition, "The photon passed through slit Y," and let z be the corresponding proposition for slit Z. Let x be the proposition, "The photon landed at point X on the screen." If the proposition "x and (y or z)" were equivalent to the proposition "(x and y) or (x and z)," then what would evolve on the screen would be a superposition of two single-slit patterns. That the double-slit pattern emerges instead reflects the nondistributive character of the proposition system.

Put more abstractly, the macroscopic observables of quantum mechanics are modelled as projection operators, whose algebraic operations form a lattice that is modular. The intersection of union of subspaces of a Hilbert space do not lie on a distributive lattice.[29]

Information and Entropy "Information" was introduced into mathematics in 1948 by Shannon [83], who used the idea that the information in a message should depend not on the message alone but also on the state of the receiver. Specifically, the information in a message, m, should be a function of the a priori probability, p_m,

[29]For more detailed explanations, see Jauch [50] or Piron [77].

of receiving that particular message, an assertion that we shall write symbolically as

$$\textbf{infor}(m) \equiv f(p_m).$$

A severe constraint is placed on the function f by requiring that the information content of independent messages be additive. Let m_1 and m_2 be two independent messages with a priori probabilities p_{m_1} and p_{m_2}. Then the joint probability is $p_{m_1}p_{m_2}$, and we have the following line of reasoning:

$$\begin{aligned}\textbf{infor}(m_1 \textbf{ and } m_2) &= \textbf{infor}(m_1) + \textbf{infor}(m_2)\\ \Rightarrow f(p_{m_1}p_{m_2}) &= f(p_{m_1}) + f(p_{m_2})\\ \Rightarrow f(p) &= -\log p.\end{aligned}$$

The negative logarithm is used so that information is a positive quantity; if the logarithm is expressed in base 2, then the unit of information is called a "bit." This definition is intuitively reasonable: less probable messages hold more information than more probable messages, and in the limit $p = 1$ the information content goes to zero.

The average, or expected, information in a set of possible messages, $\{m_1, m_2, m_3, \ldots\}$, is called the "entropy" of the set.

$$H(\{m\}) \equiv -\sum_m p_m \log p_m.$$

It is intuitively obvious, and easy to verify, that for a finite number of messages, $N = \textbf{card}[\{m\}]$, $H(\{m\})$ is maximized by making each $p_m = 1/N$.

$$H_{\max} = \log N = \log \textbf{card}[\{m\}]. \tag{2.75}$$

It is natural to ask if there is any connection between this information theoretic notion of entropy and the one that appears in thermodynamics. In fact, Shannon leaned heavily on statistical mechanics in developing his ideas, and the two concepts are virtually identical. Recall from statistical mechanics that the change in thermodynamic entropy, dS, of a system undergoing an infinitesimal isothermal expansion satisfies

$$\begin{aligned}T\,dS &= dQ = -\langle dE\rangle + d\langle E\rangle\\ &= -\frac{1}{Z}\sum_m dE_m e^{-\beta E_m} + d\langle E\rangle\\ &= \frac{1}{\beta Z} d\left(\sum_m e^{-\beta E_m}\right) + d\langle E\rangle\\ &= d\left[kT \ln Z + \langle E\rangle\right],\end{aligned}$$

where T represents temperature, $\beta = 1/kT$, and $p_m = Ze^{-\beta E_m}$. Integrate this equation with the boundary condition that $S = 0$ at $T = 0$.

$$\begin{aligned} S &= k\left[\ln Z + \beta\langle E\rangle\right] \\ &= k\left[\ln Z + \sum_m p_m \beta E_m\right] \\ &= k\left[\ln Z - \sum_m p_m \ln(Zp_m)\right] \\ &= -k\sum_m p_m \ln p_m. \end{aligned}$$

Apart from Boltzmann's constant, which only normalizes the expression and can be absorbed into the base of the logarithm, this is the average information in an observation of the system's state.

KS Entropy A partition of phase space defines a set of messages:

$$m = a \Leftrightarrow x \in a \in \alpha.$$

This message states that one observation was made on the system and it was observed to be in the cell labeled "a." To calculate its information, we must assign it an a priori probability. The "natural" one — and one that enjoys the property of being dynamically invariant — uses the phase space measure.

$$p_m \equiv \mu(a).$$

If $\mathcal{P}$ has finite measure, then upon renormalization to $\mu(\mathcal{P}) = 1$, p_m becomes a legitimate probability; if $\mathcal{P}$ has infinite measure, then we must be content with p_m as a relative probability. From this, the entropy (the information expected from a single observation) associated with the partition (or measuring device) α is

$$H(\alpha) = -\sum_{a\in\alpha} \mu(a) \log \mu(a).$$

What, then, is the expected information in the message stream: $a_0\, a_1\, a_2 \cdots a_n$? (See Eq.(2.73).) This composite message states that the state of the system, x, is in a particular cell of the partition $\bigvee_{k=0}^{n} \phi^{-k}\alpha$. The expected information in the message is therefore the entropy of this partition.

$$a_0\, a_1\, a_2 \cdots a_n \rightarrow x \in \bigcap_{k=0}^{n} \phi^{-k} a_k \in \bigvee_{k=0}^{n} \phi^{-k}\alpha$$

$$\langle \mathbf{infor}(a_0\, a_1 \cdots a_n)\rangle = H\left(\bigvee_{k=0}^{n} \phi^{-k}\alpha\right).$$

Increasing the number of observations gives us more information about the initial state — the partitions become more refined, and the initial state is determined to greater precision. The limiting *rate* at which information about the system increases with the number of observations is the crucial quantity.

$$h(\alpha, \phi) \equiv \lim_{n\to\infty} \frac{1}{n} H\left(\bigvee_{k=0}^{n} T^{-k}\alpha\right).$$

This number depends on the initial partition, α, or, equivalently, on the experimental apparatus. Nothing prevents us from choosing poorly. For example, choosing $\alpha = U$, results in zero information per observation, unless the system suddenly disappears. A better choice of apparatus will optimize $h(\alpha, \phi)$, thereby providing the maximum rate of increase of information about the system, a quantity we shall symbolize as $h(\phi)$.

$$h(\phi) \equiv \sup_{\alpha} h(\alpha, \phi).$$

The supremum is taken over all possible measurable partitions. This limiting value is called the Kolmogorov-Sinai (KS) entropy of the system.[30]

We are finally ready to give an exact meaning to "chaos":

$$\begin{array}{lll} \text{regular motion} & \textbf{means} & h(\phi) = 0, \\ \text{chaotic motion} & \textbf{means} & h(\phi) > 0. \end{array}$$

If the system evolves "regularly," then the information gain per observation approaches zero. Its future is *predictable* in the sense that as time goes by we could continue to increase the interval between observations without loss of information. This is the not the case when the motion is chaotic. There is a lower bound on the rate of gaining information. Conversely, there is an upper bound on the interval between observations that can be tolerated without losing information. This has nothing to do with quantum mechanics; it would be true even if we violated the quantum uncertainty principle. It depends only on the fact that no matter how good the measuring apparatus is, it is still necessarily finite.

The definition of KS entropy is not computationally useful. We cannot search through all possible partitions to find the supremum. Fortunately, most sensible partitions attain this limiting value and Kolmogorov's theorem tells us how to recognize them. We first must define what is meant by a "generating partition." Loosely speaking, a partition, α, is "generating for ϕ" if an infinite number of observations will determine x completely. Heuristically, this means that the size of the cells in $\bigvee_{k=0}^{n} T^{-k}\alpha$ can be made arbitrarily small by taking n large enough. A more precise way of putting it is this: if γ is the σ-algebra of sets over which μ is defined, then $\bigvee_{k=0}^{n} T^{-k}\alpha \to \gamma$ as $n \to \infty$. A partition that violates this is particularly poor in that it will have a limiting coarseness to it. The fundamental theorem that we need is then stated as follows.

[30]More exactly, of the mapping ϕ.

Theorem 2.8 (Kolmogorov's theorem) If (a) α is generating for ϕ and (b) $H(\alpha)$ is finite, then

$$h(\phi) = h(\alpha, \phi).$$

This result seems plausible. It is remarkable, nonetheless, that all generating partitions, some of which can be very coarse indeed, attain the same limiting $h(\alpha, \phi)$. If α is generating, then the original dynamical system is equivalent to shifts on strings of message symbols. This is the approach of "symbolic dynamics."

Three Examples Consider observing a harmonic oscillator at equally spaced times, say, $0, \tau, 2\tau, \ldots$, and record whether its velocity is positive or negative. If time $t = 0$ is set to correspond to maximum negative displacement of the oscillator, then the nth measurement records $\operatorname{sgn}\sin(n\omega\tau)$, where $\omega = 2\pi f$ is the angular frequency of the oscillator. Let us take the state variable, x, to be the phase of the oscillator divided by 2π; its value, mod 1, lies in the interval $U^1 = [0, 1)$ and the mapping corresponding to one sampling interval is

$$\phi : x \longmapsto x + f\tau \bmod 1.$$

Finally, the partition describing our apparatus is simply

$$\begin{aligned} \alpha &= \{a_0, a_1\}, \\ a_0 &= [0, 1/2), \\ a_1 &= [1/2, 1). \end{aligned} \tag{2.76}$$

If $x \in a_0$ the oscillator has positive velocity, and if $x \in a_1$ its velocity is negative. Designing the experiment intelligently means that we choose τ so that $f\tau$ is irrational. Then, by the "linear ergodic theorem" Theorem 2.3, α will be a generating partition, and its entropy should equal the KS entropy of the oscillator.

To evaluate this, note that each observation adds only two cells to $\bigvee_{k=0}^{n} T^{-k}\alpha$, from which it follows that

$$\textbf{card}\left[\bigvee_{k=0}^{n} T^{-k}\alpha\right] = 2(n+1).$$

By Eq.(2.75), we then have the following inequalities:

$$\begin{aligned} H\left(\bigvee_{k=0}^{n} T^{-k}\alpha\right) &\leqslant \log[2(n+1)], \\ h(\alpha, \phi) &\leqslant \lim_{n\to\infty} \frac{1}{n}\log[2(n+1)] \\ &= 0. \end{aligned}$$

Since necessarily $h(\alpha, \phi) \geqslant 0$, it follows that $h(\alpha, \phi) = 0$. Finally, since α is generating for ϕ, Kolmogorov's theorem tells us that $h(\phi) = 0$. To nobody's surprise, the harmonic oscillator undergoes regular motion. The key reason was that the number of cells in the superpartition did not increase quickly enough. Indeed, we can make the general observation that since $H \leqslant \log N$, N must grow at least exponentially fast to have $h(\phi) \neq 0$.

As another example, consider an "abstract" dynamical system defined by assuming the same partition as before, Eq. (2.76), but now we take the map to be $\phi : x \longmapsto 2x \bmod 1$. (This is an example of an irreversible, measure-preserving mapping.) Consider the first step in the evaluation of $\bigvee_{k=0}^{n} T^{-k}\alpha$. The partition $\phi^{-1}\alpha$ consists of two sets, the first of which maps into a_0 under ϕ, and the second into a_1.

$$\phi^{-1}\alpha = \{[0, \tfrac{1}{4}) \cup [\tfrac{1}{2}, \tfrac{3}{4}), [\tfrac{1}{4}, \tfrac{1}{2}) \cup [\tfrac{3}{4}, 1)\}.$$

We now "join" this with α to get

$$\alpha \vee \phi^{-1}\alpha = \{[0, \tfrac{1}{4}), [\tfrac{1}{4}, \tfrac{1}{2}), [\tfrac{1}{2}, \tfrac{3}{4}), [\tfrac{3}{4}, 1)\}.$$

Going on to subsequent steps, at the nth stage the unit interval is divided into subintervals of size $1/2^{n+1}$, giving us the following result:

$$\bigvee_{k=0}^{n} T^{-k}\alpha = \left\{\frac{1}{2^{n+1}}[m, m+1) \mid m = 0, 1, \ldots, 2^{n+1} - 1\right\}.$$

This is obviously a generating partition for ϕ, and we can proceed to evaluate the KS entropy. Since every cell in this superpartition is the same size,

$$\forall a \in \bigvee_{k=0}^{n} T^{-k}\alpha : \mu(a) = \frac{1}{2^{n+1}},$$

this becomes an easy calculation.

$$\begin{aligned} H\left(\bigvee_{k=0}^{n} T^{-k}\alpha\right) &= -\sum \mu(a) \log \mu(a) \\ &= \log[2^{n+1}] \\ &= (n+1)\log 2, \\ h(\phi) = h(\alpha, \phi) &= \lim_{n\to\infty} \frac{1}{n}(n+1)\log 2 \\ &= \log 2 \quad \text{(1 bit)}. \end{aligned}$$

What an eminently reasonable answer! It has a natural interpretation if we think of x as represented by its binary expansion. The mapping ϕ then corresponds to shifting the symbols left one position, and the partition α means that one measurement corresponds to noting whether the first digit to the right of the decimal is a 1 or a 0. A sequence of measurements just reproduces the original binary expansion. The information in a measurement that is skipped cannot be recaptured by subsequent observations: it is gone forever.

Finally, linear automorphisms of a torus provide a classic category of discrete dynamical systems studied by ergodic theorists. Suppose that $\mathcal{P}$ is a two-dimensional torus, which we consider as topologically equivalent to U^2 with identification of opposite edges.

$$\mathcal{P} \simeq U^2 = [0, 1) \times [0, 1).$$

Consider the mapping

$$\phi : \underline{x} \longmapsto \underline{\underline{A}} \cdot \underline{x} \quad \bmod U^2, \tag{2.77}$$

where $\underline{\underline{A}}$ is an integer, 2×2, unimodular matrix. Because of these conditions on $\underline{\underline{A}}$, the mapping represents a true one-to-one mapping of the torus onto itself. It is called a "cat map" for no particularly good reason but following a whimsical sketch of Arnold [7]. Calculating the entropy by means of a generating partition is more difficult than in the two previous examples; details can be found in Sinai [85].[31] The answer is

$$h(\phi) = \log \lambda_+,$$

where λ_+ is the eigenvalue of $\underline{\underline{A}}$ satisfying $\lambda_+ > 1$.

Lyapunov Exponents Although KS entropy is a well defined mathematical concept, using generating partitions to calculate its value is intractable in all but the simplest models. In addition, it assumes that all orbits are of the same character, whereas in most real problems the dynamical system will possess both regular and chaotic orbits. A different approach is needed.

Fortunately, one exists. Chaos is characterized by a phenomenon of divergence: two chaotic orbits that start out infinitesimally close to each other diverge exponentially rapidly. Consider, for example, the evolution of two infinitesimally close orbits, say, with initial conditions x and $x + \epsilon$ under the doubling map. After n iterates,

$$\begin{aligned} \phi^n(x + \epsilon) &= 2^n(x + \epsilon) \\ &= 2^n x + 2^n \epsilon \\ &= \phi^n x + 2^n \epsilon. \end{aligned}$$

[31] Although his proof may be flawed.

Thus the distance between the two orbits grows exponentially.

$$|\phi^n(x+\epsilon) - \phi^n(x)| = e^{\Gamma n}|\epsilon|, \quad \text{where } \Gamma = \log 2.$$

Consider also the behavior of two infinitesimally close orbits under the torus automorphism.

$$\phi(\underline{x}+\underline{\epsilon}) = \underline{\underline{A}}\cdot\underline{x} + \underline{\underline{A}}\cdot\underline{\epsilon} \quad (\text{mod } \mathcal{P}),$$

$$\begin{aligned}|\phi^n(\underline{x}+\underline{\epsilon}) - \phi^n(\underline{x})| &= \left|\underline{\underline{A}}^n\underline{\epsilon}\right| \\ &\approx \lambda_+^n|\underline{\epsilon}| \\ &= e^{\Gamma n}|\underline{\epsilon}|, \quad \text{where } \Gamma = \log\lambda_+.\end{aligned}$$

Of course, doing only two examples does not prove anything, but in each case infinitesimally close orbits diverge exponentially at a rate, Γ, *which is numerically equal to* $h(\phi)$, the KS entropy of the map. This suggests an alternate, more computationally tractable method for calculating $h(\phi)$.

Making a giant leap forward, we now consider the divergence of infinitesimally close orbits under iteration of a generic map, ϕ.

$$\begin{aligned}\phi(\underline{x}+\underline{\epsilon}) &= \phi(\underline{x}) + \underline{\epsilon}\cdot\underline{\underline{\partial\phi}}(\underline{x}) \\ \phi^2(\underline{x}+\underline{\epsilon}) &= \phi(\phi(\underline{x}) + \underline{\epsilon}\cdot\underline{\underline{\partial\phi}}(\underline{x})) \\ &= \phi(\phi(\underline{x})) + \underline{\epsilon}\cdot\underline{\underline{\partial\phi}}(\underline{x})\cdot\underline{\underline{\partial\phi}}(\phi(\underline{x})) \\ &\equiv \phi^2(\underline{x}) + \underline{\epsilon}\cdot\underline{\underline{\partial\phi^2}}(\underline{x}) \\ &\vdots \\ \phi^n(\underline{x}+\underline{\epsilon}) &= \phi^n(\underline{x}) + \underline{\epsilon}\cdot\underline{\underline{\partial\phi^n}}(\underline{x}),\end{aligned}$$

where the $\underline{\underline{\partial\phi^n}}$ are computed recursively.

$$\underline{\underline{\partial\phi^n}}(\underline{x}) = \underline{\underline{\partial\phi^{n-1}}}(\underline{x})\cdot\underline{\underline{\partial\phi}}(\phi^{n-1}(\underline{x})).$$

If the orbits are diverging exponentially, we can connect the rate to the norm of $\underline{\underline{\partial\phi^n}}$.

$$\begin{aligned}|\phi^n(\underline{x}+\underline{\epsilon}) - \phi^n(\underline{x})| &\stackrel{?}{=} |\epsilon|e^{\Gamma n} \\ &\approx |\epsilon|\left\|\underline{\underline{\partial\phi^n}}(\underline{x})\right\|.\end{aligned}$$

This suggests both a definition,

$$\Gamma(\underline{x}) \stackrel{?}{\equiv} \lim_{n\to\infty}\frac{1}{n}\ln\left\|\underline{\underline{\partial\phi^n}}(\underline{x})\right\|,$$

and an association, $h(\phi) \stackrel{?}{=} \Gamma$.

This cannot be correct, however. In the first place $\Gamma(\underline{x})$ depends on the initial conditions, $\underline{x}$, while $h(\phi)$ is a global number, and in the second place, the norm of $\underline{\underline{\partial\phi}}$ does not carry enough information to characterize the motion.

The correct answer is only a little more complicated [59]. Let $\lambda_1 \geqslant \lambda_2 \geqslant \cdots \geqslant \lambda_N$ be the N eigenvalues of $\underline{\underline{\partial\phi^n}}(\underline{x})$ arranged in decreasing order. The "Lyapunov exponents" of the map ϕ evaluated for the orbit passing through $\underline{x}$ are the numbers

$$\Gamma_k(\underline{x}) \equiv \lim_{n\to\infty} \frac{1}{n} \ln \lambda_k(\underline{x}).$$

The fundamental result that then connects these quantities to the KS entropy is as follows:

$$h(\phi) = \int_{\mathcal{P}} d\mu(\underline{x}) \sum_{\Gamma_k>0} \Gamma_k(\underline{x}).$$

This expresses the global quantity, KS entropy, as an average of a local function, the Lyapunov exponents. It reflects the fact that regular and chaotic motion can exist simultaneously but in different regions of phase space.[32]

Hausdorff (Fractal) Dimension Nonzero entropy is not the only signature of chaotic orbits; for example, they are associated with broad band Fourier spectra and fractal dimensions as well. The latter quantity in particular has captured people's imagination, and we shall consider it briefly before closing.

Consider the problem of operationally determining the dimension of a set of points, such as a surface of some kind. One approach, first suggested by Hausdorff, is to cover the set — or the generated points — with spheres of a given radius. If the set is infinite, the minimum number of spheres (or cubes, or ellipsoids, or whatever) with radius $\leqslant r$ needed to cover a set will grow like an inverse power of r, for small r.

$$N(r) \sim r^{-d}. \tag{2.78}$$

The exponent is the "Hausdorff dimension" of the set.

$$d = -\lim_{r\to 0} \frac{\log N(r)}{\log r}. \tag{2.79}$$

The characterization of this number as a dimension is confirmed by the observation that when the set is (an open subset of) a manifold, its value is indeed the dimension of the manifold: the Hausdorff dimension of a curve is 1, of a surface 2, of a volume 3, and so forth. To see this most trivially, consider packing a measurable subset of R^3 with small spheres of radius r. If the volume of the set is V,

$$V \approx N(r) \times \tfrac{4}{3}\pi r^3,$$

[32]For a recent review of their use in accelerator applications, see Schmidt, Willeke, and Zimmerman [82].

from which

$$N(r) \sim r^{-3},$$

so that the Hausdorff dimension is three. However, the procedure defining this number makes no assumption about the point set being sampled; it need not be a manifold, and its value need not be an integer. This can happen, for example, with samples of a chaotic orbit confined to a compact subset of phase space, for strange attractors in dissipative systems, or for basin boundaries of systems with more than one attractor. In today's terminology, a set with noninteger Hausdorff dimension is a "fractal," and its dimension is then called a "fractal dimension."

To illustrate this we consider Cantor's set, the patriarch of all fractals, devised by Cantor to demonstrate the existence of a set that had zero measure and the cardinality of the continuum. It is defined as follows. Begin with the sets

$$C_0 \equiv U^1 = [0, 1) \subset R,$$
$$S_0 \equiv \{U^1\}.$$

Notice that C_0 is an interval of the real axis, while S_0 is a set whose element is C_0. We now define a sequence of sets, $S_k, k = 0, 1, 2, \ldots$, recursively by specifying the members of S_{k+1} in terms of those of S_k. This is described most easily with a pseudoprogram:

```
S₀ ≡ {U¹};
for ( k ∈ Z⁺ ) {
  S_k ≡ ∅;
  for every [x,y) ∈ S_{k−1} :
    S_k = S_k ∪ {[x, ⅔x + ⅓y)} ∪ {[⅓x + ⅔y, y)};
}
```

In words: S_k is obtained from the intervals in S_{k-1} by a process of chopping out their middle thirds. Thus, in particular,

$$S_0 = \{[0, 1)\},$$
$$S_1 = \{[0, \tfrac{1}{3}), [\tfrac{2}{3}, 1)\},$$
$$S_2 = \{[0, \tfrac{1}{9}), [\tfrac{2}{9}, \tfrac{1}{3}), [\tfrac{2}{3}, \tfrac{7}{9}), [\tfrac{8}{9}, 1)\},$$

and so forth. The kth approximant to Cantor's set is the union of all intervals in S_k.

$$C_k \equiv \bigcup_{s \in S_k} s.$$

Finally, the Cantor set itself is the intersection of all its approximants.

$$C \equiv \bigcap_{k=0}^{\infty} C_k.$$

(This is a set-theoretic way of saying $C = \lim_{k\to\infty} C_k$.) It is easy to see that the set has measure zero, $\mu(C) = 0$, since it is covered by each approximant, C_k, and $\mu(C_k) = (2/3)^k$. Furthermore, it has the cardinality of the continuum, since C comprises all those numbers in U^1 whose expansion in base 3 contains only the digits 0 and 2. By changing each 2 to a 1, and reinterpreting the string as a *binary* expansion, we can map Cantor's set one-to-one onto U^1. Finally, its Hausdorff dimension is not an integer. To see this, note that each S_k contains 2^k intervals of radius $(1/3)^k$ and covers C. A moment's reflection is sufficient to convince that it is the smallest such set. Therefore the dimension of C can be evaluated, after taking the limit in Eq. (2.79),

$$d = -\lim_{k\to\infty} \frac{\log[2^k]}{\log[(1/3)^k]} = \log 2/\log 3. \tag{2.80}$$

Cantor's set shares one more property with other fractals: it is self-similar. Multiplication by 3 maps the lower third of Cantor's set onto the full set.

$$C' \equiv 3C \cap U^1 = C.$$

$$x \in C' \quad \text{iff} \quad x \in U^1 \quad \text{and} \quad \tfrac{1}{3}x \in C$$

This scaling property of self-similarity can be used to derive the fractal dimension in a better way, one that can be generalized to other fractal figures. Each S_k can be constructed by shrinking S_{k-1} by a factor of 3, cloning the resulting set, and translating by 2/3. Therefore, if N spheres of radius r are required to cover the Cantor set, shrinking the spheres $1/3$ will enable them to cover half the Cantor set. That is $2N(r) = N(r/3)$. Combining this with Eq. (2.78) reproduces the fractal dimension given in Eq. (2.80).

The term "fractal" was coined by Benoit Mandelbrot, the man largely responsible for their reintroduction into the modern stream of collective consciousness [66, 65]. Following his lead, researchers have found fractals arising in a wide variety of applications. In retrospect, this is not surprising. We began these lectures by talking about manifolds as the stage on which dynamical systems perform. This *idée fixe* that the background environment is smooth pervades all of physics, from introductory mechanics to general relativity. We inherited it from the Greeks, who were principally interested in studying such things as circles and triangles. The stage that they employed was the Euclidean plane, and we have basically projected their concept forward through thirty, or so, centuries. Lost somewhere in that process was a fundamental observation: circles and triangles do not exist in Nature; they were mental constructs very useful for solving a particular class of problems. Fractals, also mental constructs, are far more appropriate to describe large classes of natural and mathematical phenomena that cannot be handled by the Greek models. One need only stare at a tree to see how Nature makes use of fractal structures and to appreciate their beauty. It has been suggested that fractal figures drawn by computers are the modern equivalent of the Greeks' circles and triangles, a first step beyond those smooth models that they created and we inherited; it is conceivable that fractals may eventually lead us to new geometries; it is possible that among these we may find one that serves us better in describing the way our world works. For now, however, this is little more than a popular speculation and likely to remain so for quite some time.

2.3.5 Distributions

This book deals with dynamical systems possessing low-dimensional phase spaces. It is important at times to consider the behavior of a collection, or ensemble, of multiple instances of such a system. Imagine not a single point evolving along its orbit but an ensemble, moving like a cloud over phase space. The abstract idealization of this is to treat the collection as sampled from a distribution, $\rho(\underline{z}, t)$, defined over extended phase space. Rather than focusing on individual orbits, people attempt to analyze the evolution of this function or of its statistics. In (classical) statistical dynamical theories, ρ is considered the primitive concept and called the "state" of the system. Single orbits can be recaptured, in principle, by taking $\rho(\underline{z}, t) = \delta(\underline{z} - \underline{z}(t))$, for some orbit, $\underline{z}(t)$.

Already, however, there are difficulties: a distribution necessarily requires a *measure,* a "differential volume element," against which it is calibrated. In order to discuss the behavior of distributions we therefore must assign a measure to our phase space and examine how volume elements change with time.

Expansion Rate Equation We have encountered a number of attractors in the examples of this chapter. By its very nature, the existence of an attractor implies a shrinkage of the phase space volume occupied by an ensemble of systems: the entire basin of attraction collapses into an object of measure zero. We shall determine here the property of the vector field that makes such shrinkage possible.

Let us start from the vector field, $\underline{v}$, with usual interpretation,

$$\dot{\underline{z}} = \underline{v}(\underline{z}, t),$$

and let us select a single orbit of $\underline{v}$, say, $\underline{z}_o(t)$. We want to examine the behavior of phase space volume in the neighborhood of $\underline{z}_o$. Because of its function of serving as a reference orbit, we shall refer to $\underline{z}_o$ as a "fiducial orbit."

If N is the dimension of phase space, we construct an ensemble of N orbits "infinitesimally close," to the fiducial orbit, in a sense to be made precise shortly. The difference between one of these and the fiducial orbit will be denoted $\underline{u}$.

$$\underline{z}^k(t) = \underline{z}_o(t) + \underline{u}^k(t), k = 1, \dots, N. \tag{2.81}$$

Such a collection is sometimes called a "pencil" of orbits. An important point should be made here: $\underline{u}$ is not a geometric object; its definition is chart-dependent. In general, it does not make sense to think of "adding" or "subtracting" orbits embedded in an arbitrary phase space. It *does* make sense, however, in the tangent spaces attached to the manifold, and this is the sense in which we must interpret Eq. (2.81). This is what we mean by saying that the $\underline{u}^k$'s are "infinitesimally close" to $\underline{z}_o$: each $\underline{u}^k(t)$ belongs to the tangent space $T_{\mathbf{p}}\mathcal{P}$, where $\mathbf{p}$ is the point on $\mathcal{P}$ locally represented by $\underline{z}_o(t)$. With this in mind, we can state the added requirement that the $\underline{u}^k$'s comprise a linearly independent set.

In practical terms, what we shall do is develop orbits in the neighborhood of the fiducial orbit while retaining terms only linear in the $\underline{u}^k$'s.

$$
\begin{aligned}
\dot{\underline{z}}^k = \frac{d}{dt}\left(\underline{z} + \underline{u}^k\right) = \dot{\underline{z}} + \dot{\underline{u}}^k &= \underline{v}(\underline{z}(t) + \underline{u}^k(t), t) \\
&= \underline{v}(\underline{z}, t) + \left(\partial \underline{v}/\partial \underline{z}\right)_{(\underline{z}(t),t)} \cdot \underline{u}^k + \cdots \\
\dot{\underline{u}}^k &= \left(\partial \underline{v}/\partial \underline{z}\right)_{(\underline{z}(t),t)} \cdot \underline{u}^k \\
&\equiv \underline{\underline{A}}(t) \cdot \underline{u}^k.
\end{aligned}
\tag{2.82}
$$

This is a linear system, albeit not an autonomous one. In the event that the fiducial orbit is periodic, $\underline{\underline{A}}(t)$ is periodic, and Floquet theory can be applied to determine the stability of the orbit's neighborhood.

We shall define the phase space volume associated with the pencil as the usual one obtained by taking the determinant of the parallelepiped whose edges are the $\underline{u}^k$'s.

$$
\begin{aligned}
V(t) &= \mathbf{det}\left[\underline{u}^1(t)\underline{u}^2(t)\underline{u}^3(t)\cdots\underline{u}^N(t)\right] \\
&= \sum_{k,l,m,\ldots,n} \epsilon_{k,l,m,\ldots,n} u_k^1 u_l^2 u_m^3 \cdots u_n^N.
\end{aligned}
$$

Here, each $\underline{u}^k$ is a column in this determinant; $\epsilon_{k,l,m,\ldots,n}$ is the Levi-Civita symbol taking on the values ± 1 according to whether the ordered array, $(k, l, m, \ldots, n)$, is an even or odd permutation of the indices $1, \ldots, N$. Perpetrating a very minor abuse of terminology, we shall refer to this as the "Liouville volume element." The expansion (or contraction) rate is obtained by differentiation and application of Eq. (2.82).

$$
\begin{aligned}
\dot{V} &= \sum_{k,l,m,\ldots,n} \epsilon_{k,l,m,\ldots,n}\left[\dot{u}_k^1 u_l^2 u_m^3 \cdots u_n^N + u_k^1 \dot{u}_l^2 u_m^3 \cdots u_n^N\right. \\
&\qquad \left. + u_k^1 u_l^2 \dot{u}_m^3 \cdots u_n^N + \cdots + u_k^1 u_l^2 u_m^3 \cdots \dot{u}_n^N\right] \\
&= \sum_{i,k,l,m,\ldots,n} \epsilon_{k,l,m,\ldots,n}\left[A_{ki} u_i^1 u_l^2 u_m^3 \cdots u_n^N + u_k^1 A_{li} u_i^2 u_m^3 \cdots u_n^N\right. \\
&\qquad \left. + u_k^1 u_l^2 A_{mi} u_i^3 \cdots u_n^N + \cdots + u_k^1 u_l^2 u_m^3 \cdots A_{ni} u_i^N\right] \\
&= \sum_{r,i,k,l,m,\ldots,n} \left[\epsilon_{rlm\cdots n}\delta_{ik} + \epsilon_{krm\cdots n}\delta_{il} + \epsilon_{klr\cdots n}\delta_{im}\right. \\
&\qquad \left. + \cdots + \epsilon_{klm\ldots r}\delta_{in}\right] A_{ri} u_k^1 u_l^2 u_m^3 \cdots u_n^N.
\end{aligned}
$$

We now use a combinatorial lemma whose proof, though simple, is tedious and left to the reader.

$$\epsilon_{rlm\cdots n}\delta_{ik} + \epsilon_{krm\cdots n}\delta_{il} + \epsilon_{klr\cdots n}\delta_{im} + \cdots + \epsilon_{klm\cdots r}\delta_{in} = \epsilon_{klm\cdots n}\delta_{ir}. \tag{2.83}$$

The final result is written as follows:

$$\begin{aligned} \dot{V} &= \sum_{r,i,k,l,m,\ldots,n} (\delta_{ir}A_{ri})(\epsilon_{klm\cdots n}u_k^1 u_l^2 u_m^3 \cdots u_n^N) \\ &= \mathrm{Tr}\left[\underline{\underline{A}}\right] \mathbf{det}\left[\underline{u}^1\underline{u}^2\underline{u}^3 \cdots \underline{u}^N\right] \\ \dot{V}(t) &= \mathrm{Tr}\left[\underline{\underline{A}}(t)\right] V(t). \end{aligned} \tag{2.84}$$

Recall that

$$\underline{\underline{A}}(t) = \left(\partial \underline{v}/\partial \underline{z}\right)_{(\underline{z}(t),t)},$$

so that $\mathrm{Tr}\left[\underline{\underline{A}}(t)\right]$ is the divergence of the vector field along the fiducial orbit.

$$\begin{aligned} \mathrm{Tr}\left[\underline{\underline{A}}(t)\right] &= \sum_k \left(\partial v_k/\partial z_k\right)_{(\underline{z}(t),t)} \\ &= (\underline{\partial} \cdot \underline{v})_{(\underline{z}(t),t)}, \\ \dot{V}(t) &= (\underline{\partial} \cdot \underline{v})_{(\underline{z}(t),t)} V(t). \end{aligned} \tag{2.85}$$

Regardless of the nonlinearities of the dynamical system, the differential volume element associated with a pencil of orbits is itself a (one-dimensional) linear system with coefficient equal to the divergence of the vector field along the fiducial orbit. The solution, of course, is

$$V(t) = V(0) \exp\left(\int_0^t d\tau (\underline{\partial} \cdot \underline{v})_{(\underline{z}(\tau),\tau)}\right).$$

Whether a region of phase space is contracting or expanding depends only on the sign of $\underline{\partial} \cdot \underline{v}$ within that region. In particular, if $\underline{\partial} \cdot \underline{v} = 0$ everywhere, then the system is "conservative," in the sense of conserving phase space volume, *not* energy. Attractors can exist only in regions for which $\underline{\partial} \cdot \underline{v} < 0$; if there are none, then the system cannot possess an attractor.

Boltzmann and Vlasov Equations Let us return to the ensemble with which we began this section and to its underlying distribution, which we express as a real-valued function on extended phase space, say, $\rho(\underline{z}, t)$, locally. To determine the behavior of ρ we idealize the ensemble as a "fluid" whose density is ρ and whose current is $\rho\underline{v}$. Since systems are neither created nor destroyed by their dynamics, the number of systems in the ensemble is conserved and the fluid must obey a continuity equation of the form

$$\underline{\partial} \cdot (\rho \underline{v}) + \frac{\partial \rho}{\partial t} = 0,$$

which upon expansion and a little rearrangement yields the following:

$$\dot{\rho} = \underline{v} \cdot \underline{\partial} \rho + \frac{\partial \rho}{\partial t} = -\rho \underline{\partial} \cdot \underline{v}. \tag{2.86}$$

This is the "collisionless Boltzmann equation" describing evolution of a distribution function representing an ensemble of independent systems. To evaluate $\rho(\underline{z}, t)$ one integrates — perhaps numerically — Eq. (2.86) along an orbit. Let $\underline{z}_o$ be the point satisfying $\phi_{t t_o}^{-1} : \underline{z} \rightarrow \underline{z}_o$.

$$\begin{aligned} \rho(\underline{z}, t) &= \rho(\underline{z}_o, t_o) + \int_{t_o}^{t} d\tau \dot{\rho}(\phi_{\tau t_o}(\underline{z}_o), \tau) \\ &= \rho(\underline{z}_o, t_o) - \int_{t_o}^{t} d\tau (\rho \underline{\partial} \cdot \underline{v})_{(\phi_{\tau t_o}(\underline{z}_o), \tau)}. \end{aligned}$$

Matters simplify considerably when the dynamic is conservative, in which case the equation is called the "Vlasov equation." If $\underline{\partial} \cdot \underline{v} = 0$, then Eq. (2.86) implies immediately that $\dot{\rho} = 0$ as well, so that ρ is a dynamical invariant. Solving for the orbit histories is sufficient to track the evolution of ρ.

$$\rho(\underline{z}, t) = \rho(\underline{z}_o, t_o).$$

This has the additional, beautiful consequence that *level surfaces of* ρ *propagate according to the time-evolution operator*. Let $S(t, x) \equiv \{\underline{z} \mid \rho(\underline{z}, t) = x\}$. Then conservative dynamics possess the property

$$\phi_{t t_o} : S(t_o, x) \longrightarrow S(t, x), \quad \forall x.$$

The task of tracking the contour map of ρ over time is the same as that of following these orbits.

The two problems are, in fact, logically equivalent: not only does knowing the orbits allow us to calculate ρ but a procedure that calculates the time evolution of ρ would also find orbit histories when run on an initial distribution sufficiently concentrated. Thus there is nothing to be gained, in general, in attempting to calculate ρ in detail. Rather, scientists work with less information and study various statistics of the distribution, notably moments or spectral components.

A distribution is in equilibrium if its value does not change over time at any point in phase space: $\partial \rho / \partial t = 0$. Eq. (2.86) then implies that

$$\begin{aligned} \underline{v} \cdot \underline{\partial} \rho &= -\rho \underline{\partial} \cdot \underline{v}, && \text{for dissipative systems} \\ &= 0, && \text{for conservative systems.} \end{aligned}$$

In its turn, this implies that if the system is conservative ($\underline{\partial} \cdot \underline{v} = 0$), then the Lie derivative of ρ must vanish. This means geometrically that the level surfaces, $S(t, x)$, need only be *tangential to the orbits' paths* for ρ to be in equilibrium. There is no such simple interpretation for dissipative systems, however.

Equilibrium distributions for conservative systems possessing chaotic orbits are constrained by the requirement that ρ take on a single value along an entire orbit. The fractal nature of the sets of orbit samples is in contrast with our requirement that ρ be a smooth function. The only way out of this dilemma is to let ρ be constant over the closure of any chaotic orbit, which is some measurable region of phase space. Even so, one is not assured that this is sufficient for an equilibrium density to exist. Certainly, unless ρ is constant everywhere within a chaotic layer, it must have some interesting, nonanalytic properties on the fractal boundaries between chaotic and nonchaotic regions.

Recurrence Conservative dynamical systems on compact phase spaces possess a "recurrence" property that, loosely stated, says that any region of the phase space, no matter how small, intersects orbits that eventually return to the region. Mathematicians call such orbits "nonwandering." The early proofs of this property were written by Poincaré and Carathéodory, and subsequent treatments follow their line of reasoning. The recurrence theorem was one of the earliest fruits of what is called "ergodic theory," whose original motivation was the quest to prove the ergodic hypothesis of Boltzmann, a cornerstone of classical statistical mechanics. Ergodic theory is now a mathematical discipline in its own right, dealing with the properties of measure-preserving transformations.

We have already introduced a specific "differential volume," the Liouville volume, but the development of this section is more general and requires only the existence of some measure function, $\mu : \mathcal{B}(\mathcal{P}) \longrightarrow R$, which maps the "measurable" subsets of the phase space, $\mathcal{P}$, into the reals. That is, $B \in \mathcal{B}$ means that $B \subset \mathcal{P}$ and B is "measurable," it is in the domain of μ. I am being somewhat circular, because I do not want to enter into a long introduction to measure theory, Borel sets, and all that. The distinction "measurable" would not be at all necessary were it not for the inconvenient fact that nonmeasurable sets exist — at least they do for those whose mathematical faith includes the Axiom of Choice in its strongest form.[33] For our purposes, we can naively think of $\mu(B)$ as the volume of a subset $B \subset \mathcal{P}$ without qualms and move on.

The recurrence theorem is usually expressed in terms of mappings rather than flows. Let $\phi : \mathcal{P} \longrightarrow \mathcal{P}$ be a measure-preserving mapping, by which we mean that

$$\mu(\phi^{-1}B) = \mu(B), \tag{2.87}$$

[33]For a particularly disturbing application of this inconvenient "fact" read the Banach-Tarski "paradox," preferably by candlelight on some dark and lonely night, during a thunderstorm, with the wind howling and the rain rattling on windows suggestive of another sort of howling and rattling, which are, perhaps, reminders of mortal finitude.

for all $B \in \mathcal{B}(\mathcal{P})$. ϕ^{-1} need not exist in a pointwise sense; in this section, and in this section only, we shall not restrict ourselves to invertible maps. Rather, $\phi^{-1}B$ is defined as the preimage of B under ϕ.

$$\phi^{-1}B \equiv \{\mathbf{p} \in \mathcal{P} \mid \phi(\mathbf{p}) \in B\}.$$

Similarly, the symbol ϕB represents the image of B under ϕ.

$$\phi B \equiv \{\phi(\mathbf{p}) \mid \mathbf{p} \in B\}.$$

By repeating this, we define a doubly infinite chain of sets, each of which maps into the next under ϕ.

$$\begin{aligned} \mathcal{T} &\equiv \ldots, \phi^{-3}B, \phi^{-2}B, \phi^{-1}B, \phi^{0}B = B, \phi^{1}B, \phi^{2}B, \phi^{3}B, \ldots \\ \phi^{-n}B &\equiv \{\mathbf{p} \mid \phi(\mathbf{p}) \in \phi^{-n+1}B\} = \phi^{-1}\phi^{-n+1}B, \quad n = 2, 3, \ldots \\ \phi^{n}B &\equiv \{\phi(\mathbf{p}) \mid \mathbf{p} \in \phi^{n-1}B\} = \phi\phi^{n-1}B, \quad n = 2, 3, \ldots. \end{aligned}$$

The theorem is stated as follows.

Theorem 2.9 (Recurrence) Let $\phi : \mathcal{P} \rightarrow \mathcal{P}$ be a measure-preserving mapping, and let $B \in \mathcal{B}(\mathcal{P})$ be a subset of $\mathcal{P}$ with nonzero measure. If $\mu(\mathcal{P})$ is finite, then there is a point $\mathbf{p} \in B$ and a positive integer, $n \in Z^{+}$, such that $\phi^{n}(\mathbf{p}) \in B$.

The connection with dynamics occurs when we identify ϕ as a time-evolution, period-advance, or Poincaré mapping. That $\mu(\mathcal{P})$ is finite at first appears to restrict the theorem to special cases, such as the rigid body problem, in which $\mathcal{P}$ is compact. But $\mathcal{P}$ need not be the *entire* phase space of the system: any invariant subspace will do, and the measure, μ, can be restricted to it. For example, if the energy of a system is conserved, then μ can act on a level set of the energy function, which surface is frequently compact.

Proof (Recurrence theorem): Since $\mathcal{P}$ has finite measure and $\mu(B) \neq 0$, we can define $\sigma \equiv \mu(B)/\mu(\mathcal{P})$ and be assured that $0 < \sigma \leq 1$. Consider the subchain of $\mathcal{T}$ given by the sets, $B, \phi^{-1}B$, $\phi^{-2}B$, $\phi^{-3}B$, ... Because ϕ preserves measure at most $1/\sigma$ of these can be mutually disjoint. For some $0 < k, l < 1 + 1/\sigma$, $k \neq l$, we therefore must have

$$0 < \mu(\phi^{-k}B \cap \phi^{-l}B), \tag{2.88}$$

otherwise we would have

$$\mu(\mathcal{P}) < \mu\left(\bigcup_{m=0}^{\lfloor 1+1/\sigma \rfloor} \phi^{-m}B\right),$$

which would be impossible, as the union must be a subset of $\mathcal{P}$. We can "reduce" Eq. (2.88) one step by using the identity $\phi^{-1}(A \cap B) = \phi^{-1}A \cap \phi^{-1}B$.

$$\begin{aligned} 0 &< \mu(\phi^{-k}B \cap \phi^{-l}B) \\ &= \mu(\phi^{-1}\phi^{-k+1}B \cap \varphi^{-1}\phi^{-l+1}B) \\ &= \mu(\phi^{-1}(\phi^{-k+1}B \cap \phi^{-l+1}B)) \\ &= \mu(\phi^{-k+1}B \cap \phi^{-l+1}B). \end{aligned}$$

Let us assume, without loss of generality, that $l < k$, and define the *positive* integer $n \equiv k - l$. Performing this step l times leads to the inequality

$$0 < \mu(B \cap \phi^{-n}B).$$

Since the measure is nonzero, the intersection itself must be nonvoid, which means that for some $\mathbf{p} \in \mathcal{P}$ we must have $\mathbf{p} \in B$ *and* $\mathbf{p} \in \phi^{-n}B$. Put another way,

$$\exists \mathbf{p} \in B, \exists n \in Z^+ : \phi^n(\mathbf{p}) \in B,$$

which is the statement of recurrence: there is some point in B whose orbit returns to B. □

The proof provides an upper bound for the "Poincaré recurrence time," as no more than $\lfloor 1 + 1/\sigma \rfloor$ iterates of the map are required to assure that at least two members of $\mathcal{T}$ overlap. A number of texts choose to prove the recurrence theorem using forward, rather then backward, mappings and write

$$\mu(\phi B) = \mu(B), \tag{2.89}$$

as the definition of "measure-preserving" rather than Eq. (2.87). We have gone through the proof in the way that we have in order to employ the *weakest* assumptions possible. Because the proposition "$B = \phi\phi^{-1}B$" is a set-theoretic identity but "$B = \phi^{-1}\phi B$" is *not*, it is possible to derive Eq. (2.87) from Eq. (2.89), but not vice versa. The former is therefore a weaker assumption and holds for noninvertible, measure-preserving maps — such as the famous $\phi : [0, 1) \to [0, 1), x \mapsto 2x \bmod 1$ — while the latter does not.

The recurrence theorem does *not* state that the same orbit need return indefinitely many times, nor that every orbit through B returns to B. Indeed, the phase space of a simple pendulum provides a trivial counterexample to this. Take B to be a small neighborhood of any point on the separatrix, and choose ϕ to be a flow map for time sufficiently large that $B \cap \phi B = \varnothing$. The orbit on the separatrix does not return to B. Of course, it is isolated in this property, as it must be. The recurrence theorem assures that any nonrecurring orbit must be arbitrarily close to an infinite family of recurring ones. This means that no finite computational algorithm can determine with certainty that an arbitrary point in phase space lies on a nonrecurring orbit if in fact it "does"; without extra information, the proposition is not decidable by finite means.

PROBLEMS

2.1 Confirm Eqs. (2.7).

2.2 What is the form of $\exp(t\underline{\underline{A}})$ when $\underline{\underline{A}}$ is a single, $n \times n$ Jordan block?

2.3 Sketch all topologically distinct volume-preserving flows in a small neighborhood of a fixed point of a two-dimensional dynamical system.

2.4 Starting from Eq. (2.9), show that

$$\frac{\partial}{\partial I} = \frac{1}{2I}\left(a\frac{\partial}{\partial a} + a^*\frac{\partial}{\partial a^*}\right)$$

$$\frac{\partial}{\partial \varphi} = i\left(a^*\frac{\partial}{\partial a^*} - a\frac{\partial}{\partial a}\right)$$

Use this to extend Eq. (2.57).

2.5 Suppose a harmonic oscillator is kicked randomly but periodically. Express the limiting amplitude growth rate in terms of the autocorrelation of the kicks.

2.6 Verify "Sundman's transformation" of Kepler motion into simple harmonic motion. Work in the complex plane: letting $z \equiv x + iy$, write the (suitably normalized) equation of motion,

$$\ddot{z} = -z/r^3, \quad \text{with } r^2 = zz^*.$$

Now define a new "coordinate," u, and a new "time," s, according to

$$u^2 = z$$
$$r\,ds = dt,$$

and show that the equation of motion for $u(s)$ is,

$$u'' - \tfrac{1}{2}hu = 0,$$

where h is the energy of the orbit,

$$h = \tfrac{1}{2}|\dot{z}|^2 - 1/r.$$

Why does *not* this transformation make a Kepler dynamical system topologically equivalent to the harmonic oscillator?

2.7 Accepting Lemma 2.1 as correct, prove Lemma 2.2.

2.8 Suppose that the two eigenvalues of Hill's equation are equal, that is, $\mathrm{Tr}[W(2\pi)] = \pm 2$. Show that there is one nontrivial periodic solution $p(\theta)$ with period either 2π or 4π and that any solution $y(\theta)$, which is linearly independent of p, must satisfy

$$y(\theta + 2\pi) = \rho y(\theta) + \gamma p(\theta),$$

where ρ is the eigenvalue, and γ is a constant.

2.9 Show that β defined in Eq. (2.28) obeys the differential equation

$$\tfrac{1}{2}\beta\beta'' - \tfrac{1}{4}\beta'^2 + K\beta^2 = 1,$$

where prime indicates d/ds, and K is defined in Eq. (2.26).

2.10 Particles are transported from one circular accelerator, A, to another, B, across a beamline. Let θ_A be the point of attachment to A, and similarly for θ_B. Making the beamline "linearly matched" to A and B requires that its transfer map matrix, $\underline{\underline{M}}$, turn eigenvectors of $\underline{\underline{M}}_{2\pi}(\theta_A)$ into eigenvectors of $\underline{\underline{M}}_{2\pi}(\theta_B)$. That is, let the eigenvectors of $\underline{\underline{M}}_{2\pi}(\theta_A)$ and $\underline{\underline{M}}_{2\pi}(\theta_B)$ be collected into sets U_A and U_B, respectively. Then, $\underline{\underline{M}}$ is linearly matched only if $\underline{\underline{M}} \cdot \underline{u} \in U_B$, for all $\underline{u} \in U_A$. Show that, in an n degree of freedom phase space, the set of all linearly matched matrices is a manifold each connected piece of which is topologically equivalent to an n-dimensional torus. Remark on the similarities and differences of your argument with singular value decomposition.

2.11 Using Eqs. (2.30) and (2.22), write an explicit representation of $\underline{\underline{M}}(\theta_2, \theta_1)$ using the lattice functions at θ_1 and θ_2.

2.12 Repeat the discussion of accelerator lattice functions but starting from the representation

$$\begin{pmatrix} x \\ p \end{pmatrix} \propto \begin{pmatrix} i + \alpha \\ -\gamma \end{pmatrix}.$$

What would be the expressions corresponding to Eq. (2.32)?

2.13 Show that the parameter f, which appears in the thin lens model of Eq. (2.25), is indeed the focal length of the lens.

2.14 Calculate the tune, ν, corresponding to one FODO cell of the thin lens model, showing that it satisfies $\sin \pi\nu = l/2f$. Show that we must have $\nu < 1/2$.

2.15 Calculate the lattice function β at the location of the quadrupoles (or lenses) of the thin lens FODO lattice. Show that their product, $\beta_F\beta_D$, does not depend on l.

2.16 The resonance condition of Eq. (2.34) remains the same under the change $\underline{m} \longmapsto k\underline{m}$, $n \longmapsto kn$. For a two-dimensional problem, explain the difference in the placement of resonant orbits depending on whether m_1, m_2, and n are or are not co-prime. (Hint: Complete solution requires material from Chapter 4.)

2.17 Write a representation for a vector field that possesses a *repelling* periodic orbit.

2.18 Compare the small γ estimate of the radius of the Van der Pol limit cycle with numerical computations.

2.19 Confirm that the two off-origin fixed points of the Lorenz dynamic are point attractors when

$$1 < \rho < \frac{\sigma(\sigma + \beta + 3)}{\sigma - \beta - 1}.$$

Sketch the paths taken by the eigenvalues of the linearized flow in crossing the bifurcation.

2.20 Verify that λ, appearing in the Hénon map of Eq. (2.52), can be transformed away by a change of scale.

2.21 Show that the normal form Hénon map of Eq. (2.51) possesses both a hyperbolic and an elliptic fixed point when $-(1 - b)^2/4 < a < 3(1 - b)^2/4$.

2.22 The mappings $f(x) = -x$ and $f(x) = 1/x$ both possess the remarkable property that *all* orbits in their domain are periodic with prime period 2: that is, $f(f(x)) = x$. Is there any function possessing the same property for period 3? (Comment on the connection with Sharkovsky's theorem.)

2.23 Using bitmap graphics, draw the set of starting points in the complex plane for which Newton's method fails to find n^{th} roots of unity. In particular, find all complex z for which the iterative scheme $z \leftarrow (2z^3 + 1)/3z^2$ fails to converge to one of the three cube roots of unity: 1 and $e^{\pm 2\pi i/3}$. The roots of unity in this example are attractors of Newton's map. The set of all points in the complex plane that converge to one of them is called its "basin of attraction." The points for which Newton's method fail are on the boundary between adjoining basins; hence the term "fractal basin boundary."

2.24 Verify explicitly that for linear systems, with matrix $\underline{\underline{A}}$,

$$e^{t\mathbf{V}}\underline{z} = e^{t\underline{\underline{A}}} \cdot \underline{z}.$$

2.25 Repeat the calculation of Eq. (2.66) for $e^{t\mathbf{V}}a$ and $e^{t\mathbf{V}}a^*$, with the representation of $\mathbf{V}$ given at the end of Eq. (2.57).

2.26 Sketch orbits of the vector field over R^2,

$$\mathbf{V} = (-x + x^2 - p^2)\frac{\partial}{\partial x} + (-p + 2xp)\frac{\partial}{\partial p}.$$

In particular, show that their paths are circles. Write the simplest algorithm you can for finding $\underline{z}(t)$ given $\underline{z}(0)$.

2.27 Using the concept of winding number, "prove" the fundamental theorem of algebra: Every polynomial of degree n has n solutions (counting according to multiplicities).

2.28 Show that any commutator algebra is a Lie algebra.

2.29 Show that (a) linear dynamics are closed under the Lie bracket and (b) their matrix commutator algebra is an antimorphism of their Lie bracket algebra. That is, show that if $\underline{a}$ and $\underline{b}$ are linear dynamics with matrix representations $\underline{\underline{A}}$ and $\underline{\underline{B}}$, then their Lie bracket, $[\underline{a}, \underline{b}]$, is a linear dynamic with matrix $-[\underline{\underline{A}}, \underline{\underline{B}}]$.

2.30 Confirm that $\mathbf{W} = [\mathbf{U}, \mathbf{V}]$ is indeed a vector field by showing it is coordinate independent. (Equivalently, show that the local representative components of $\mathbf{W}$ transform correctly.)

2.31 Interpret the Lie bracket in an extended phase space and separate the *time* sector.

2.32 Let $f(z)$ and $g(z)$ be solutions of the "generalized Hill's equation,"

$$y''(z) + p(z)y'(z) + q(z)y(z) = 0,$$

where p and q are arbitrary, not necessarily periodic. Show that the bracket (aka Wronskian),

$$w(z) = [f, g]_{(z)} = f(z)g'(z) - f'(z)g(z),$$

evolves according to

$$w(z) = w(z_0)\exp\left[-\int_{z_0}^{z} p(u)du\right].$$

2.33 Explain the difficulty in classifying linear systems, all of whose eigenvalues are purely imaginary. Give an example of two such systems that possess the same eigenvalues but are not topologically equivalent.

2.34 Show that $\phi^{-1}(A \cap B) = \phi^{-1}(A) \cap \phi^{-1}(B)$.

2.35 Show that the proposition "$U = \phi\phi^{-1}U$" is a set-theoretic identity but "$U = \phi^{-1}\phi U$" is not.

2.36 Confirm Eq. (2.74) for the lattice algebra of partitions.

2.37 Let us define a "Koch operation" on a line segment in the following way: (a) divide the line segment into thirds, (b) construct an equilateral triangle on the middle third, and (c) remove its base, the intersection of the triangle with the original segment. The initial line segment is thus replaced by four connected line segments, each one-third the length of the original. Now consider starting from a given polygon and recursively generating an infinite family of polygons by, at each step of the recursion, performing a Koch operation on every segment of the preceding polygon. Use the self-similar, scaling method to calculate the fractal dimension of the "limiting curve."

2.38 Let $f_1, f_2, f_3 : U^2 \longrightarrow U^2$, be three contractive mappings of the unit square into itself, defined as follows:

$$f_1 : (x, y) \longmapsto (x/2, y/2),$$
$$f_2 : (x, y) \longmapsto ((x + 1)/2, y/2),$$
$$f_3 : (x, y) \longmapsto (x/2, (y + 1)/2).$$

Construct a stochastic process on U^1 by randomly choosing at each step from f_1, f_2, and f_3. That is, $z_0, z_1, z_2, \ldots$ will be a random sequence of points in U^2, and for all k, $\text{Prob}[z_{k+1} = f_n(z_k)] = 1/3, n = 1, 2, 3$. Beginning anywhere in U^2, plot an orbit of this process. The strange attractor that results has an obvious relation to Pascal's triangle mod 2. What is its Hausdorff dimension?

2.39 Show that the property of linear independence of a pencil of orbits is conserved.

2.40 Find the regions of phase space that are contracting for (a) the Lorenz equation, (b) the Van der Pol oscillator, and (c) Duffing's equation.

2.41 Prove, using the expansion rate equation, Eq. (2.84), that for any matrix, $\underline{\underline{A}}$

$$\det\left[e^{\underline{\underline{A}}}\right] = e^{\mathrm{Tr}(\underline{\underline{A}})}.$$

2.42 Which observables $\rho : \mathcal{P} \longrightarrow R$ can serve as an equilibrium distribution of some vector field? Given the equilibrium distribution, is the vector field unique, apart from trivial multiplication by a scalar?

2.43 Let $\mathcal{P} = R$ and consider the Gaussian distribution,

$$\rho(x) = \exp(-x^2/2\sigma^2)/\sqrt{2\pi}\sigma.$$

Suppose that a stream of particles are emitted with positive velocity from a source at $x = -\sigma$ and that a force field $F(x)$ is established in the region $x \in \Sigma = (-\sigma, \sigma)$. For what $F(x)$ will the equilibrium distribution of particles be proportional to $\rho(x)$ within Σ? (For the criminally ambitious: build a mechanical model that could be a high school physics demo.)

2.44 Show that at a fixed point, $\underline{z}_o$, of an autonomous system a distribution function decays (or grows) exponentially,

$$\rho(\underline{z}_o, t) = \rho(\underline{z}_o, 0)e^{-\Gamma t},$$

and find the decay rate, Γ.

2.45 Demonstrate that it is possible to derive Eq. (2.87) from Eq. (2.89), but not vice versa.

2.46 Is it possible for the set of recurrent orbits associated with a neighborhood of phase space to be countable?

2.47 Prove the combinatorial lemma in Eq. (2.83).

2.48 The projection of a (non-relativistic) Kepler orbit from phase space into the position sector lies on a conic section: an ellipse if closed, a hyperbola (generically) if open. Hamilton was the first to notice that its projection into the momentum sector always lies on a circle. Show that this is so. Overlay sketches of the two projections, and relate the eccentricity of the spatial conic to the geometry of the momentum circle.

CHAPTER 3

Hamiltonian Dynamics

> Human beings are incredibly conceited fellows, too prompt to admire their own work, and to indulge in wishful thinking! ... We select experimental results that appear to us as logically connected together, and we ignore many facts that do not fit into our "logic." This rather artificial procedure is our own invention and we are so proud of it that we insist its results should be considered as "laws of nature." We also select a few laws as of superior importance and we call them "principles."
>
> — Leon Brillouin
> *Scientific Uncertainty, and Information*

Among all the possible dynamics, Hamiltonian systems hold a special place in the arsenal of physics. A goodly fraction of beam physics is concerned with Hamiltonian models whose Poincaré maps are essential for understanding orbit stability in accelerators.[1] The subject is deep, and our treatment will be "intermediate," rather than "advanced": neither as tutorial and complete as Goldstein [41] nor as mathematically rigorous as Abraham and Marsden [1], but with enough ideas in place to serve as stepping stones between these two excellent sources. It is regrettable that much will be left out — we do not examine Noether's theorem, the KAM theorem, or Hamilton-Jacobi theory,[2] for example — but choices had to be made.

The first section is given to reviewing the most basic notions of Hamiltonian dynamics from the viewpoint established in the preceding chapter. Of special importance is the further extension of the Lie algebras of commutators and brackets

[1] Which is not the same as saying that orbit stability in accelerators is understandable.

[2] {See Chapter 6, however.}

into the Hamiltonian domain. Transformations of Hamiltonian systems are presented next, followed by a treatment of action-angle coordinates and their connection with invariant tori. The action-angle formalism will be especially useful in the later study of Hamiltonian normal forms and resonances. This chapter ends with an introduction to two Hamiltonian models used in accelerator physics.

3.1 HAMILTON'S EQUATION

A Hamiltonian dynamic is linearly related to a gradient field. Let $H : \mathcal{P} \rightarrow R$ be an observable. Components of the gradient field, $\mathbf{V}$, associated with H are written locally as follows:

$$\underline{v} = \frac{\partial H}{\partial \underline{z}} \quad \text{(gradient field components).} \tag{3.1}$$

If the phase space has *even* dimension, say, $2N$, then we can derive a Hamiltonian vector field from this by introducing the matrix, $\underline{\underline{J}}$, written in block form,

$$\underline{\underline{J}} \equiv \begin{pmatrix} \underline{\underline{0}} & \underline{\underline{1}} \\ -\underline{\underline{1}} & \underline{\underline{0}} \end{pmatrix}, \tag{3.2}$$

where $\underline{\underline{0}}$ and $\underline{\underline{1}}$ are, respectively, the $N \times N$ null and identity matrices. We note here for future reference two important properties: $\underline{\underline{J}}$ is

$$\begin{aligned} &\text{(a) skew-symmetric:} \quad && \underline{\underline{J}}^T = -\underline{\underline{J}}, \quad \text{and} \\ &\text{(b) idempotent:} && \underline{\underline{J}}^2 = -\underline{\underline{1}},\ \underline{\underline{J}}^4 = \underline{\underline{1}}. \end{aligned} \tag{3.3}$$

A Hamiltonian field is then obtained from a gradient one by premultiplication with $\underline{\underline{J}}$, so that in place of Eq. (3.1) we have,

$$\underline{v} = \underline{\underline{J}} \cdot \frac{\partial H}{\partial \underline{z}} \quad \text{(Hamiltonian field components).} \tag{3.4}$$

Given an observable H, the Hamiltonian vector field generated by H is symbolized as X_H; we shall also use boldface capitals, as in Chapter 1. Using the antisymmetry of $\underline{\underline{J}}$ and Eq. (2.56) X_H is evaluated as follows:

$$X_H = \mathbf{V} = \underline{v} \cdot \underline{\partial} = -\underline{\partial} H \cdot \underline{\underline{J}} \cdot \underline{\partial}.$$

X_H is a symbol more traditional with mathematicians; if we separate $\underline{z}$ into "position" and "momentum" sectors, the components take on a form more readily recognized by physicists.

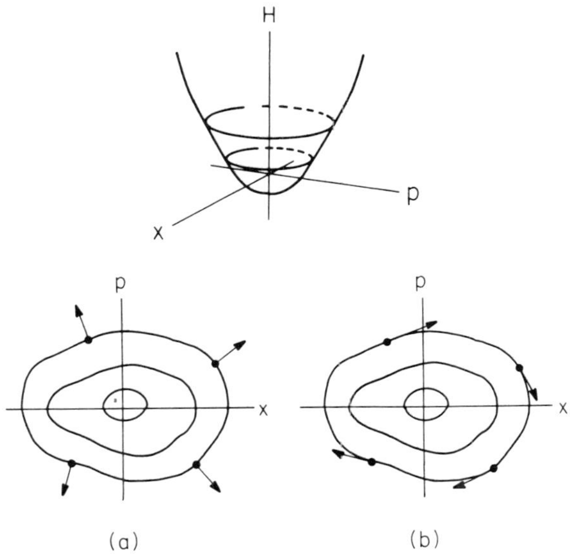

FIGURE 3.1 Rotations change a gradient field to a Hamiltonian one.

$$\underline{z} = \begin{pmatrix} \underline{x} \\ \underline{p} \end{pmatrix},$$

$$\dot{\underline{z}} = \underline{v}(\underline{z}, t) = \underline{\underline{J}} \cdot \partial H / \partial \underline{z} = \begin{pmatrix} \underline{\underline{0}} & \underline{\underline{1}} \\ -\underline{\underline{1}} & \underline{\underline{0}} \end{pmatrix} \cdot \begin{pmatrix} \partial H / \partial \underline{x} \\ \partial H / \partial \underline{p} \end{pmatrix},$$

$$\dot{\underline{x}} = \partial H / \partial \underline{p}, \quad \dot{\underline{p}} = -\partial H / \partial \underline{x}.$$

When the vector field is represented in this form, each (x_k, p_k) pair are said to comprise a "conjugate" pair of coordinates; that is, x_k is "conjugate" to p_k and vice versa. We shall also consider Eq. (3.4) to be the "canonical form" of Hamilton's equations; we shall see shortly that other forms are possible.

One can heuristically think of $\underline{\underline{J}}$ as performing a $\pi/2$ rotation of the gradient field.[3] (See Figure 3.1.) Indeed, by the above identities, $\underline{\underline{J}}$ is a member of the orthogonal group, O^{2N} — that is, $\underline{\underline{J}}^T \cdot \underline{\underline{J}} = \underline{\underline{1}}$ — and $\underline{\underline{J}}^4 = \underline{\underline{1}}$. Furthermore, the Lie derivative of H vanishes,

$$\underline{v} \cdot \underline{\partial} H = \frac{\partial H}{\partial \underline{z}} \cdot \underline{\underline{J}}^T \cdot \frac{\partial H}{\partial \underline{z}} = -\frac{\partial H}{\partial \underline{z}} \cdot \underline{\underline{J}} \cdot \frac{\partial H}{\partial \underline{z}} = 0,$$

by the skew-symmetry of $\underline{\underline{J}}$. Besides implying that $\dot{H} = \partial H / \partial t$, a result of interest in itself, this means that the gradient field and the Hamiltonian field are orthogonal, in a dot-product sense. In two dimensions this rotation picture is complete as well as

[3]Although the concept of "angle" has no home in a space that does not possess a metric.

descriptive, since

$$\underline{\underline{J}} = \begin{pmatrix} 0 & 1 \\ -1 & 0 \end{pmatrix} = \begin{pmatrix} \cos(\pi/2) & \sin(\pi/2) \\ -\sin(\pi/2) & \cos(\pi/2) \end{pmatrix}.$$

A second, immediate property that follows from skew-symmetry is that Hamiltonian fields (whether autonomous or not) must be conservative.

$$\begin{aligned} \underline{\partial} \cdot \underline{v} &= \frac{\partial}{\partial \underline{z}} \cdot \left(\underline{\underline{J}} \cdot \frac{\partial H}{\partial \underline{z}} \right) \\ &= \mathrm{Tr}\left[\underline{\underline{J}} \cdot \frac{\partial^2 H}{\partial \underline{z} \partial \underline{z}} \right] \\ &= 0. \end{aligned}$$

This means that *Hamiltonian systems cannot possess attractors*, an attribute that makes them much more difficult to study than their dissipative cousins.[4]

Linear Hamiltonian Systems Hamiltonian functions that are quadratic polynomials in the coordinates lead to linear dynamics.

$$\begin{aligned} H &\equiv \underline{z} \cdot \underline{\underline{B}} \cdot \underline{z}, \\ \underline{v} &= \underline{\underline{J}} \cdot \frac{\partial}{\partial \underline{z}} (\underline{z} \cdot \underline{\underline{B}} \cdot \underline{z}) \\ &= (\underline{\underline{J}} \cdot (\underline{\underline{B}} + \underline{\underline{B}}^T)) \cdot \underline{z}. \end{aligned}$$

Conversely, the linear system, $\underline{v} = \underline{\underline{A}} \cdot \underline{z}$, is Hamiltonian provided $\underline{\underline{A}} = \underline{\underline{J}}\,\underline{\underline{S}}$, where $\underline{\underline{S}}$ is a symmetric matrix. Using Eqs. (3.3) we can solve for $\underline{\underline{S}} = -\underline{\underline{J}}\,\underline{\underline{A}}$ and the condition $\underline{\underline{S}} = \underline{\underline{S}}^T$ can then be written

$$\underline{\underline{J}}\,\underline{\underline{A}} + \underline{\underline{A}}^T \underline{\underline{J}} = \underline{\underline{0}}. \tag{3.5}$$

This is the necessary and sufficient condition on the matrix of a linear system in order that it be Hamiltonian.

The most important member of this class is, of course, the harmonic oscillator. The phase space for an ensemble of N independent harmonic oscillators is $\mathcal{P} = R^{2N}$,

[4]There is a structure, called a "vague attractor" by some, which we shall encounter later as a "separatrix," but this is not a true attractor in the sense of absorbing regions of phase space. Attractors pull in not only phase space but researchers as well: the fact that one can start anywhere in phase space and eventually land on an "interesting" structure has tremendous appeal. Among the more sophistically inclined, it leads to speculations about self-organization, the origins of life, and other alluring topics.

and its Hamiltonian is written as follows:

$$H = \sum_{k=1}^{N} \left(\frac{p_k^2}{2m_k} + \frac{1}{2} m_k \omega_k^2 x_k^2 \right). \tag{3.6}$$

Applying Hamilton's equation, Eq. (3.4), provides the individual component equations,

$$\begin{aligned} \dot{x}_k &= \frac{\partial H}{\partial p_k} = p_k/m_k, \\ \dot{p}_k &= -\frac{\partial H}{\partial x_k} = -m_k \omega_k^2 x_k, \end{aligned}$$

exactly as in Eq. (2.2) with $\Gamma \equiv 0$. Note in passing that the level surfaces of this Hamiltonian are ellipsoids, in keeping with the origin being an elliptic fixed point.

3.1.1 Brackets

In Chapter 1 we associated a vector field with a linear operator on the vector space of observables, the Lie derivative for observables. The commutator of these operators, upon evaluation in a local representation, led to the notion of the Lie bracket. We shall show here that if two vector fields, **U** and **V**, are Hamiltonian, then so is their commutator, $\mathbf{W} = [\mathbf{U}, \mathbf{V}]$. Furthermore, the Hamiltonian, W, is a skew-symmetric, bilinear function of U and V, which, not to anyone's surprise, obeys the Jacobi identity.

Poisson Bracket We express these facts in the form of a theorem.

Theorem 3.1 Let $\underline{u} = \underline{\underline{J}} \cdot \underline{\partial} U$ and $\underline{v} = \underline{\underline{J}} \cdot \underline{\partial} V$ be the local components of two Hamiltonian vector fields with Lie bracket, $\underline{w} \equiv [\underline{u}, \underline{v}]$. Then there is an observable, W, such that $\underline{w} = \underline{\underline{J}} \cdot \underline{\partial} W$. That is, $\mathbf{W} = \underline{w} \cdot \underline{\partial}$ is a Hamiltonian vector field generated by the Hamiltonian, W.

Proof: We begin with the expression in Eq. (2.68),

$$[\underline{u}, \underline{v}] = \underline{u} \cdot \underline{\underline{\partial v}} - \underline{v} \cdot \underline{\underline{\partial u}},$$

where $\underline{u} = \underline{\underline{J}} \cdot \underline{\partial} U$ and $\underline{v} = \underline{\underline{J}} \cdot \underline{\partial} V$. First, let us manipulate $\underline{\underline{\partial v}}$, using component notation for clarity (Einstein summation convention assumed).

$$\begin{aligned} (\underline{\underline{\partial v}})_{ik} &= \partial_i J_{km} \partial_m V \\ &= J_{km} \partial_i \partial_m V \\ &= \partial_i \partial_m V J_{mk}^T \\ &= -\partial_i \partial_m V J_{mk}. \end{aligned}$$

All indices are now in the appropriate location so that we can write this in matrix form.

$$\underline{\underline{\partial v}} = -(\underline{\underline{\partial\partial V}}) \cdot \underline{\underline{J}}$$

We now substitute this into the expression for Lie's bracket, first pausing to note that, by Eq. (3.3), $\underline{\underline{J}} \cdot \underline{\partial} U = -\underline{\partial} U \cdot \underline{\underline{J}}$.

$$\begin{aligned}[\underline{u}, \underline{v}] &= (-\underline{\partial} U \cdot \underline{\underline{J}}) \cdot (-\underline{\underline{\partial\partial}} V \cdot \underline{\underline{J}}) - (U \leftrightarrow V) \\ &= (-\underline{\underline{J}}) \cdot (\underline{\underline{\partial\partial}} V \cdot (-\underline{\underline{J}}) \cdot \underline{\partial} U) - (U \leftrightarrow V) \\ &= \underline{\underline{J}} \cdot (\underline{\underline{\partial\partial}} V \cdot \underline{\underline{J}} \cdot \underline{\partial} U - \underline{\underline{\partial\partial}} U \cdot \underline{\underline{J}} \cdot \underline{\partial} V).\end{aligned}$$

Now it again becomes temporarily convenient to introduce indices. The kth component of the last expression is written

$$\begin{aligned}&(\underline{\underline{J}})_{km}(\partial_m \underline{\partial} V \cdot \underline{\underline{J}} \cdot \underline{\partial} U - \partial_m \underline{\partial} U \cdot \underline{\underline{J}} \cdot \underline{\partial} V) \\ &\quad = (\underline{\underline{J}})_{km}(\partial_m \underline{\partial} V \cdot \underline{\underline{J}} \cdot \underline{\partial} U + \underline{\partial} V \cdot \underline{\underline{J}} \cdot \partial_m \underline{\partial} U) \\ &\quad = (\underline{\underline{J}})_{km} \partial_m (\underline{\partial} V \cdot \underline{\underline{J}} \cdot \underline{\partial} U).\end{aligned}$$

We can now replace the matrix notation and collect the pieces.

$$\begin{aligned}\underline{w} \equiv [\underline{u}, \underline{v}] &= \underline{\underline{J}} \cdot \underline{\partial} W, \quad \text{where} \\ W &= (\underline{\partial} V \cdot \underline{\underline{J}} \cdot \underline{\partial} U).\end{aligned} \tag{3.7}$$

The components of $\underline{w}$ are those of a Hamiltonian vector field with Hamiltonian W. □

We have shown that the set of Hamiltonian vector fields is closed under commutation. Because $\underline{\underline{J}}$ is skew-symmetric, W is a bilinear, skew-symmetric function of the observables U and V. We shall anticipate the fact that this bilinear operation obeys the Jacobi identity as well and overload the bracket notation one last time, extending its application to observables.

$$[f, g] \equiv \underline{\partial} f \cdot \underline{\underline{J}} \cdot \underline{\partial} g \rightsquigarrow \frac{\partial f}{\partial \underline{x}} \cdot \frac{\partial g}{\partial \underline{p}} - \frac{\partial g}{\partial \underline{x}} \cdot \frac{\partial f}{\partial \underline{p}}. \tag{3.8}$$

This operation is called the "Poisson bracket" of f with g.

The above derivation establishes an *antimorphism* that is most easily expressed using the mathematicians' symbolism: what we have demonstrated is that

$$\begin{aligned}[X_U, X_V] &= [\underline{u}, \underline{v}] \cdot \underline{\partial} \\ &= (\underline{\underline{J}} \cdot \underline{\partial}(-[U, V])) \cdot \underline{\partial} \\ &= X_{-[U,V]}.\end{aligned} \tag{3.9}$$

By repeated use of this identity, and the fact that the vector field operators themselves obey the Jacobi identity, it is easy to show that

$$X_{[U,[V,W]]+[V,[W,U]]+[W,[U,V]]} = 0.$$

However, we cannot deduce from this alone that the Poisson bracket obeys the Jacobi identity and therefore generates a Lie algebra. That final calculation is left as an exercise for the reader. Its completion validates the following theorem.

Theorem 3.2 The Poisson bracket is a Lie algebraic operation over the vector space of observables.

Another important connection exists between Poisson brackets and Hamiltonian vector fields. Let f and H be any two observables, and consider the Lie derivative of f associated with the Hamiltonian vector field $\mathbf{V} = X_H$. Applying Eq. (2.59), we have

$$\dot{f} = X_H f = \mathbf{V}f = \underline{v} \cdot \underline{\partial} f = -\underline{\partial} H \cdot \underline{\underline{J}} \cdot \underline{\partial} f = [f, H], \tag{3.10}$$

with the natural generalization to extended observables,

$$\dot{f} = [f, H] + \frac{\partial f}{\partial t}.$$

Applying this to the coordinates produces the Hamiltonian form of Eq. (2.60),

$$\underline{v} = [\underline{z}, H], \tag{3.11}$$

which is interpreted component-wise.

Acting as a linear operator on the space of observables, the action of a Hamiltonian vector field is, within a sign, to perform a Poisson bracket. In the language of abstract Lie theory, this means that X_H acts as the "adjoint operator" associated with H. (Dragt [30] introduced the notation $:H:$ into the accelerator physics literature in order to indicate this; Dragt's $:H:$ is equivalent to the vector field, $:H: = -X_H = -\mathbf{V}$.)

3.1.2 Correspondence Principle

The morphisms between vector field commutators, Lie brackets, and Poisson brackets (for Hamiltonian dynamics) are reminiscent of that between classical observables and the linear operators of quantum theory. In our introductory courses to quantum mechanics we were taught Dirac's quantization procedure, which postulated a morphism between the Lie algebra of Poisson brackets and the commutator algebra of Hermitian operators on a Hilbert space [29]. What is frequently not mentioned in these courses is that this procedure fails for any system more complicated than harmonic oscillators.[5] Loosely speaking, this is the content of a theorem, due to

[5]More precisely, it is impossible to extend it to observables that are cubic or higher order polynomials in x and p.

Groenwold [42] and Van Hove [48], which deserves to be more widely known. Let $f \longmapsto \bar{f}$ be a correspondence between classical observables of a phase space and the linear operators of a Hilbert space. We shall assume this mapping obeys the desired morphism, which we shall call the "naive correspondence principle" (NCP).

$$\begin{aligned} [\bar{f},\bar{g}] &= i\overline{[f,g]}, \\ \bar{1} &= \text{identity}. \end{aligned} \tag{3.12}$$

The brackets on the left represent commutators of linear operators while those on the right represent Poisson brackets between classical observables. We shall show that this equality cannot be satisfied for polynomials of degree larger than two.

To save space, we shall accept the following result: the operators associated with second order polynomials are determined by the NCP conditions to be

$$\overline{x^2} = \bar{x}^2, \quad \overline{p^2} = \bar{p}^2, \quad \overline{px} = \overline{xp} = \tfrac{1}{2}(\bar{p}\bar{x} + \bar{x}\bar{p}). \tag{3.13}$$

That this is so can be shown by the methods we shall apply below.

We now try to extend NCP to third degree polynomials. Without loss of generality, we shall write

$$\overline{x^3} = \bar{x}^3 + R,$$

where R is a "residual" operator that we must determine. In fact, we shall show that our expectation, $R = 0$, is correct. First, from $[x^3, x] = 0$ and the NCP it follows immediately that $[R, \bar{x}] = 0$. R also commutes with $\bar{p}$, since NCP provides the following implications:

$$\begin{aligned} [x^3, p] = 3x^2 \Rightarrow 3i\bar{x}^2 &= [\bar{x}^3 + R, \bar{p}] \\ &= 3i\bar{x}^2 + [R, \bar{p}] \\ \Rightarrow [R, \bar{p}] &= 0. \end{aligned}$$

Since R commutes with both $\bar{x}$ and $\bar{p}$, it must commute with $\overline{xp}$:

$$[R, \overline{xp}] = \frac{1}{2}[R, \bar{x}\bar{p} + \bar{p}\bar{x}] = 0.$$

NCP then provides the following implication:

$$[x^3, xp] = 3x^3 \Rightarrow [\bar{x}^3 + R, \overline{xp}] = [\bar{x}^3, \overline{xp}] = 3i\bar{x}^3 + 3iR.$$

But, from Eq. (3.13), $[\bar{x}^3, \overline{xp}] = \frac{1}{2}(\bar{x}[\bar{x}^3, \bar{p}] + [\bar{x}^3, \bar{p}]\bar{x}) = 3i\bar{x}^3$. Thus we must have $R = 0$ and $\overline{x^3} = \bar{x}^3$. A parallel line of reasoning leads to the conclusion that $\overline{p^3} = \bar{p}^3$.

We now consider the monomials px^2 and p^2x. Our new results, combined with NCP and the Poisson bracket, $[x^3, p^2] = 6px^2$, lead to the following:

$$\begin{aligned} 6i\overline{px^2} &= [\overline{x^3}, \overline{p^2}] = [\bar{x}^3, \bar{p}^2] \\ &= \bar{p}[\bar{x}^3, \bar{p}] + [\bar{x}^3, \bar{p}]\bar{p} \\ &= 3i\bar{p}\bar{x}^2 + 3i\bar{x}^2\bar{p} \\ \overline{px^2} &= \tfrac{1}{2}(\bar{p}\bar{x}^2 + \bar{x}^2\bar{p}). \end{aligned}$$

A similar calculation leads to the result

$$\overline{p^2x} = \tfrac{1}{2}(\bar{p}^2\bar{x} + \bar{x}\bar{p}^2).$$

And now we are in trouble, because the Poisson bracket identity,

$$\tfrac{1}{9}[x^3, p^3] = \tfrac{1}{3}[px^2, p^2x] \quad (= x^2p^2),$$

taken with the NCP must imply that,

$$\tfrac{1}{9}[\bar{x}^3, \bar{p}^3] = \tfrac{1}{3}[\overline{px^2}, \overline{p^2x}].$$

However, a rather tedious but straightforward calculation — which is left to the reader — using our previously generated identities shows that

$$\begin{aligned} \tfrac{1}{9}[\bar{x}^3, \bar{p}^3] &= i(\bar{x}^2\bar{p}^2 - \tfrac{2}{3}) + 2\bar{x}\bar{p} \\ \tfrac{1}{3}[\overline{px^2}, \overline{p^2x}] &= i(\bar{x}^2\bar{p}^2 - \tfrac{1}{3}) + 2\bar{x}\bar{p}. \end{aligned}$$

So there is a serious discrepancy: $2 \neq 1$. Nor can we take refuge in ignoring the third degree: taking commutators of higher degree polynomials with each other and with linear polynomials eventually leads to polynomials of the third degree. There is no avoiding them.

Theorem 3.3 (Groenwold, Van Hove; paraphrased) The morphism postulated by Dirac's correspondence principle, as expressed in Eq. (3.12) cannot be extended consistently to polynomials beyond second degree.

The reason the morphism works in classical mechanics is that $X_1 = \mathbf{0}$; the reason it fails in quantum mechanics is that $\overline{1} = \mathbf{1}$. More detailed treatments can be found in Abraham and Marsden [1] and in Guillemin and Sternberg [44].

3.1.3 Lagrangian Dynamics

Hamiltonian dynamics would not have deserved the attention they have enjoyed were it not that so many classes of physical systems can be expressed naturally

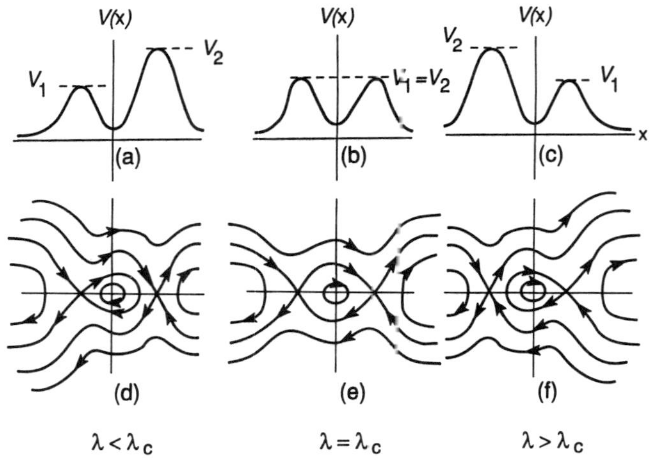

FIGURE 3.2 Saddleswitch is a global bifurcation that car occur when the Hamiltonian takes the same value at two or more hyperbolic critical orbits.

in Hamiltonian form. Indeed, one of physics' deepest articles of faith is that the "fundamental" laws of nature can be written efficiently using either Hamiltonians or Lagrangians.

Newtonian Dynamics, and the Saddleswitch For instance, Newton's law governing the motion of a nonrelativistic particle can be written in Hamiltonian form when the force on the particle is conservative, so that it is expressible as the (negative) gradient of a "potential." To establish this, we need only combine Newton's law,

$$\dot{\underline{p}} = \underline{F} = -\underline{\partial}V,$$

with the (nonrelativistic) definition of momentum, $\underline{p} = m\dot{\underline{x}}$. The two of these are expressible in the form of Eq. (3.4) provided we identify $\mathcal{P} = R^6 = \{(\underline{x}, \underline{p})\}$ and

$$H(\underline{x}; \underline{p}; t) = \frac{1}{2m}\underline{p} \cdot \underline{p} + V(\underline{x}, t), \tag{3.14}$$

the total energy of the particle.

If we introduce a control parameter into the problem and remove the time dependence, this Hamiltonian provides a simple model of a global bifurcation, called the "saddleswitch" bifurcation. This is illustrated in Figure 3.2 for $\mathcal{P} = R^2$.Suppose $V(x; \lambda)$ to be an autonomous doubly peaked potential with control parameter λ. Let $x_1(\lambda)$ and $x_2(\lambda)$ be the position coordinates of the hyperbolic fixed points at the peak values of V, $V_1(\lambda) = V(x_1(\lambda); \lambda)$ and $V_2(\lambda) = V(x_2(\lambda); \lambda)$, and imagine that at a critical value, say, $\lambda = \lambda_c$, $V_1 - V_2$ changes sign. The situation is shown in Figure 3.2(a–c). Below each of these diagrams is sketched the flow in R^2. The saddleswitch bifurcation that takes place in Figure 3.2(e) is characterized by having two branches of the separatrix merge and then separate in a different configuration. It is a *global* rather than a *local* bifurcation: the linearized dynamics around the fixed points remain nonsingular and vary smoothly throughout the entire process. It is impossible to tell that a bifurcation has occurred by making localized observations in their neighborhoods; it reveals itself only upon taking a global view.

Equation (3.14) did not contain velocity-dependent forces, but we can extend the formalism by allowing V to depend on $\underline{p}$ as well as $\underline{x}$. The inescapable price for doing this is to alter the definition of "momentum."

$$H = \frac{1}{2m}\underline{p}\cdot\underline{p} + V(\underline{x},\underline{p},t),$$

$$\dot{\underline{x}} = \partial H/\partial\underline{p} = \underline{p}/m + \partial V/\partial\underline{p}, \quad \text{or}$$

$$\underline{p} = m\dot{\underline{x}} - m\partial V/\partial\underline{p}, \quad \text{and}$$

$$\dot{\underline{p}} = -\partial H/\partial\underline{x} = -\underline{\partial}V = \underline{F}.$$

In another section we shall examine the only example of velocity-dependent forces that concerns us and the one dynamic of greatest relevance to the material in later chapters: that of a relativistic, charged particle in an electromagnetic environment. But first we shall review the connections between Newtonian, Lagrangian, and Hamiltonian dynamical systems (of finite dimension). That we can go through these arguments so blithely centuries after the fact should belie neither their original brilliance nor the labor of mathematicians in laying foundations to support them.

From Newtonian Dynamics to Lagrangians Newtonian mechanics proceeded under the assumption that one knew the forces acting on all the "particles" of a system — typically the solar system, in Newton's time — and the objective was to calculate their orbits. Lagrange went on to consider systems for which we do not know *a priori* all the forces, such as rigid bodies coupled with joints, rods, chains, pulleys, frictionless surfaces, and the like. The characteristic of these systems is that their "particles" do not move independently of each other. Rather, their orbits are coupled via "constraints" imposed by the joints, rods, surfaces, and so forth. The forces that act on the "particles" of these systems are split into two categories: (a) the known "external" forces and (b) the unknown forces that maintain the constraints.

In the case of integrable, or "holonomic," constraints, the spatial position of the kth "particle," $\vec{r}_k$, can be written as a function of configuration coordinates, $\underline{x}$, formally expressed as $\vec{r}_k(\underline{x},t)$. In this section we shall use the arrow notation to denote components of the ordinary three-vectors of Newtonian mechanics. These are not logically different from the components of vectors defined over manifolds, which we have been symbolizing with underlines, but it will be convenient to make the distinction here. Unlike the individual "particle" coordinates, $\vec{r}_k$, the "generalized" coordinates, $\underline{x}$, *refer to the system as a whole*. They shall fill the coordinate charts of its configuration manifold.

The differential forms indicating the coupling between "particle" coordinates are written

$$d\vec{r}_k = (\partial\vec{r}_k/\partial\underline{x})\cdot d\underline{x} + (\partial\vec{r}_k/\partial t)\,dt. \tag{3.15}$$

"Nonholonomic," or nonintegrable, constraints are expressed via differential form equations of this type without the assumption of an existing $\vec{r}_k(\underline{x}, t)$; the partial derivatives are simply the arbitrary coefficients of the differentials $d\underline{x}$ and dt. From Eq. (3.15) we obtain immediately the relationship between the velocity of a "particle" and the system's "velocity" coordinates, $\dot{\underline{x}}$,

$$\vec{v}_k(\underline{x}, \dot{\underline{x}}, t) = (\partial\vec{r}_k/\partial\underline{x}) \cdot \dot{\underline{x}} + \partial\vec{r}_k/\partial t.$$

This already leads to an important observation: $\dot{\underline{x}}$ appears only once and linearly, as both $\partial\vec{r}_k/\partial\underline{x}$ and $\partial\vec{r}_k/\partial t$ are independent of it. Picking off the coefficient, we see that

$$\partial\vec{v}_k/\partial\dot{\underline{x}} = \partial\vec{r}_k/\partial\underline{x}. \tag{3.16}$$

To eliminate the unknown forces of constraint from the dynamical equations, we invoke "D'Alembert's principle": the forces of constraint do no net work on the system — or put another way, they cannot alter the system's total energy — when the system undergoes a "virtual displacement," $\delta\vec{r}_k \equiv d\vec{r}_k - (\partial\vec{r}_k/\partial t)\,dt$. The equation expressing D'Alembert's principle is

$$\sum_k \vec{F}_k^{(c)} \cdot \delta\vec{r}_k = 0, \tag{3.17}$$

where $\vec{F}_k^{(c)}$ is the total constraint force acting on "particle" k. In the case of continuous bodies, k would be considered a continuous variable, and sums over k become integrals. We shall not discuss these matters in detail. D'Alembert's principle would not be satisfied if $d\vec{r}_k$ were used rather than $\delta\vec{r}_k$, as constraint forces certainly change the energy of the system when they arise from driven supports. Virtual displacements freeze the system at an instant of time and move its "particles" consistent with the system's constraints, but *not with its explicit, nonautonomous time dependence*.

Eq. (3.17) is not a physical law but *an assumption about the system under consideration*. It would be violated, for example, in the case of an object moving along a surface with friction (whatever that is); in a system of four "particles" at the vertices of a tetrahedron — two of whose edges are springs, the other two, rigid rods — the springs must be considered as exerting "external" forces, *not* constraint forces.

We now combine Eq. (3.17) with Newton's law, $\vec{F}_k - \dot{\vec{p}}_k = 0$, to eliminate the constraint forces from consideration.

$$\sum_k (\vec{F}_k - \dot{\vec{p}}_k) \cdot \delta\vec{r}_k = \sum_k (\vec{F}_k^{(c)} + \vec{F}_k^{(e)} - \dot{\vec{p}}_k) \cdot \delta\vec{r}_k = \sum_k (\vec{F}_k^{(e)} - \dot{\vec{p}}_k) \cdot \delta\vec{r}_k = 0.$$

Obviously, this cannot be used to equate $\vec{F}_k^{(e)}$ and $\dot{\vec{p}}_k$, as the components of $\delta\vec{r}_k$ are not linearly independent but related by the constraint equations. We obviate that by substituting from Eq. (3.15), with $dt \equiv 0$:

$$\sum_k (\vec{F}_k^{(e)} - \dot{\vec{p}}_k) \cdot (\partial\vec{r}_k/\partial\underline{x}) \cdot d\underline{x} = 0.$$

The components of $d\underline{x}$ *are* linearly independent, so we can conclude that

$$\sum_k \vec{\mathrm{F}}_k^{(e)} \cdot (\partial \vec{\mathrm{r}}_k / \partial \underline{x}) = \sum_k \dot{\vec{\mathrm{p}}}_k \cdot (\partial \vec{\mathrm{r}}_k / \partial \underline{x}). \tag{3.18}$$

The expression on the left-hand side of this equation is called the "generalized force."

$$\underline{Q} \equiv \sum_k \vec{\mathrm{F}}_k^{(e)} \cdot (\partial \vec{\mathrm{r}}_k / \partial \underline{x}).$$

In the most general case, this must be evaluated by actually carrying out the sum (or integral, for nondiscrete objects) over all the "particles" in the object. If the forces are derived from a potential function, V, then $\underline{Q}$ is found more easily by using the chain rule:

$$\underline{Q} = -\sum_k (\partial V / \partial \vec{\mathrm{r}}_k) \cdot (\partial \vec{\mathrm{r}}_k / \partial \underline{x}) = -\partial V / \partial \underline{x}. \tag{3.19}$$

All reference to individual "particles" has been eliminated from this expression; $\underline{Q}$ is the (negative) gradient of the potential energy of the object as a whole with respect to its system coordinates. We see in this a step toward establishing a "phase space" for the object.

We manipulate the right-hand side of Eq. (3.18) as follows:

$$\dot{\vec{\mathrm{p}}}_k \cdot (\partial \vec{\mathrm{r}}_k / \partial \underline{x}) = \frac{d}{dt}(\vec{\mathrm{p}}_k \cdot (\partial \vec{\mathrm{r}}_k / \partial \underline{x})) - \vec{\mathrm{p}}_k \cdot \frac{d}{dt}(\partial \vec{\mathrm{r}}_k / \partial \underline{x}).$$

In the first term we substitute from Eq. (3.16), while in the second we commute d/dt with $\partial/\partial \underline{x}$, so that the right-hand side of Eq. (3.18) becomes

$$\sum_k \left(\frac{d}{dt}(\vec{\mathrm{p}}_k \cdot (\partial \vec{\mathrm{v}}_k / \partial \dot{\underline{x}})) - \vec{\mathrm{p}}_k \cdot (\partial \vec{\mathrm{v}}_k / \partial \underline{x}) \right).$$

The terms that appear in the parentheses are both of the form $\vec{\mathrm{p}}_k \cdot \partial \vec{\mathrm{v}}_k$, where symbolically ∂ stands for either $\partial/\partial \underline{x}$ or $\partial/\partial \dot{\underline{x}}$. Nonrelativistically we can identify

$$\begin{aligned} \vec{\mathrm{p}}_k \cdot \partial \vec{\mathrm{v}}_k &= m_k \vec{\mathrm{v}}_k \cdot \partial \vec{\mathrm{v}}_k \\ &= \partial \left(\frac{1}{2} m_k \vec{\mathrm{v}}_k^2 \right) \\ &= \partial T_k, \end{aligned}$$

where T_k is the kinetic energy of the kth "particle." The summation over k then results in the kinetic energy of the object as a whole. Putting the pieces together results in the following replacement for Eq. (3.18):

$$\underline{Q} = \frac{d}{dt}\frac{\partial T}{\partial \underline{\dot{x}}} - \frac{\partial T}{\partial \underline{x}}. \tag{3.20}$$

All references to individual particles have been eliminated; everything is now written in system coordinates, $\underline{x}$ and $\underline{\dot{x}}$.

A final step is taken when the dynamics are conservative, so that we can use Eq. (3.19). By defining the "Lagrangian" function,

$$L \equiv T - V,$$

and assuming that $\partial V/\partial \underline{\dot{x}} = 0$, Eq. (3.20) is rewritten in its final form:

$$\frac{d}{dt}\frac{\partial L}{\partial \underline{\dot{x}}} - \frac{\partial L}{\partial \underline{x}} = 0.$$

This is the "Euler-Lagrange equation" describing the behavior of a system in terms of system coordinates.

Calculus of Variations It is well known that Lagranians can be interpreted as the integrand of an optimization problem. Scanned quickly, the argument goes as follows. Define the "action integral,"

$$A \equiv \int_{t_1}^{t_2} dt\, L(\underline{x}(t), \underline{\dot{x}}(t), t).$$

We seek an "orbit," $\underline{x}(t)$, such that A is stationary against its variation.[6] Such an orbit will be called a "critical orbit" of L. Let $\underline{x}(t) \rightarrow \underline{x}(t) + \underline{u}(t)$ and $\underline{\dot{x}}(t) \rightarrow \underline{\dot{x}}(t) + \underline{\dot{u}}(t)$, where $\underline{u}$ is arbitrary, except that: (a) it is at least as smooth as $\underline{x}$, as a function of t, and (b) it vanishes at the endpoints, t_1 and t_2. Our requirement is that the variation in A be second order in $\underline{u}$; its calculation proceeds via an integration by parts.

$$\begin{aligned}
&\int_{t_1}^{t_2} dt\, L(\underline{x} + \underline{u}, \underline{\dot{x}} + \underline{\dot{u}}, t) - \int_{t_1}^{t_2} dt\, L(\underline{x}, \underline{\dot{x}}, t) + O(\underline{u}^2) \\
&= \int_{t_1}^{t_2} dt\, ((\partial L/\partial \underline{x}) \cdot \underline{u} + (\partial L/\partial \underline{\dot{x}}) \cdot \underline{\dot{u}}) \\
&= \int_{t_1}^{t_2} dt \frac{d}{dt}\left(\frac{\partial L}{\partial \underline{\dot{x}}} \cdot \underline{u}\right) - \int_{t_1}^{t_2} dt \left(\frac{d}{dt}\frac{\partial L}{\partial \underline{\dot{x}}} - \frac{\partial L}{\partial \underline{x}}\right) \cdot \underline{u}.
\end{aligned} \tag{3.21}$$

The first integral vanishes, by virtue of the boundary conditions, $\underline{u}(t_1) = \underline{u}(t_2) = \underline{0}$; the requirement that the second vanish for all allowed $\underline{u}$ reproduces the Euler-Lagrange equation. By itself, this does *not* mean that A takes on an extremum value at $\underline{x}$, any more than the vanishing of a function's derivative means that its value is a

[6]In the modern language of functional derivatives, $\delta A/\delta \underline{x}(t) = 0$.

local minimum; other conditions, involving second derivatives of the integrand, must be satisfied for that to be true; for an explanation of these matters, see Ewing [33].

From Lagrangians to Hamiltonians The Euler-Lagrange equation has the disadvantage of leading to second order differential equations, since $\partial L/\partial \underline{\dot{x}}$ typically contains $\underline{\ddot{x}}$. This makes them unsuitable for our picture of a dynamic as a vector field over a (phase space) manifold. We get around this by defining a "momentum," $\underline{p}$, which is a function of $\underline{x}$ and $\underline{\dot{x}}$, so that $\underline{\dot{p}}$ involves $\underline{\ddot{x}}$. Inverting these relations and replacing $\underline{\dot{x}}$ and $\underline{\ddot{x}}$ everywhere by the appropriate functions of $\underline{p}$ and $\underline{\dot{p}}$ would then produce first order differential equations, returning us to the phase space paradigm. Fortunately, there is no need to go through such arduous manipulations explicitly. The "Legendre transformation" provides a most elegant and efficient method for doing them, the result of which is a Hamiltonian dynamic equivalent to the Lagrangian system from which it springs.[7]

Starting from the Lagrangian, $L(\underline{x}, \underline{\dot{x}}, t)$, we write the differential form,

$$dL = (\partial L/\partial \underline{x}) \cdot d\underline{x} + (\partial L/\partial \underline{\dot{x}}) \cdot d\underline{\dot{x}} + (\partial L/\partial t)\, dt.$$

(Legitimately, this should be interpreted as *defining* the partial derivative symbols that appear in the parentheses.) We now define "momentum" to be $\underline{p} \equiv \partial L/\partial \underline{\dot{x}}$, so that the Euler-Lagrange equation implies that $\underline{\dot{p}} = \partial L/\partial \underline{x}$. Making these substitutions provides the following:

$$dL = \underline{\dot{p}} \cdot d\underline{x} + \underline{p} \cdot d\underline{\dot{x}} + (\partial L/\partial t)\, dt. \tag{3.22}$$

Now define H, which will turn out to be our Hamiltonian, as follows:

$$H = \underline{p} \cdot \underline{\dot{x}} - L. \tag{3.23}$$

We write the differential form, dH, and substitute from Eq. (3.22) for dL.

$$\begin{aligned} dH &= \underline{\dot{x}} \cdot d\underline{p} + \underline{p} \cdot d\underline{\dot{x}} - dL \\ &= \underline{\dot{x}} \cdot d\underline{p} - \underline{\dot{p}} \cdot d\underline{x} - (\partial L/\partial t)\, dt. \end{aligned}$$

The key result of this is that $d\underline{\dot{x}}$ has been swallowed. We now read and identify the coefficients.

$$\begin{aligned} \underline{\dot{x}} &= \left(\frac{\partial H}{\partial \underline{p}}\right)_{\underline{x},t}, \\ \underline{\dot{p}} &= -\left(\frac{\partial H}{\partial \underline{x}}\right)_{\underline{p},t}, \\ \frac{\partial H}{\partial t} &= -\frac{\partial L}{\partial t}. \end{aligned}$$

[7] See Arnold [5] for a very nice presentation of the geometric motivation behind Legendre's transformation.

The first two of these are indeed Hamilton's equations for $\dot{\underline{x}}$ and $\dot{\underline{p}}$, confirming that H is the system's Hamiltonian function; the third is thrown in for completeness.

The frequent association of a Hamiltonian with the total energy occurs when (a) the kinetic energy, T, is a quadratic form in $\dot{\underline{x}}$, and (b) the potential, V, is independent of $\dot{\underline{x}}$. Let

$$T = \tfrac{1}{2}\dot{\underline{x}} \cdot \underline{\underline{\mathcal{M}}}(\underline{x}, t) \cdot \dot{\underline{x}},$$

where $\underline{\underline{\mathcal{M}}}$ is the "mass matrix," which without loss of generality we can assume to be symmetric. Then the momentum is evaluated,

$$\underline{p} = \partial L/\partial \dot{\underline{x}} = \partial T/\partial \dot{\underline{x}} = \underline{\underline{\mathcal{M}}} \cdot \dot{\underline{x}},$$

because of the second assumption. Solve for $\dot{\underline{x}}$ and substitute above, keeping in mind that $\underline{\underline{\mathcal{M}}}$ is symmetric, to get

$$T = \tfrac{1}{2}\underline{p} \cdot \underline{\underline{\mathcal{M}}}^{-1} \cdot \underline{p}.$$

Finally, we write the Hamiltonian,

$$\begin{aligned} H &= \underline{p} \cdot \dot{\underline{x}} - L = \underline{p} \cdot \underline{\underline{\mathcal{M}}}^{-1} \cdot \underline{p} - T + V \\ &= 2T - T - V = T + V, \end{aligned}$$

which is indeed the total energy of the system. For a one-particle system with no constraints, this brings us back to Eq. (3.14).

3.1.4 Examples

Before moving on we shall pause to examine two simple models: the brachistochrone and the rotating pendulum.

Brachistochrone The associations between the calculus of variations, the Euler-Lagrange equation, and Hamiltonians mean that one should be able to attack (some) optimization problems using Hamilton's equation. As an example, consider the classic brachistochrone problem. The physical situation is that of a bead on a wire imbedded in a constant gravity field. (See Figure 3.3.) For what curve $y(x)$ representing the shape of the wire will the transit time from the origin, $(0, 0)$, to a predetermined point, $(a, b = y(a))$, be mimimum? The problem is formulated as minimization of an integral:

$$\text{minimize over paths} \quad \int_{\text{path}} dt = \int_{\text{path}} \frac{ds}{v} = \int_0^a dx \sqrt{\frac{1 + y'^2}{2gy}},$$

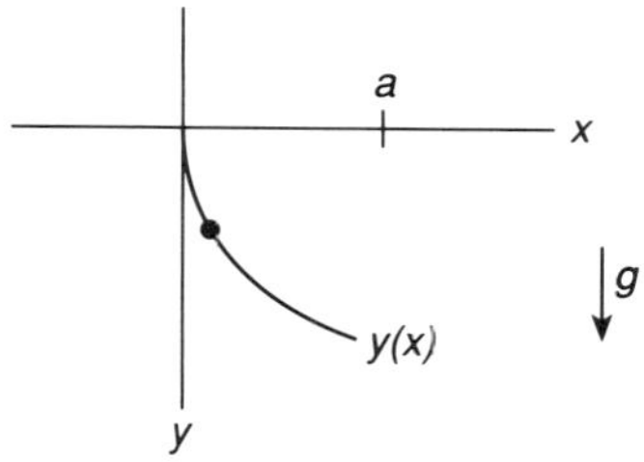

FIGURE 3.3 The brachistochrone problem. For what shape is the transit time minimum?

where we have used conservation of energy to write $mv^2/2 = mgy$, and we have chosen y to be decreasing downward. The symbolism has changed slightly from Eq. (3.21): here x, not t, is our orbit parameter, primes have replaced dots to represent differentiation, and the the phase space coordinates are (y, y'). Ignoring the factor $\sqrt{2g}$, we can take the Lagrangian to be

$$L(y, y') = \sqrt{\frac{1 + y'^2}{y}},$$

and the Euler-Lagrange equation becomes

$$\frac{d}{dx}\left(\frac{\partial L}{\partial y'}\right) - \frac{\partial L}{\partial y} = 0.$$

Rather than use the Lagrangian, we shall analyze the problem in its "phase space" by constructing its associated Hamiltonian. First define the "momentum":

$$p \equiv \frac{\partial L}{\partial y'} = \frac{y'}{\sqrt{y(1 + y'^2)}}.$$

Inverting this, we obtain y' as a function of p and y.

$$y'^2 = \frac{p^2 y}{1 - p^2 y}.$$

Multiply by p^2 and take the square root to get

$$py' = p^2 \sqrt{\frac{y}{1 - p^2 y}}.$$

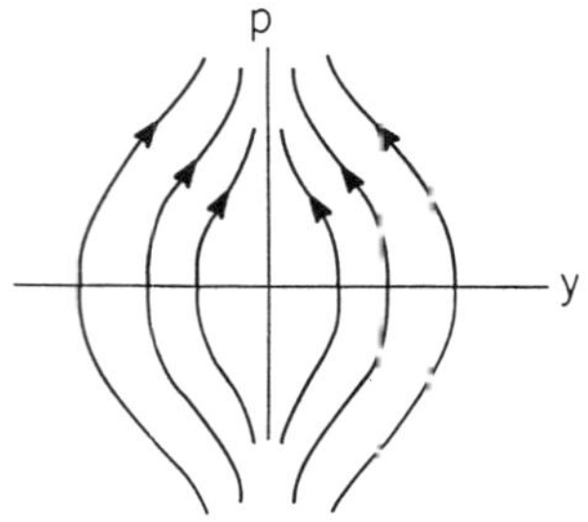

FIGURE 3.4 Flow diagram of the phase space associated with the brachistochrone problem.

Taking the positive square root is legitimate, as $py' \geqslant 0$. If we also use the obvious fact that $y \geqslant 0$, then we can write a Hamiltonian for this system as follows:

$$\begin{aligned} H = py' - L &= p^2\sqrt{\frac{y}{1-p^2y}} - \sqrt{\frac{1}{y(1-p^2y)}} \\ &= \frac{1}{\sqrt{y(1-p^2y)}}[p^2y-1] \\ &= -\sqrt{1/y - p^2}. \end{aligned} \tag{3.24}$$

Figure 3.4 shows a flow diagram for this system, including the unphysical region $y < 0$. Each curve corresponds to a level set, $H = \text{const}$. We now can write Hamilton's equations of motion.

$$\begin{aligned} y' &= \frac{\partial H}{\partial p} = -\frac{p}{H}, \\ p' &= -\frac{\partial H}{\partial y} = -\frac{1}{2Hy^2}. \end{aligned} \tag{3.25}$$

As is well known, the solution to the original problem is that the bead travels along a cycloid, which is the path followed by a point on the edge of a circle that rolls without slipping along a straight line. We shall confirm this result. The parametric equations of the cycloid are written

$$\begin{aligned} x &= R(\phi - \sin\phi), \\ y &= R(1 - \cos\phi), \end{aligned}$$

where ϕ is the roll angle, and R is the radius of the circle. From these we obtain easily that

$$y' = R\sin\phi/y.$$

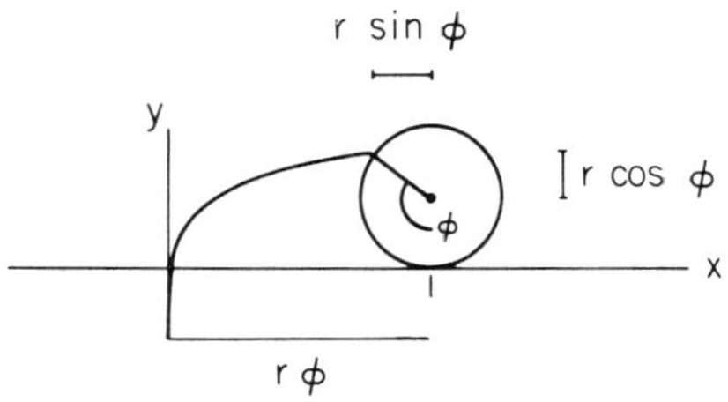

FIGURE 3.5 The solution to the brachistochrone problem is a cycloid.

Now a little manipulation yields the following result:

$$\begin{aligned} y(1 + y'^2) &= y\left(1 + \frac{R^2 \sin^2 \phi}{y^2}\right) \\ &= \frac{R^2}{y}\left(\frac{y^2}{R^2} + \sin^2 \phi\right) \\ &= \frac{R^2}{y}\left((1 - \cos\phi)^2 + \sin^2 \phi\right) \\ &= \frac{2R^2}{y}(1 - \cos\phi) \\ &= 2R. \end{aligned}$$

However, combining Eq. (3.25) with Eq. (3.24) gives us an identical result,

$$y(1 + y'^2) = y(1 + p^2/H^2) = y(1/H^2 y) = 1/H^2,$$

provided that we identify the constant of motion: $R \equiv 1/2H^2$.

In fact, we can go further. The cycloid describes not only the spatial path of the bead but also its phase space history: if we roll the circle according to $\dot{\phi} =$ constant, then the orbit of the marker (i.e., its time history) indeed corresponds to that of the bead on the wire. Proving this is left as an exercise.

Rotating Pendulum Think of the pendulum as a bead on a frictionless ring, as shown in Figure 3.6, and assume that the ring is driven to rotate azimuthally at constant angular velocity, ω. Using polar coordinates, the kinetic energy of the system is

$$\begin{aligned} T &= \tfrac{1}{2}m(\dot{r}^2 + r^2\dot{\theta}^2 + r^2 \sin^2 \theta \dot{\phi}^2) \\ &= \tfrac{1}{2}ml^2(\dot{\theta}^2 + \omega^2 \sin^2 \theta), \end{aligned}$$

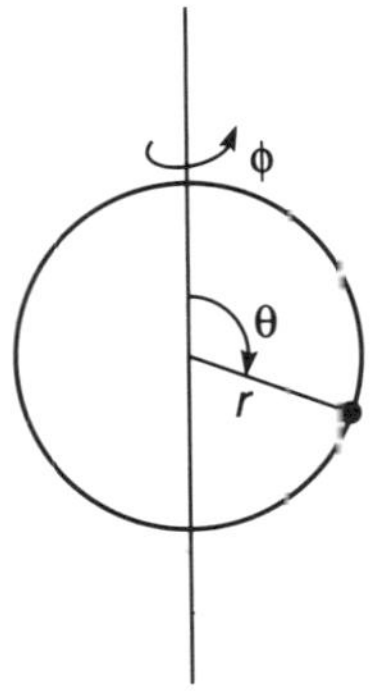

FIGURE 3.6 Bead on a rotating ring.

where in the last line we have substituted the constraints, $r \equiv l$ and, suitably defining time zero, $\phi \equiv \omega t$. The potential energy is written

$$V = mgr\cos\theta = mgl\cos\theta.$$

We calculate successively the Lagrangian, the momentum conjugate to θ, and the Hamiltonian.

$$\begin{aligned} L = T - V &= \tfrac{1}{2}ml^2(\dot\theta^2 + \omega^2\sin^2\theta) - mgl\cos\theta, \\ p = \partial L/\partial\dot\theta &= ml^2\dot\theta, \\ H = p\dot\theta - L &= \frac{p^2}{2ml^2} + mgl\cos\theta - \tfrac{1}{2}ml^2\omega^2\sin^2\theta. \end{aligned} \tag{3.26}$$

Notice that H is *not* the total energy of the system, $H \neq T + V$, because T is not a quadratic form in the system velocity coordinates. Of course, in the limit $\omega \longrightarrow 0$ we recover Eq. (1.2) after two trivial changes: (a) using the symbol p for the generalized momentum, which is not the angular momentum when $\omega \neq 0$, and (b) changing the angular coordinate, θ, to $\pi - \theta$.

There is one subtle point worth mentioning. Because ϕ is no longer a dynamical coordinate, its value being now a linear function of time, the domain of the coordinate θ is $[0, 2\pi)$ rather than the usual half-range, $[0, \pi)$, one expects for polar coordinates. The phase space manifold is thus fully cylindrical.

The vector field is obtained from Hamilton's equation.

$$\underline{v} = \begin{pmatrix}\dot\theta \\ \dot p\end{pmatrix} = \begin{pmatrix}\partial H/\partial p \\ -\partial H/\partial\theta\end{pmatrix} = \begin{pmatrix} p/ml^2 \\ ml^2(\omega^2\cos\theta + g/l)\sin\theta\end{pmatrix}.$$

As usual, the equation for $\dot\theta$ just repeats the definition of the canonical momentum, p; it does not yield any new information. Our first priority is to find the fixed points, at which $\underline{v} = \underline{0}$. There are four of these, located at

$$p_o = 0 \quad \textbf{and} \quad \begin{cases} \theta_o = 0 \text{ or } \pi, \quad \textbf{or} & : \text{primary} \\ \cos\theta_o = -\omega_o^2/\omega^2 & : \text{secondary} \end{cases},$$

where we have reintroduced $\omega_o^2 = g/l$, the small oscillation frequency of a simple pendulum. The locations of the two primary fixed points are identical to those of a simple pendulum; the secondary ones exist only if ω is sufficiently large: we must have $\omega^2 > \omega_o^2$ in order to satisfy $|\cos\theta_o| < 1$. Thus ω_o is the "critical value" of this problem's only control parameter, ω.

To sketch the phase space flow, it is convenient to write the Hamiltonian using a set of scaled, dimensionless variables. To this end, let

$$\begin{aligned} h &\equiv H/ml^2\omega^2, \\ \kappa &\equiv \omega_o^2/\omega^2, \\ \xi &\equiv p/ml^2\omega, \\ \eta &\equiv \cos\theta. \end{aligned} \tag{3.27}$$

Despite the seemingly "unnatural" choice of using ω to set the scale rather than ω_o, so that the scale of the variables themselves varies with the strength of the control parameter, this does turn out to be slightly more convenient. Making these substitutions in Eq. (3.26) yields

$$h = \tfrac{1}{2}\xi^2 + \kappa\eta - \tfrac{1}{2}(1 - \eta^2), \tag{3.28}$$

which, upon a little rearrangement, produces the following:

$$\xi^2 + (\eta + \kappa)^2 = 2h + (1 + \kappa^2).$$

Because H is conserved, orbits consist of the level sets of H, which are identical to the level sets of h. On this chart these are simply circles with centers at $\xi = 0, \eta = -\kappa$, as illustrated in Figure 3.7 for the two cases $\omega < \omega_o$ $(1 < \kappa)$ and $\omega_o < \omega$ $(\kappa < 1)$. Orbits are displayed using both the new and old coordinates. For $1 < \kappa$ the only unstable fixed point occurs at $\eta = 1$; for $\kappa < 1$ the fixed point at $\eta = -1$ becomes unstable and splits off two stable fixed points corresponding to $\theta = \pm\cos^{-1}(-\kappa)$.

To see this more clearly, we shall examine the bifurcation in differential regions about $\theta_o = 0$ and $\theta_o = \pi$. Expanding everything and keeping only second order terms, we have, near $\theta_o = 0$,

$$\begin{aligned} \eta &\equiv \cos(\theta_o + \delta\theta) \approx 1 - \tfrac{1}{2}(\delta\theta)^2, \\ \eta^2 &\approx 1 - (\delta\theta)^2, \\ \xi &\approx \delta\xi, \\ h &\approx h_o + \tfrac{1}{2}(\delta\xi)^2 - \tfrac{1}{2}(\delta\theta)^2 - \tfrac{1}{2}\kappa(\delta\theta)^2 \\ &= h_o + \tfrac{1}{2}(\delta\xi)^2 - \tfrac{1}{2}(1 + \kappa)(\delta\theta)^2. \end{aligned}$$

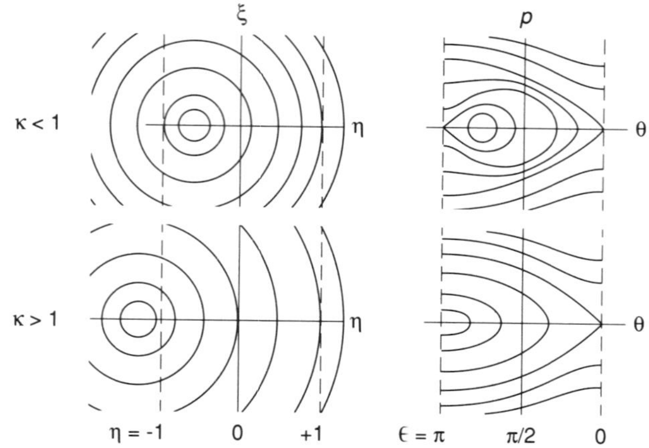

FIGURE 3.7 In appropriately scaled and distorted variables, orbits of the "bead on a rotating ring" become circles.

The level sets of this quadratic form are hyperbolas, so that the fixed point is hyperbolic (unstable) regardless of the value of κ (which is always positive). On the other hand, doing a similar expansion in the vicinity of $(\xi, \theta)_o = (0, \pi)$ produces the following result:

$$\begin{aligned} \eta &\equiv \cos(\pi + \delta\theta) = -\cos\delta\theta \\ &\approx -1 + \tfrac{1}{2}(\delta\theta)^2, \\ \eta^2 &\approx 1 - (\delta\theta)^2, \\ \xi &\approx \delta\xi, \\ h &\approx h_o + \tfrac{1}{2}(\delta\xi)^2 - \tfrac{1}{2}(\delta\theta)^2 + \tfrac{1}{2}\kappa(\delta\theta)^2 \\ &= h_o + \tfrac{1}{2}(\delta\xi)^2 + \tfrac{1}{2}(\kappa - 1)(\delta\theta)^2. \end{aligned}$$

This is very different. For $1 < \kappa$ the fixed point is elliptic, while for $\kappa < 1$ it is hyperbolic. The local flows on either side of the bifurcation are sketched in Figure 3.8 along with the trajectories of the fixed points as a function of κ. This latter picture is called a "bifurcation diagram," and this particular bifurcation is called a "pitchfork bifurcation," because of its resemblance to that exceedingly useful instrument. Its characteristic signature includes (a) a stable fixed point becoming unstable and (b) the creation of two new stable fixed points.

It is left to the reader to complete this analysis by expanding h about the two fixed points $(\xi, \theta)_o = (0, \pm\cos^{-1}(-\kappa))$.

3.1.5 Electrodynamics

The one example of velocity-dependent forces relevant to our studies involves motion of a charged particle in a classical, background, electromagnetic field; it is also the

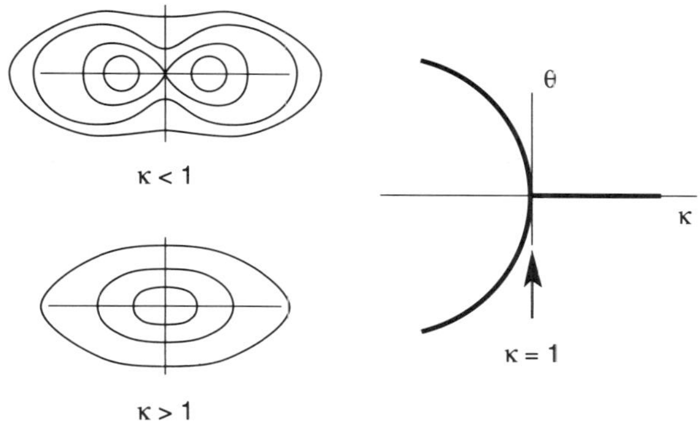

FIGURE 3.8 Pitchfork bifurcation of the principal fixed point.

only relativistic system that we shall consider. The Hamiltonian for this dynamic can be motivated heuristically by (a) starting from the relativistic dispersion relation,

$$E = \sqrt{\underline{p}^2 c^2 + m^2 c^4},$$

(b) performing the "minimal coupling" substitutions,

$$\begin{aligned} \underline{p} &\rightarrow \underline{p} - e\underline{A}, \quad \text{and} \\ E &\rightarrow E - e\phi, \end{aligned}$$

where $\underline{A}$ and ϕ are the gauge fields producing electromagnetic interactions, and (c) identifying E with the Hamiltonian function, H:

$$H = \sqrt{(\underline{p} - e\underline{A})^2 c^2 + m^2 c^4} + e\phi. \tag{3.29}$$

Lorentz Force Of course, merely writing this does not mean it is correct. We shall validate this expression by showing that it indeed leads to the relativistic Lorentz force law on charged particles. Applying Hamilton's equations to Eq. (3.29) first yields

$$\dot{\underline{x}} = \partial H / \partial \underline{p} = (\underline{p} - e\underline{A})c^2 / \sqrt{(\underline{p} - e\underline{A})^2 c^2 + m^2 c^4}. \tag{3.30}$$

We introduce a new dynamical variable,

$$\underline{\mathbf{p}} \equiv \underline{p} - e\underline{A}, \tag{3.31}$$

so that Eq. (3.30) takes the form

$$\dot{\underline{x}} = \underline{\mathbf{p}} c^2 / \sqrt{\underline{\mathbf{p}}^2 c^2 + m^2 c^4}. \tag{3.32}$$

But this means that, although $\underline{p}$ is the *conjugate momentum* associated with $\underline{x}$, $\underline{\mathbf{p}}$ must be the *kinematic* momentum of the particle. To see this, consider that

$$1/\gamma^2 = 1 - \underline{\dot{x}}^2/c^2 = 1 - \underline{\mathbf{p}}^2c^2/(\underline{\mathbf{p}}^2c^2 + m^2c^4) = m^2c^4/(\underline{\mathbf{p}}^2c^2 + m^2c^4).$$

This allows us to identify the kinematic energy of the particle,

$$\gamma mc^2 = \sqrt{\underline{\mathbf{p}}^2c^2 + m^2c^4},$$

and it also enables us to rewrite Eq. (3.32) in the familiar form

$$\underline{\mathbf{p}} = \gamma m\underline{\dot{x}}, \tag{3.33}$$

completing the identification of $\underline{\mathbf{p}}$, not $\underline{p}$, with the kinematic momentum.

The momentum sector of Hamilton's equation gives us $\underline{\dot{p}}$.

$$\underline{\dot{p}} = -\partial H/\partial\underline{x} = e\underline{\dot{x}} \cdot \partial\underline{A}/\partial\underline{x} - e\,\partial\phi/\partial\underline{x}. \tag{3.34}$$

Here we have used Eq. (3.32) to simplify the expression after differentiating. Now notice that, since $\underline{B} = \mathbf{curl}\,\underline{A}$, a simple vector identity allows the replacement

$$\underline{\dot{x}} \cdot \partial\underline{A}/\partial\underline{x} = \underline{\dot{x}} \times \underline{B} + \left(\underline{\dot{x}} \cdot \frac{\partial}{\partial\underline{x}}\right)\underline{A}. \tag{3.35}$$

Before making this substitution, however, we introduce the "total" time derivative of $\underline{A}$ along the orbit (and notice that this is *not* the Lie derivative of the vector field, $\underline{A}$).

$$\underline{\dot{A}} = \left(\underline{\dot{x}} \cdot \frac{\partial}{\partial\underline{x}}\right)\underline{A} + \partial\underline{A}/\partial t. \tag{3.36}$$

Making the replacements from Eq. (3.36) and Eq. (3.35) into Eq. (3.34) brings us close to the final result.

$$\underline{\dot{p}} = e\underline{\dot{x}} \times \underline{B} + e\underline{\dot{A}} - e\,\partial\underline{A}/\partial t - e\,\partial\phi/\partial\underline{x}.$$

Now we rearrange terms slightly, substitute the kinematic momentum from Eq. (3.31), recognize that the combination $-\partial\phi/\partial\underline{x} - \partial\underline{A}/\partial t$ is the electric field, $\underline{E}$, and obtain

$$\underline{\dot{\mathbf{p}}} = e(\underline{E} + \underline{\dot{x}} \times \underline{B}),$$

which is the correct, relativistic, Lorentz force on a charged particle in an electromagnetic field.

Covariant Form: Lagrangian It does not, however, look particularly relativistic. A more manifestly invariant expression can be obtained by writing the Lagrangian corresponding to the Hamiltonian in Eq. (3.29).

$$L = \underline{p} \cdot \underline{\dot{x}} - H.$$

We eliminate $\underline{p}$ from the first term by using Eq. (3.33).

$$\begin{aligned}\underline{p} \cdot \underline{\dot{x}} &= (\underline{p} - e\underline{A}) \cdot \underline{\dot{x}} + e\underline{A} \cdot \underline{\dot{x}} \\ &= \gamma m \underline{\dot{x}}^2 + e\underline{A} \cdot \underline{\dot{x}}.\end{aligned}$$

Substituting this above and using

$$H = \gamma mc^2 + e\phi,$$

we are provided with the following:

$$\begin{aligned}L &= \gamma m \underline{\dot{x}}^2 + e\underline{A} \cdot \underline{\dot{x}} - \gamma mc^2 - e\phi \\ &= -\gamma mc^2(1 - \underline{\dot{x}}^2/c^2) + e\underline{A} \cdot \underline{\dot{x}} - e\phi \\ &= -\frac{1}{\gamma} mc^2 + e\underline{A} \cdot \underline{\dot{x}} - e\phi.\end{aligned}$$

To write this (relativistically) invariantly, multiply by the differential, dt, use $\underline{\dot{x}}dt = d\underline{x}$, and recognize that $dt/\gamma = d\tau$, the relativistic "proper time" of the particle.

$$L\,dt = \underline{p} \cdot d\underline{x} - H\,dt = -mc^2 d\tau + e\underline{A} \cdot d\underline{x} - e\phi\,dt.$$

The second and third terms can be combined into a four-vector dot product. In the (rationalized mks) gauge that we have been using,

$$\begin{aligned}A^\mu &= (\phi, c\underline{A}), \text{ and} \\ dx^\mu &= (cdt, d\underline{x}).\end{aligned}$$

This gives us our final expression,

$$L\,dt = -mc^2 d\tau - \frac{e}{c} A^\mu dx_\mu,$$

where the Einstein summation convention has been assumed.

3.2 TRANSFORMATIONS

We began this chapter by writing a Hamiltonian vector field in the form of Eq. (3.4). This is indeed the canonical form in which it is traditionally presented. It is not, however, the *only* possible representation: the canonical form does not necessarily survive a transformation to another coordinate chart. This section is devoted to the transformation properties of Hamiltonian dynamics.

3.2.1 General Theory

Scaling and Offset The most trivial transformations involve altering the Hamiltonian directly either by multiplication or addition with a function of the orbit parameter, "time." Addition, say, $H \longrightarrow H + g(t)$, leaves the vector field invariant, as it depends only on the gradient of H with respect to dynamical coordinates. Multiplication, $H \longrightarrow g(t)H$, is more interesting; by Hamilton's equation, Eq. (3.4), this has the effect of scaling the vector field by $g(t)$. By Eq. (2.53), if $g(t)$ is monotonic, this rescaling can be interpreted as a "time" distortion (see the discussion following Eq. (2.53)).

When the multiplier is a constant, say, $g(t) = \lambda$, there is an alternative interpretation. Suppose that $\underline{z} = (\underline{x}, \underline{p})$ are the local coordinates for a Hamiltonian, H. Then $\underline{z}' = (\alpha_1 \underline{x}, \alpha_2 \underline{p})$, with $\alpha_1 \alpha_2 = \lambda$, would be local coordinates whose evolution is governed by λH with *no* time distortion:

$$d(\alpha_1 \underline{x})/dt = \partial(\lambda H)/\partial(\alpha_2 \underline{p}) \quad \text{and} \quad d(\alpha_2 \underline{p})/dt = -\partial(\lambda H)/\partial(\alpha_1 \underline{x}). \tag{3.37}$$

We shall find a use for this when we consider isolated resonance models in the next chapter.

Space-Time Interchange We now treat a rather unusual transformation that is commonly used in accelerator physics. Begin with a Hamiltonian, $H(\underline{x}, \underline{p}, t)$, written here with the "position" and "momentum" sectors separated. Suppose that over a phase space region of interest the orbits are such that one of the spatial coordinates, say, x_1, is monotonic in t. Then x_1 could be used to parametrize orbits in place of t, for each point on an orbit would correspond to one and only one value of x_1. We would think of t as a "dynamical" coordinate, rather than the orbit parameter, and write its equation of motion,

$$\frac{dt}{dx_1} = \left(\frac{dx_1}{dt}\right)^{-1} = \left(\frac{\partial H}{\partial p_1}\right)^{-1}. \tag{3.38}$$

We write the differential form of H,

$$dH = \frac{\partial H}{\partial \underline{x}} \cdot d\underline{x} + \frac{\partial H}{\partial \underline{p}} \cdot d\underline{p} + \frac{\partial H}{\partial t} dt, \tag{3.39}$$

which can be solved for dp_1 provided that $\partial H/\partial p_1$ does not vanish.

$$\begin{aligned} dp_1 &= \left(\frac{\partial H}{\partial p_1}\right)^{-1} \left[dH - \frac{\partial H}{\partial \underline{x}} \cdot d\underline{x} - \sum_{k=2}^{N} \frac{\partial H}{\partial p_k} \cdot dp_k - \frac{\partial H}{\partial t} dt \right] \\ &= \frac{\partial p_1}{\partial H} dH + \frac{\partial p_1}{\partial \underline{x}} \cdot d\underline{x} + \sum_{k=2}^{N} \frac{\partial p_1}{\partial p_k} dp_k + \frac{\partial p_1}{\partial t} dt. \end{aligned} \tag{3.40}$$

The last expression is what one would obtain by inverting H to solve for p_1, but as always it is best to think of the differential forms as defining the partial derivatives. We have immediately that $(\partial H/\partial p_1)^{-1} = \partial p_1/\partial H$, so that Eq. (3.38) becomes

$$\frac{dt}{dx_1} = \frac{\partial p_1}{\partial H}. \tag{3.41}$$

This looks exactly like the differential equation one would obtain by treating x_1 as the orbit parameter, t as a dynamical "position" coordinate, p_1 as the Hamiltonian function, and H as the "momentum" coordinate conjugate to t. To test this further, we evaluate dH/dx_1.

$$\frac{dH}{dx_1} = \frac{dH}{dt}\frac{dt}{dx_1} = \frac{\partial H}{\partial t}\frac{\partial p_1}{\partial H}.$$

The temptation is strong to "cancel ∂H" and write this as $\partial p_1/\partial t$, but this would be wrong.[8] Again we use Eq. (3.39) and Eq. (3.40), *zeroing all differentials except dH and dt.*

$$\begin{aligned} dp_1 = 0 &= \frac{\partial p_1}{\partial H}\,dH + \frac{\partial p_1}{\partial t}\,dt \\ &= \left(\frac{\partial p_1}{\partial H}\frac{\partial H}{\partial t} + \frac{\partial p_1}{\partial t}\right)dt. \end{aligned}$$

Making this substitution above, we obtain

$$\frac{dH}{dx_1} = -\frac{\partial p_1}{\partial t}, \tag{3.42}$$

which, combined with Eq. (3.41), reinforces the idea that p_1 must be the new Hamiltonian.

However, this does not quite work. Pressing forward, we calculate the vector field component, dx_k/dx_1, with $k \neq 1$.

$$\frac{dx_k}{dx_1} = \frac{dx_k}{dt}\frac{dt}{dx_1} = \frac{\partial H}{\partial p_k}\frac{\partial p_1}{\partial H}.$$

Now use Eqs. (3.39) and (3.40), this time zeroing all differentials except dp_k and dH.

$$\begin{aligned} dp_1 = 0 &= \frac{\partial p_1}{\partial H}\,dH + \frac{\partial p_1}{\partial p_k}\,dp_k \\ &= \left(\frac{\partial p_1}{\partial H}\frac{\partial H}{\partial p_k} + \frac{\partial p_1}{\partial p_k}\right)dp_k. \end{aligned}$$

[8]Not only mathematically wrong: it produces the wrong answer, which is worse.

From this we obtain

$$\frac{dx_k}{dx_1} = -\frac{\partial p_1}{\partial p_k}, \tag{3.43}$$

and a parallel computation yields

$$\frac{dp_k}{dx_1} = \frac{\partial p_1}{\partial x_k}. \tag{3.44}$$

These look like Hamilton's equation, with Hamiltonian p_1, except for the sign. We are confronted with a choice: either (a) interchange the roles of "position" and "momentum" for the remaining coördinates, or (b) use $-p_1$ as the new Hamiltonian. Normally, the latter is chosen, in which case the "momentum" conjugate to t is not H but $-H$. Equations (3.41), (3.42), (3.43), and (3.44) are then written as follows:

$$\begin{aligned} dt/dx_1 &= \partial(-p_1)/\partial(-H), \\ d(-H)/dx_1 &= -\partial(-p_1)/\partial t, \\ dx_k/dx_1 &= \partial(-p_1)/\partial p_k, \\ dp_k/dx_1 &= -\partial(-p_1)/\partial x_k, \end{aligned} \tag{3.45}$$

the last two holding for $k \neq 1$. These are now in the correct form, and the conversion to employing x_1 as an orbit parameter is complete.

Phase Space Transformations The way in which a coordinate transformation alters the components of a vector field was written in Eq. (2.54). Applying this to Hamiltonian fields leads to the following development:

$$\begin{aligned} \underline{v}' &= (\partial \underline{z}'/\partial \underline{z}) \cdot \underline{v} \\ &= (\partial \underline{z}'/\partial \underline{z}) \cdot \underline{\underline{J}} \cdot (\partial H/\partial \underline{z}) \\ &= [(\partial \underline{z}'/\partial \underline{z}) \cdot \underline{\underline{J}} \cdot (\partial \underline{z}'/\partial \underline{z})^T](\partial H/\partial \underline{z}') \\ &\equiv \underline{\underline{J}}'(\underline{z}') \cdot (\partial H/\partial \underline{z}'). \end{aligned} \tag{3.46}$$

The new representation is still linearly related to a gradient but the matrix, $\underline{\underline{J}}'(\underline{z}')$, need not be constant: it may depend on the state. Note that $\underline{\underline{J}}'$ is still skew-symmetric, but it need not be anti-idempotent. If the transformation is designed so that $\underline{\underline{J}}'(\underline{z}') = \underline{\underline{J}}$, that is, if the Jacobian matrix, $\underline{\underline{M}} = \partial \underline{z}'/\partial \underline{z}$, satisfies

$$\underline{\underline{M}} \cdot \underline{\underline{J}} \cdot \underline{\underline{M}}^T = \underline{\underline{J}}, \quad \forall \underline{z}, \tag{3.47}$$

then the form of the equations of motion is exactly the same on both charts. In such a case $\underline{\underline{M}}$ is called a "symplectic" matrix, and by an informal extension of the terminology, the transformation itself is usually called a "symplectic transformation." Let us emphasize, however: *it is not at all necessary that a transformation be symplectic to be "legitimate."* Nonsymplectic transformations not only are allowed but can even be useful.

It is easy to show that the set of $N \times N$ matrices that obey Eq. (3.47) is a group, the "symplectic group" of order $n = N/2$, symbolized Sp(n). Furthermore, it follows readily from the chain rule that symplectic transformations belong to a group under composition. Now, because $\underline{\underline{J}}$ is antisymmetric, Eq. (3.47) contains $(N^2 - N)/2$ independent scalar equations, which its matrix elements must satisfy. The number of matrix elements that can be independently assigned values is therefore $N^2 - (N^2 - N)/2 = N(N+1)/2 = n(2n+1)$. Alternatively, all matrix elements of a symplectic matrix can be specified as functions of $n(2n+1)$ arguments. The Jacobian of a symplectic transformation is actually a matrix function over (a local region of) phase space and therefore produces a continuum of symplectic matrices. The $n(2n+1)$ arguments needed to specify each member of this continuum can be viewed as the coordinates of a manifold, which can be identified with the group itself. Put in one sentence: the symplectic group Sp(n) is a manifold of dimension $n(2n+1)$. In particular, the group manifolds Sp(1) and Sp(2) are, respectively, 3- and 10-dimensional.

One of the simpler consequences of Eq. (3.47) is that symplectic transformations preserve Liouville measure. This follows from (a) the fact that $\underline{\underline{M}}$ is the Jacobian of the transformation and (b) the multiplicative morphism properties of determinants.

$$\det\left[\underline{\underline{J}}\right] = \det\left[\underline{\underline{M}} \cdot \underline{\underline{J}} \cdot \underline{\underline{M}}^T\right] = \left(\det\left[\underline{\underline{M}}\right]\right)^2 \det\left[\underline{\underline{J}}\right].$$

As $\det\left[\underline{\underline{J}}\right]$ is manifestly nonzero, this leaves two possibilities: $\det\left[\underline{\underline{M}}\right] = \pm 1$. However, a deep theorem states that the manifold of a symplectic group, Sp(n), is connected.[9] This makes it impossible for the determinant to be negative — since any symplectic $\underline{\underline{M}}$ must be connected to the identity matrix via a path through the group — so only $\det\left[\underline{\underline{M}}\right] = 1$ is allowed.

An important category consists of symplectic, linear transformations of the coordinates. We write $\underline{z}' = \underline{\underline{M}} \cdot \underline{z}$ and express $\underline{\underline{M}}$ in $N \times N$ block form as follows:

$$\underline{\underline{M}} \equiv \begin{pmatrix} \underline{\underline{M}}_{11} & \underline{\underline{M}}_{12} \\ \underline{\underline{M}}_{21} & \underline{\underline{M}}_{22} \end{pmatrix}, \tag{3.48}$$

$$\underline{\underline{M}}\,\underline{\underline{J}}\,\underline{\underline{M}}^T = \begin{pmatrix} \underline{\underline{A}} & \underline{\underline{B}} \\ -\underline{\underline{B}}^T & \underline{\underline{C}} \end{pmatrix} \quad \text{where,}$$

$$\begin{aligned} \underline{\underline{A}} &= \underline{\underline{M}}_{11}\underline{\underline{M}}_{12}{}^T - \underline{\underline{M}}_{12}\underline{\underline{M}}_{11}{}^T, \\ \underline{\underline{B}} &= \underline{\underline{M}}_{11}\underline{\underline{M}}_{22}{}^T - \underline{\underline{M}}_{12}\underline{\underline{M}}_{21}{}^T, \\ \underline{\underline{C}} &= \underline{\underline{M}}_{21}\underline{\underline{M}}_{22}{}^T - \underline{\underline{M}}_{22}\underline{\underline{M}}_{21}{}^T. \end{aligned} \tag{3.49}$$

[9] A proof can be found in Siegel [84].

The necessary and sufficient conditions that make this a symplectic transformation are:

$$\begin{aligned}&\text{(a) both } \underline{\underline{M}}_{11}\underline{\underline{M}}_{12}^{\;T} \text{ and } \underline{\underline{M}}_{21}\underline{\underline{M}}_{22}^{\;T} \text{ must be symmetric matrices, and}\\ &\text{(b) } \underline{\underline{M}}_{11}\underline{\underline{M}}_{22}^{\;T} - \underline{\underline{M}}_{12}\underline{\underline{M}}_{21}^{\;T} = \underline{\underline{1}}.\end{aligned} \tag{3.50}$$

A special subclass of these do not mix the position and momentum sectors: $\underline{\underline{M}}_{12} = \underline{\underline{M}}_{21} = \underline{\underline{0}}$. The two conditions then collapse to the single requirement $\underline{\underline{M}}_{11}\underline{\underline{M}}_{22}^{T} = \underline{\underline{1}}$. In particular, if this is just a scaling transformation—so that $\underline{\underline{M}}_{11} = \lambda\underline{\underline{1}}$, $\underline{\underline{M}}_{22} = \mu\underline{\underline{1}}$—then symplecticity requires that $\lambda = 1/\mu$. An important example of this type is the scaling transformation that puts the harmonic oscillator Hamiltonian, Eq. (3.6), into its "normal" form. Take $(\underline{\underline{M}}_{11})_{ij} = \delta_{ij}\sqrt{m_j\omega_j}$ and $(\underline{\underline{M}}_{22})_{ij} = \delta_{ij}/\sqrt{m_j\omega_j}$, so that

$$\begin{aligned} H &= \sum_k \left(p_k^2/2m_k + \frac{1}{2}m_k\omega_k^2 x_k^2 \right)\\ &= \sum_k \frac{1}{2}\omega_k \left((p_k/\sqrt{m_k\omega_k})^2 + (\sqrt{m_k\omega_k}x_k)^2 \right)\\ &= \sum_k \frac{1}{2}\omega_k \left(p_k'^2 + x_k'^2 \right). \end{aligned} \tag{3.51}$$

An even more useful chart uses the polar, or "action-angle," coordinates introduced in Eq. (2.9).

$$\begin{aligned} x_k' &= \sqrt{2I_k}\sin\varphi_k,\\ p_k' &= \sqrt{2I_k}\cos\varphi_k,\\ \underline{\underline{M}}|_{k\text{th block}} &= \partial(x_k' p_k')/\partial(\varphi_k I_k)\\ &= \begin{pmatrix} p_k' & x_k'/2I_k \\ -x_k' & p_k'/2I_k \end{pmatrix}. \end{aligned} \tag{3.52}$$

The factor of $\sqrt{2}$ is introduced to make this transformation symplectic; it is a trivial exercise, left to the reader, to show that the symplectic condition, Eq. (3.47), is indeed satisfied. It can be motivated more heuristically by requiring invariance of the phase space volume element, which symplecticity guarantees.

$$dx'\,dp' = r\,dr\,d\varphi = d(\tfrac{1}{2}r^2)d\varphi = dI\,d\varphi.$$

From Eq. (3.51), the Hamiltonian, expressed in these polar coordinates, is written

$$H = \sum_k \omega_k I_k \equiv \underline{\omega}\cdot\underline{I}, \tag{3.53}$$

in agreement with Eq. (2.10). Because the transformation is symplectic, the components of the vector field — also referred to as the equations of motion — are written in canonical form.

$$\dot{\underline{I}} = -\partial H/\partial \underline{\varphi} = \underline{0},$$
$$\dot{\underline{\varphi}} = \partial H/\partial \underline{I} = \underline{\omega}\ . \tag{3.54}$$

Notice the agreement with Eq. (2.12).

The transformation to the eigencoordinates, a and a^*, introduced in Eq. (2.8) and Eq. (2.9), is almost, but not quite, symplectic. We relax our implicit assumption that $\underline{\underline{M}}$ is a real matrix and introduce the transformation

$$a_k \equiv \frac{1}{\sqrt{2}}(x'_k + ip'_k) = i\sqrt{I_k}\, e^{-i\varphi_k}, \quad a^*_k \equiv \frac{1}{\sqrt{2}}(x'_k - ip'_k) = -i\sqrt{I_k}\, e^{i\varphi_k},$$

in terms of which the Hamiltonian in Eq. (3.51) or Eq. (3.53) is written

$$H = \sum_k \omega_k a^*_k a_k. \tag{3.55}$$

The Jacobian matrix of this transformation is

$$\underline{\underline{M}}|_{k\text{th block}} = \partial(a_k a^*_k)/\partial(x'_k p'_k) = \sqrt{1/2}\begin{pmatrix} 1 & i \\ 1 & -i \end{pmatrix},$$

from which we see immediately that $\underline{\underline{J}}' = \underline{\underline{M}}\,\underline{\underline{J}}\,\underline{\underline{M}}^T = -i\underline{\underline{J}}$. Eq. (2.54) then yields the equation of motion,

$$\begin{pmatrix} \dot{a}_k \\ \dot{a}^*_k \end{pmatrix} = -i\underline{\underline{J}} \cdot \begin{pmatrix} \partial H/\partial a_k \\ \partial H/\partial a^*_k \end{pmatrix},$$

which upon separation becomes

$$\dot{a}_k = -i\partial H/\partial a^*_k = -i\omega_k a_k\,, \quad \dot{a}^*_k = i\partial H/\partial a_k = i\omega_k a^*_k.$$

All of which is in agreement with the discussion in the previous chapter (e.g., see Eq. (2.57)).

Furthermore, let H be *any* Hamiltonian function expressed on the complex $(\underline{a}, \underline{a}^*)$ chart. The vector field is written on this chart as follows:

$$X_H = \underline{v} \cdot \underline{\partial} = -\underline{\partial} H \cdot \underline{\underline{J}}' \cdot \underline{\partial} = i\underline{\partial} H \cdot \underline{\underline{J}} \cdot \underline{\partial} = i\sum_k \left(\frac{\partial H}{\partial a_k}\frac{\partial}{\partial a^*_k} - \frac{\partial H}{\partial a^*_k}\frac{\partial}{\partial a_k} \right). \tag{3.56}$$

These manipulations with a and a^* may seem uncomfortably formal. To make them more palatable, consider the following argument. Let α be an arbitrary complex number and define

$$a \equiv \sqrt{1/2}\,(\alpha x + ip/\alpha) \quad \text{and} \quad a^* = \sqrt{1/2}\,(\alpha x - ip/\alpha).$$

The differentials (more correctly: differential forms), $\alpha\,dx$ and dp/α, are written in terms of da and da^* as follows:

$$\begin{aligned} \alpha\,dx &= \sqrt{1/2}\,(da + da^*), \\ dp/\alpha &= -i\sqrt{1/2}\,(da - da^*). \end{aligned}$$

Let H be an arbitrary Hamiltonian and write the differential (form), dH,

$$\begin{aligned} dH &= H_{,x}\,dx + H_{,p}\,dp \\ &= (H_{,x}/\alpha)\alpha\,dx + (\alpha H_{,p})\,dp/\alpha \\ &= \sqrt{1/2}\,[(H_{,x}/\alpha - i\alpha H_{,p})\,da + (H_{,x}/\alpha + i\alpha H_{,p})\,da^*\,] \\ &\equiv H_{,a}\,da + H_{,a^*}\,da^*, \end{aligned}$$

where $H_{,x} \equiv (\partial H/\partial x)_p$ and $H_{,p} \equiv (\partial H/\partial p)_x$. The equation of motion for a is then written

$$\begin{aligned} \dot{a} &= \sqrt{1/2}\,(\alpha\dot{x} + i\dot{p}/\alpha) \\ &= \sqrt{1/2}\,(\alpha H_{,p} - iH_{,x}/\alpha) \\ &= -i\sqrt{1/2}\,(H_{,x}/\alpha + i\alpha H_{,p}) \\ &= -iH_{,a^*} \equiv -i\partial H/\partial a^*. \end{aligned}$$

Taking the complex conjugate, or going through a similar sequence of manipulations, yields the partner equation. This uses the exact meaning of "$\partial H/\partial a^*$," which is *the coefficient of da^* in the differential form dH*. It is easier to understand this definition than to write something like $(\partial H/\partial a^*)_a$, as it is confusing to think of "varying a^* while holding a fixed."

Rotating Pendulum, Revisited As an example, let us apply a few of these transformations to the scaled Hamiltonian of the "rotating pendulum" model, which was introduced in Eq. (3.28) and which we repeat here for convenience.

$$h = \tfrac{1}{2}\xi^2 + \kappa\eta - \tfrac{1}{2}(1 - \eta^2).$$

We shall write the equations of motion on this chart and obtain the frequency of small oscillation about the elliptic fixed point at $(\xi, \eta)_o = (0, -\kappa)$. The original Hamiltonian corresponded to a chart in which the equations of motion were in canonical form, and the transformations to these coordinates were given by Eq. (3.27), which we also repeat for convenience.

$$\begin{aligned} h &\equiv H/ml^2\omega^2, \\ \kappa &\equiv \omega_o^2/\omega^2, \\ \xi &\equiv p/ml^2\omega, \\ \eta &\equiv \cos\theta. \end{aligned}$$

We interpret the rescaling of the Hamiltonian function as a change in the "time" coordinate, as given by Eq. (2.53).

$$t \rightarrow \tau \quad \text{where} \quad \frac{dt}{d\tau} = \frac{1}{ml^2\omega^2} \equiv \lambda. \tag{3.57}$$

The transformation in dynamical coordinates is compensated by a change in the symplectic form, according to Eq. (3.46).

$$\begin{aligned} \underline{\underline{M}} &= \partial(\eta,\xi)/\partial(\theta,p) = \begin{pmatrix} -\sin\theta & 0 \\ 0 & \lambda\omega \end{pmatrix}, \\ \underline{\underline{J}}' = \underline{\underline{M}} \cdot \underline{\underline{J}} \cdot \underline{\underline{M}}^T &= -\lambda\omega\sin\theta\underline{\underline{J}} \\ &= -\lambda\omega\sqrt{1-\eta^2}\underline{\underline{J}}. \end{aligned}$$

The correct equations of motion on this chart are

$$\frac{d}{d\tau}\begin{pmatrix} \xi \\ \eta+\kappa \end{pmatrix} = -\lambda\omega\sqrt{1-\eta^2}\underline{\underline{J}} \cdot \begin{pmatrix} \partial h/\partial\xi \\ \partial h/\partial\eta \end{pmatrix} = \lambda\omega\sqrt{1-\eta^2}\begin{pmatrix} -(\eta+\kappa) \\ \xi \end{pmatrix},$$

where we have made use of the fact that κ is constant.

We now can return to our original clock by making use of Eq. (3.57).

$$\frac{d}{dt}\begin{pmatrix} \xi \\ \eta+\kappa \end{pmatrix} = \frac{1}{\lambda}\frac{d}{d\tau}\begin{pmatrix} \xi \\ \eta-\kappa \end{pmatrix} = \omega\sqrt{1-\eta^2}\begin{pmatrix} -(\eta+\kappa) \\ \xi \end{pmatrix}.$$

Small amplitude motion about the fixed point is then obtained by linearization, which amounts to no more than making the simple replacement $\sqrt{1-\eta^2} \approx \sqrt{1-\kappa^2}$. The frequency of small oscillations is then given by Ω, where

$$\Omega^2 = \omega^2(1-\kappa^2) = \omega^2 - \omega_o^2.$$

Time Evolution Is Symplectic We have seen already that Liouville measure is conserved by Hamiltonian flows and that this is also a property of symplectic transformations. Except for two-dimensional phase spaces, this is not enough by itself to establish that the time-evolution maps of Hamiltonian vector fields are symplectic. Nonetheless they are, and we shall now prove this rather strong proposition, which is important enough to be labeled a theorem.

Theorem 3.4 (Symplectic evolution) Let $\phi_{t_2 t_1} : \underline{z} \longmapsto \underline{z}'$ be a time-evolution map of a Hamiltonian dynamic, and let $\underline{\underline{M}}(\underline{z}; t_2, t_1) = \partial \underline{z}'/\partial \underline{z}$ be its Jacobian. Then $\underline{\underline{M}}$ is a symplectic matrix.

The proof written below follows a line of reasoning found in Dragt [30] and, most probably, others. It is facilitated by two preliminary lemmas, which together establish the theorem up to terms of $O\left((t_2 - t_1)^2\right)$.

Lemma 3.1 Let $\underline{v}(\underline{z}, t)$ (canonically)[10] represent a Hamiltonian vector field. Then $\underline{\underline{\partial v}} \cdot \underline{\underline{J}} + \underline{\underline{J}} \cdot \underline{\underline{\partial v}}^T = \underline{\underline{\mathbf{0}}}$.

Proof: From Eq. (3.4), $\underline{v} = \underline{\underline{J}} \cdot \underline{\partial H} = -\underline{\partial H} \cdot \underline{\underline{J}}$, $\underline{\underline{\partial v}} = -\underline{\underline{\partial\partial H}} \cdot \underline{\underline{J}}$, and $\underline{\underline{\partial v}}^T = \underline{\underline{J}} \cdot \underline{\underline{\partial\partial H}}$. Then we simply calculate

$$\begin{aligned} \underline{\underline{\partial v}} \cdot \underline{\underline{J}} + \underline{\underline{J}} \cdot \underline{\underline{\partial v}}^T &= (-\underline{\underline{\partial\partial H}} \cdot \underline{\underline{J}}) \cdot \underline{\underline{J}} + J \cdot (\underline{\underline{J}} \cdot \underline{\underline{\partial\partial H}}) \\ &= \underline{\underline{\partial\partial H}} - \underline{\underline{\partial\partial H}} = \underline{\underline{\mathbf{0}}}. \end{aligned}$$

□

This result should be compared with Eq. (3.5), which is a special case. The second lemma follows immediately from this.

Lemma 3.2 Let $\underline{\underline{M}}$ be defined as above. Then there exists a matrix $\underline{\underline{D}}$ such that: (a) $\underline{\underline{M}}(\underline{z}; t + \Delta t, t) = \underline{\underline{D}} + O(\Delta t^2)$, and (b) $\underline{\underline{D}}\,\underline{\underline{J}}\,\underline{\underline{D}}^T = \underline{\underline{J}} + O(\Delta t^2)$.

Proof: The time-evolution map is $\phi_{t+\Delta t, t} : \underline{z} \longmapsto \underline{z}'$, where

$$\underline{z}' = \underline{z} + \Delta t\, \underline{v}(\underline{z}, t) + O(\Delta t^2).$$

The Jacobian is

$$\underline{\underline{M}}(\underline{z}, t) = \partial \underline{z}'/\partial \underline{z} = \underline{\underline{\mathbf{1}}} + \Delta t\, \underline{\underline{\partial v}}(\underline{z}, t) + O(\Delta t^2).$$

Using the previous lemma, we have

$$\begin{aligned} \underline{\underline{M}}\,\underline{\underline{J}}\,\underline{\underline{M}}^T &= \underline{\underline{J}} + \Delta t\, (\underline{\underline{\partial v}}\,\underline{\underline{J}} + \underline{\underline{J}}\,\underline{\underline{\partial v}}^T) + O(\Delta t^2) \\ &= \underline{\underline{J}} + O(\Delta t^2). \end{aligned}$$

[10] See the statement of Darboux's theorem below.

Now identify

$$\underline{\underline{D}}(\underline{z}; t + \Delta t, t) \equiv \underline{\underline{1}} + \Delta t \, \underline{\underline{\partial v}}(\underline{z}, t). \qquad \square$$

The strategy for proving the main theorem is to partition the time interval, say, $(t, t + \tau)$, into a large number, N, of subintervals and apply Lemma 3.2 to each one.

Proof (Symplectic evolution): Define the monotonic, arithmetic sequence,

$$\begin{aligned} t_0 &= t < t_1 < t_2 < t_3 < \cdots < t_{N-1} < t_N = t + \tau, \\ t_{k+1} &= t_k + \Delta t, \quad k = 0, \ldots, N-1, \\ \Delta t &= \tau/N. \end{aligned}$$

The Chapman-Kolmogorov equation,[11] Eq. (1.4), allows us to evaluate the time-evolution map as a convolution over these subintervals. Let $(\underline{z}_0, \underline{z}_1, \underline{z}_2, \ldots, \underline{z}_N)$ be the sequence of points along the orbit corresponding to the endpoints of the subintervals; that is,

$$\begin{aligned} \underline{z}' &= \phi_{t+\tau\, t}(\underline{z}), \quad \underline{z}_0 = \underline{z}, \quad \text{and} \quad \underline{z}_N = \underline{z}', \\ \forall k = 0, \ldots, N-1 : \underline{z}_{k+1} &= \phi_{t_{k+1} t_k}(\underline{z}_k). \end{aligned}$$

Correspondingly, the chain rule provides that the overall Jacobian be calculated as a product of the component, subinterval Jacobians. To ease the notation, we denote these as $\underline{\underline{M}}_k \equiv \underline{\underline{M}}(\underline{z}_k; t_{k+1}, t_k)$.

$$\underline{\underline{M}}(\underline{z}; t + \tau, t) = \underline{\underline{M}}_N \underline{\underline{M}}_{N-1} \underline{\underline{M}}_{N-2} \cdots \underline{\underline{M}}_2 \underline{\underline{M}}_1 = \prod_k^{(dec)} \underline{\underline{M}}_k.$$

The superscript *dec* (*inc*) means that the terms are to be arranged in decreasing (increasing) order. From this, and Lemma 3.2, we obtain the final result.

$$\begin{aligned} \underline{\underline{M}}\,\underline{\underline{J}}\,\underline{\underline{M}}^T &= \left(\prod_k^{(dec)} \underline{\underline{M}}_k\right) \underline{\underline{J}} \left(\prod_k^{(inc)} \underline{\underline{M}}_k^T\right) \\ &= \left(\prod_k^{(dec)} \underline{\underline{D}}_k + O(\Delta t^2)\right) \underline{\underline{J}} \left(\prod_k^{(inc)} \underline{\underline{D}}_k^T + O(\Delta t^2)\right) \\ &= \left(\prod_k^{(dec)} \underline{\underline{D}}_k\right) \underline{\underline{J}} \left(\prod_k^{(inc)} \underline{\underline{D}}_k^T\right) + O(\Delta t^2) \\ &= \underline{\underline{J}} + (N + 1)\, O((\tau/N)^2). \end{aligned}$$

[11] And how do we know that *it* is satisfied?

In going from the third line to the fourth, we use the second half of Lemma 3.2 to reduce the product, successively replacing each $\underline{\underline{D}}_k \underline{\underline{J}} \underline{\underline{D}}_k^T$ with $\underline{\underline{J}}$. According to the lemma, each step adds an error of $O(\Delta t^2) = O((\tau/N)^2)$; since there are N such steps, the final line results. For large N, the residual term is $O(\tau^2/N)$ and therefore can be made arbitrarily small by partitioning into a sufficiently large number of subintervals. This means that, in fact, $\underline{\underline{M}}\,\underline{\underline{J}}\,\underline{\underline{M}}^T = \underline{\underline{J}}$, which was to be proven. □

If you are unhappy with the technique of this proof, then congratulations on your discernment. By all means: worry about it, go over it repeatedly, lose sleep wondering if a swindle has not been perpetrated. In fact, another proof is available using the calculus of differential forms, which we shall touch upon briefly in the last chapter.

That time-evolution maps are symplectic is the classical analog to the quantum mechanical observation that time-evolution operators of Hamiltonian systems are unitary.

Darboux's Theorem We have not yet examined how Poisson brackets transform under an arbitrary change of coordinates. Using Eq. (3.46) gives us the following development:

$$\begin{aligned}
[f, g] &= \partial f/\partial \underline{z} \cdot \underline{\underline{J}} \cdot \partial g/\partial \underline{z} \\
&= (\partial f/\partial \underline{z}') \cdot [(\partial \underline{z}'/\partial \underline{z}) \cdot \underline{\underline{J}} \cdot (\partial \underline{z}'/\partial \underline{z})^T] \cdot (\partial g/\partial \underline{z}') \\
&= (\partial f/\partial \underline{z}') \cdot \underline{\underline{J}}'(\underline{z}') \cdot (\partial g/\partial \underline{z}').
\end{aligned} \tag{3.58}$$

These equations attest to the fact, already implicit in the (anti)morphism expressed by Eq. (3.9), that the Poisson bracket is a scalar object, a geometric invariant, so that $\dot{\underline{z}} = X_H \underline{z} = [\underline{z}, H]$ (as in Eq. (3.11)) is the correct "covariant" form of Hamilton's equations of motion. (See Eq. (2.60).) In this respect, $\underline{\underline{J}}$ behaves much like a metric. For example, the inner product construction in an n-dimensional real vector space is defined, in the usual way,

$$(\underline{x}, \underline{y}) = \sum_i x_i y_i = \underline{x} \cdot \underline{y}. \tag{3.59}$$

If we confine ourselves to charts related by orthogonal transformations, then this defines a legitimate geometric object, an invariant of the group $O(n)$.

$$\forall \underline{\underline{M}} \in O(n) : \underline{x}' = \underline{\underline{M}}\,\underline{x} \quad \textbf{and} \quad \underline{y}' = \underline{\underline{M}}\,\underline{y} \Rightarrow \underline{x}' \cdot \underline{y}' = \underline{x} \cdot \underline{\underline{M}}^T \underline{\underline{M}} \cdot \underline{y} = \underline{x} \cdot \underline{y}.$$

However, if we enlarge the allowed transformations and let $\underline{\underline{M}}$ be arbitrary, $\underline{\underline{M}} \in GL(n)$, then Eq. (3.59) no longer defines a geometric scalar: it is not invariant under $GL(n)$. To correct it we must introduce a new object, the "metric tensor," $\underline{\underline{g}}$, with transformation laws appropriate to reestablish the invariance.

$$\begin{aligned}(\underline{x},\underline{y}) &\equiv \underline{x}\cdot\underline{\underline{g}}\cdot\underline{y},\\ (\underline{x}',\underline{y}') &= \underline{x}'\cdot\underline{\underline{g}}'\cdot\underline{y}'\\ &= \underline{x}\cdot\underline{\underline{M}}^T\underline{\underline{g}}'\underline{\underline{M}}\cdot\underline{y}\\ &= \underline{x}\cdot\underline{\underline{g}}\cdot\underline{y}\\ &= (\underline{x},\underline{y}).\end{aligned}$$

This reestablishes the quadratic form $(\underline{x},\underline{y})$ as a geometric (scalar) object, a "two-form," at least under $GL(n)$. If it is symmetric and positive definite, so that $\underline{\underline{g}}$ is a symmetric, positive definite matrix, then matrix theory tells us that the eigenvalues of $\underline{\underline{g}}$ are positive reals and that $\underline{\underline{g}}$ is diagonalizable. That being the case, there will be some chart on which the form of the inner product reverts back to being the dot product of Eq. (3.59). That is, for some $\underline{\underline{M}} \in GL(n)$ we shall have $\underline{\underline{M}}^T\cdot\underline{\underline{g}}\cdot\underline{\underline{M}} = \underline{\underline{1}}$. This statement is a sort of "representation theorem": any symmetric, positive definite two-form (i.e., an inner product) must look like an ordinary dot product on some chart.

A similar, but deeper, result holds for the symplectic form. Suppose that a vector field is given to us as in Eq. (3.46), $\underline{v}(\underline{z},t) = \underline{\underline{J}}(\underline{z})\cdot\underline{\partial}H(\underline{z},t)$. To identify this as a Hamiltonian vector field, it is not enough that $\underline{\underline{J}}$ be antisymmetric. $\underline{\underline{J}}$ also appears in the definition of the Poisson bracket, via Eq. (3.58), and *we must insist that this bracket be a Lie algebraic operation,* which means that $\underline{\underline{J}}$ must satisfy an extra condition in order that the Jacobi identity,

$$[f,[g,h]] + [g,[h,f]] + [h,[f,g]] = 0,$$

be satisfied. It was left as a tedious but straightforward exercise to confirm that this is true in the case that $\underline{\underline{J}}$ was a constant, antisymmetric matrix. In the case that $\underline{\underline{J}}$ is *not* constant, the extra terms that appear are

$$-\sum_{k,i,n,m}(J_{mk}\partial_k J_{in} + J_{ik}\partial_k J_{nm} + J_{nk}\partial_k J_{mi})\,\partial_i f\,\partial_n g\,\partial_m h.$$

Since this must vanish for all test functions, it is required that

$$\sum_k J_{mk}\partial_k J_{in} + J_{ik}\partial_k J_{nm} + J_{nk}\partial_k J_{mi} = 0 \qquad (3.60)$$

for all m, i, n.

This condition is written more succinctly in terms of the inverse, $\underline{\underline{J}}^{-1}$. From $\underline{\underline{J}}^{-1}\underline{\underline{J}} = \underline{\underline{\mathbf{1}}}$ we easily obtain the matrix identity,

$$\partial_i\underline{\underline{J}} = -\underline{\underline{J}}(\partial_i\underline{\underline{J}}^{-1})\underline{\underline{J}}.$$

Substituting this into Eq. (3.60) provides,

$$\sum_{k,s,t} J_{mk} J_{is} (\partial_k \underline{\underline{J}}^{-1}_{st}) J_{tn} + J_{ik} J_{ns} (\partial_k \underline{\underline{J}}^{-1}_{st}) J_{tm} + J_{nk} J_{ms} (\partial_k \underline{\underline{J}}^{-1}_{st}) J_{ti} = 0.$$

We now multiply by $\underline{\underline{J}}^{-1}_{rm} \underline{\underline{J}}^{-1}_{ui} \underline{\underline{J}}^{-1}_{nw}$ and sum over r, u, and w; since $\underline{\underline{J}}$ is antisymmetric, the final result is

$$\forall r, u, w: \quad \partial_r \underline{\underline{J}}^{-1}_{uw} + \partial_u \underline{\underline{J}}^{-1}_{wr} + \partial_w \underline{\underline{J}}^{-1}_{ru} = 0. \tag{3.61}$$

This elegant expression suggests that $\underline{\underline{J}}^{-1}$ is more fundamental than $\underline{\underline{J}}$; this is in fact the case when Hamiltonian dynamics is expressed in the advanced language of differentiable manifolds: the components of $\underline{\underline{J}}^{-1}$ appear as coefficients in an algebraic structure defining a "symplectic manifold."

It turns out that this extra condition, Eq. (3.61), is precisely what is needed to assure that $\underline{v}(\underline{z}, t) = \underline{\underline{J}}(\underline{z}) \cdot \underline{\partial} H(\underline{z}, t)$ has all the desirable properties of a Hamiltonian vector field. A natural question now arises (actually, should have arisen long ago in the mind of an alert student): Given $\underline{v}$, does there exist a coordinate chart on which $\underline{v}$ assumes canonical form, as in Eq. (3.4), with a constant $\underline{\underline{J}}$, as given by Eq. (3.2)? The answer is yes, at least locally. This beautiful proposition is the conclusion of a remarkable theorem attributed to Darboux, which we merely state without proof.

Theorem 3.5 (Darboux; paraphrased) Let $\underline{v}$ be a vector field given locally as

$$\underline{v}(\underline{z}, t) = \underline{\underline{J}}(\underline{z}) \cdot \partial H(\underline{z}, t) / \partial \underline{z},$$

where $\underline{\underline{J}}$ is skew-symmetric and obeys Eq.(3.61). Then there is a chart on which $\underline{\underline{J}}$ locally assumes the canonical form of Eq.(3.2), and $\underline{v}$ is represented in canonical form.

In fact, this is *not* the usual way of stating Darboux's theorem, whose subject matter concerns "closed two-forms" defined over manifolds and makes no reference to specific vector fields. We have paraphrased it to conform to the "intermediate" language of this text.

3.2.2 Generating Functions

By far, the most popular technique for creating symplectic transformations is to use the machinery of generating functions. The method is so ingrained that transformations obtained in this way are called "canonical," and many (most?) physicists make no logical distinction between "canonical" and "symplectic" transformations. We shall review the generating function techniques and show that they do, in fact, generate symplectic transformations.

The procedure is usually motivated from a Lagrangian viewpoint. Let $L(\underline{x}, \dot{\underline{x}}, t)$ and $L'(\underline{x}', \dot{\underline{x}}', t')$ be two Lagrangian functions expressed on different (tangent bundle) coordinate charts for extended phase space. If the difference, $L\,dt - L'\,dt'$, is a total differential, that is,

$$L\,dt - L'\,dt' = dS, \tag{3.62}$$

then the Lagrangians describe the same dynamic; that is, the same orbits satisfy the Euler-Lagrange equations on both coordinate charts. We attempt to support this statement by exploiting the connection with the calculus of variations: the Euler-Lagrange equation is satisfied by the critical orbits of $\int L\,dt$, that is, those orbits that make the value of $\int L\,dt$ stationary against orbit variations that vanish at the endpoints. Let Γ represent a generic orbit (not necessarily satisfying the Euler-Lagrange equations) passing through the boundary points at t_1 and t_2. If Eq. (3.62) is satisfied, then

$$\int_\Gamma L\,dt - \int_\Gamma L'\,dt' = \int_\Gamma dS = S\,|_{t_1}^{t_2}$$

Now, the argument goes, $S\,|_{t_1}^{t_2}$ retains its value as Γ ranges over the allowed orbit variations, since they all vanish at the endpoints. Therefore the difference between the two integrals on the left-hand side is simply a constant, so that an orbit that is critical for one must be critical for the other as well.

There may be a weakness in this line of reasoning: the calculus of variations considers orbit variations that vanish at the endpoints in the "position" sector, but not the "velocity," or "momentum," sector. (See Eq. (3.21), for example.) Thus, for example, in the velocity representation (tangent bundle chart), $S\,|_{t_1} = S(\underline{x}(t_1), \dot{\underline{x}}(t_1), t_1)$ need not remain constant under allowed variations because of its dependence on $\dot{\underline{x}}$. Thus we must restrict orbit variations further in order for the argument to hold.[12] Notwithstanding the validity of this particular argument, it is nonetheless correct that canonical transformations are indeed symplectic and therefore preserve the form of Hamilton's equations of motion, a proposition we shall confirm in the following paragraphs.

The Four Types We take the differential form equation, Eq. (3.62), as our starting point and write it in terms of Hamiltonians, using Eq. (3.23).

$$dS = (\underline{p} \cdot d\underline{x} - H dt) - (\underline{p}' \cdot d\underline{x}' - H' dt'). \tag{3.63}$$

We shall assume throughout this section, without further comment, that $dt = dt'$. From this we immediately obtain the transformation equations, in their usual "mixed"

[12]It turns out that this point is discussed in I.M. Gelfand and S.V. Fomin, *Calculus of Variations* (translated by Richard A. Silverman), Prentice-Hall, Inc., Englewood Cliffs, N.J.(1963). My thanks to Francois Ostiguy for introducing me to this text.

representation,

$$
\begin{aligned}
\underline{p} &= (\partial S/\partial \underline{x})_{\underline{x}',t},\\
\underline{p}' &= -(\partial S/\partial \underline{x}')_{\underline{x},t},\\
H' &= H + (\partial S/\partial t)_{\underline{x},\underline{x}'}.
\end{aligned}
\tag{3.64}
$$

In this picture, we represent S as a function of $\underline{x}$, $\underline{x}'$, and t. This is traditionally called a "Type 1" representation, and symbolized with a subscript, as $S_1(\underline{x}, \underline{x}', t)$. Other representations are possible, and even useful. Employing $d(\underline{p}' \cdot \underline{x}') = d\underline{p}' \cdot \underline{x}' + \underline{p}' \cdot d\underline{x}'$ in Eq. (3.63),

$$
d(S_1 + \underline{p}' \cdot \underline{x}') \equiv dS_2 = \underline{p} \cdot d\underline{x} + \underline{x}' \cdot d\underline{p}' + (H' - H)\, dt, \tag{3.65}
$$

from which

$$
\begin{aligned}
\underline{p} &= (\partial S_2/\partial \underline{x})_{\underline{p}',t},\\
\underline{x}' &= (\partial S_2/\partial \underline{p}')_{\underline{x},t},
\end{aligned}
\tag{3.66}
$$

the relation between H and H' being the same as before, but with S_2 replacing S_1. This is the so-called "Type 2" representation, $S_2(\underline{x}, \underline{p}', t)$. The other two possibilities are

$$
\begin{aligned}
d(S_1 - \underline{p} \cdot \underline{x}) &\equiv dS_3 = -\underline{x} \cdot d\underline{p} - \underline{p}' \cdot d\underline{x}' + (H' - H)\, dt,\\
d(S_1 - \underline{p} \cdot \underline{x} + \underline{p}' \cdot \underline{x}') &\equiv dS_4 = -\underline{x} \cdot d\underline{p} + \underline{x}' \cdot d\underline{p}' + (H' - H)\, dt,
\end{aligned}
$$

whose mixed equations of transformation, omitting those for H and H', are as follows:

$$
\begin{aligned}
\underline{x} &= -(\partial S_3/\partial \underline{p})_{\underline{x}',t},\\
\underline{p}' &= -(\partial S_3/\partial \underline{x}')_{\underline{p},t},\\
\underline{x} &= -(\partial S_4/\partial \underline{p})_{\underline{p}',t},\\
\underline{x}' &= (\partial S_4/\partial \underline{p}')_{\underline{p},t}.
\end{aligned}
\tag{3.67}
$$

These types are all different ways of representing, in mixed form, the *same* transformation, as they all can be derived from one another; each one is simply more convenient in the context of a different set of problems

The symplectic transformations we have considered so far can all be recast into the language of generating functions. For example, the transformations in Eq. (3.48), with $\underline{\underline{M}}_{12} = \underline{\underline{M}}_{21} = \underline{\underline{O}}$, arise from a generating function,

$$
S_2(\underline{x}, \underline{p}') = \underline{p}' \cdot \underline{\underline{M}}_{11} \cdot \underline{x}, \tag{3.68}
$$

while those of Eq. (3.52) can be obtained from the generating function,

$$S_1(\underline{x}', \underline{\varphi}) = \sum_k \tfrac{1}{2} x_k'^2 \cot \varphi_k. \tag{3.69}$$

Symplectic Condition We shall now verify the following:

Theorem 3.6 Canonical transformations are symplectic. That is, the Jacobian matrix of a transformation produced by a generating function, as given in Eq.(3.63) or any of its variants, is a symplectic matrix.

We shall consider only the Type 2 representation explicitly, the others being equivalent. Accordingly, suppose that a transformation, $\underline{z} \longmapsto \underline{z}'$, is generated by $S(\underline{x}, \underline{p}'; t)$ according to the equations

$$\underline{x}' = (\partial S/\partial \underline{p}') \quad \text{and} \quad \underline{p} = (\partial S/\partial \underline{x}).$$

The notation is liable to become cumbersome, so we shall simplify it before beginning by writing derivatives as comma-prefixed subscripts:

$$S_{,\underline{p}'} \equiv \frac{\partial S}{\partial \underline{p}'}, \quad S_{,\underline{x}\underline{p}'} \equiv \frac{\partial^2 S}{\partial \underline{x} \partial \underline{p}'},$$

and so forth. In these terms, the equations of transformation are written

$$\underline{x}' = S_{,\underline{p}'} \quad \text{and} \quad \underline{p} = S_{,\underline{x}}. \tag{3.70}$$

Proof: The differential forms of Eqs. (3.70) are

$$\begin{aligned} d\underline{x}' &= d\underline{x} \cdot S_{,\underline{x}\underline{p}'} + d\underline{p}' \cdot S_{,\underline{p}'\underline{p}'} + dt S_{,t\underline{p}'} \\ &= S_{,\underline{p}'\underline{x}} \cdot d\underline{x} + S_{,\underline{p}'\underline{p}'} \cdot d\underline{p}' + S_{,\underline{p}'t} dt, \\ d\underline{p} &= d\underline{x} \cdot S_{,\underline{x}\underline{x}} + d\underline{p}' \cdot S_{,\underline{p}'\underline{x}} + dt S_{,t\underline{x}} \\ &= S_{,\underline{x}\underline{x}} \cdot d\underline{x} + S_{,\underline{x}\underline{p}'} \cdot d\underline{p}' + S_{,\underline{x}t} dt. \end{aligned} \tag{3.71}$$

In what follows, $dt = 0$ throughout; we are not considering "contact transformations" on extended phase space here. We seek to show that $\underline{\underline{M}}\,\underline{\underline{J}}\,\underline{\underline{M}}^T = \underline{\underline{J}}$, where $\underline{\underline{M}}$ is the matrix (function), here written in block form,

$$\begin{aligned} \underline{\underline{M}} = \partial \underline{z}'/\partial \underline{z} &= \begin{pmatrix} (\partial \underline{x}'/\partial \underline{x})_{\underline{p}} & (\partial \underline{x}'/\partial \underline{p})_{\underline{x}} \\ (\partial \underline{p}'/\partial \underline{x})_{\underline{p}} & (\partial \underline{p}'/\partial \underline{p})_{\underline{x}} \end{pmatrix} \\ &\equiv \begin{pmatrix} \underline{\underline{M}}_{11} & \underline{\underline{M}}_{12} \\ \underline{\underline{M}}_{21} & \underline{\underline{M}}_{22} \end{pmatrix}, \end{aligned}$$

as in Eq. (3.48). Begin by setting $d\underline{p} = 0$ in Eqs. (3.71).

$$\begin{aligned} d\underline{p} = \underline{0} \Rightarrow S_{,\underline{x}\underline{x}} \cdot d\underline{x} &= -S_{,\underline{x}\underline{p}'} \cdot d\underline{p}' \\ \Rightarrow \qquad d\underline{p}' &= -(S_{,\underline{x}\underline{p}'})^{-1} \cdot S_{,\underline{x}\underline{x}} \cdot d\underline{x}. \end{aligned} \tag{3.72}$$

This provides the block,

$$\underline{\underline{M}}_{21} = -(S_{,\underline{x}\underline{p}'})^{-1} \cdot S_{,\underline{x}\underline{x}}.$$

Furthermore, substituting the form $d\underline{p}'$, from Eq. (3.72), back into the form $d\underline{x}'$, in Eq. (3.71), provides us with the identity

$$d\underline{p} = \underline{0} \Rightarrow d\underline{x}' = \left(S_{,\underline{p}'\underline{x}} - S_{,\underline{p}'\underline{p}'} \cdot (S_{,\underline{x}\underline{p}'})^{-1} \cdot S_{,\underline{x}\underline{x}}\right) \cdot d\underline{x},$$

from which

$$\underline{\underline{M}}_{11} = S_{,\underline{p}'\underline{x}} - S_{,\underline{p}'\underline{p}'} \cdot (S_{,\underline{x}\underline{p}'})^{-1} \cdot S_{,\underline{x}\underline{x}}.$$

The two remaining blocks can be calculated much more simply by setting $d\underline{x} = \underline{0}$ in Eq. (3.71).

$$\begin{aligned} d\underline{x} = \underline{0} \Rightarrow d\underline{x}' &= S_{,\underline{p}'\underline{p}'} \cdot d\underline{p}', \\ d\underline{p} &= S_{,\underline{x}\underline{p}'} \cdot d\underline{p}' \\ d\underline{p}' &= (S_{,\underline{x}\underline{p}'})^{-1} \cdot d\underline{p} \end{aligned}$$

From these we immediately obtain the two remaining blocks.

$$\begin{aligned} \underline{\underline{M}}_{22} &= (S_{,\underline{x}\underline{p}'})^{-1}, \\ \underline{\underline{M}}_{12} &= S_{,\underline{p}'\underline{p}'} \cdot (S_{,\underline{x}\underline{p}'})^{-1}. \end{aligned}$$

Having written $\underline{\underline{M}}$ in terms of the generating function, we are finally ready to test the symplectic condition. We shall apply the conditions written in Eq. (3.50): that $\underline{\underline{M}}_{11}\underline{\underline{M}}_{12}^{T}$ and $\underline{\underline{M}}_{21}\underline{\underline{M}}_{22}^{T}$ be symmetric, and that $\underline{\underline{M}}_{11}\underline{\underline{M}}_{22}^{T} - \underline{\underline{M}}_{12}\underline{\underline{M}}_{21}^{T} = \underline{\underline{1}}$. Proceeding with the most arduous calculation first,

$$\begin{aligned} \underline{\underline{M}}_{11}\underline{\underline{M}}_{12}^{T} &= \left(S_{,\underline{p}'\underline{x}} - S_{,\underline{p}'\underline{p}'} \cdot (S_{,\underline{x}\underline{p}'})^{-1} \cdot S_{,\underline{x}\underline{x}}\right) \cdot \left(S_{,\underline{p}'\underline{p}'} \cdot (S_{,\underline{x}\underline{p}'})^{-1}\right)^{T} \\ &= S_{\underline{p}'\underline{p}'} - S_{,\underline{p}'\underline{p}'} \cdot (S_{,\underline{x}\underline{p}'})^{-1} \cdot S_{,\underline{x}\underline{x}} \cdot (S_{,\underline{p}'\underline{x}})^{-1} \cdot S_{,\underline{p}'\underline{p}'}. \end{aligned}$$

Clearly this is symmetric. Similarly, we calculate

$$\underline{\underline{M}}_{21}\underline{\underline{M}}_{22}^{T} = -\left((S_{,\underline{x}\underline{p}'})^{-1} \cdot S_{,\underline{x}\underline{x}}\right) \cdot \left((S_{,\underline{x}\underline{p}'})^{-1}\right)^{T},$$

which also is symmetric. Finally,

$$\begin{aligned}\underline{\underline{M}}_{11}\underline{\underline{M}}_{22}{}^T - \underline{\underline{M}}_{12}\underline{\underline{M}}_{21}{}^T &= \left(S_{,\underline{p}'\underline{x}} - S_{,\underline{p}'\underline{p}'} \cdot (S_{,\underline{x}\underline{p}'})^{-1} \cdot S_{,\underline{x}\underline{x}}\right) \cdot (S_{,\underline{p}'\underline{x}})^{-1} \\ &\quad + \left(S_{,\underline{p}'\underline{p}'} \cdot (S_{,\underline{x}\underline{p}'})^{-1}\right) \cdot \left(S_{,\underline{x}\underline{x}} \cdot (S_{,\underline{p}'\underline{x}})^{-1}\right) \\ &= \underline{\underline{1}},\end{aligned}$$

and the demonstration of symplecticity is complete. □

Given the complete machinery of differential forms and exterior calculus, this theorem could be proved in a few lines (see the epilogue). However, the statement is important enough that it should be understood even before becoming immersed in that formalism.

3.2.3 Symplectic Numerical Integration

Most problems cannot be solved "in closed form"; generally, one must rely on numerical analysis to obtain approximate solutions. A variety of numerical methods exist for approximating solutions of differential equations: Runge-Kutta, predictor-corrector, and the like. We think of these methods as producing approximations to a system's exact time-evolution map. In a strategy finding its roots in Riemann integration, the time interval over which the system is to evolve, say, (t_i, t_f), is partitioned into a set of N subintervals,

$$\{(t_k, t_{k+1}), \mid k = 0, \ldots, N-1,\ t_0 = t_i,\ t_N = t_f\}.$$

The partitions are devised so that $\Delta t = \max(t_{k+1} - t_k)$ approaches zero as N approaches infinity; typically, all subintervals are the same size, so that $t_{k+1} = t_k + \Delta t$. The numerical procedure is then used to approximate the time-evolution maps, $\phi_{t_{k+1}t_k}$, sequentially, and the Chapman-Kolmogorov equation is invoked (usually unconsciously) on the fly to catenate these together into $\phi_{t_f t_i}$. This is fundamentally the strategy behind the "tracking programs" used by accelerator physicists. When dealing with Hamiltonian systems, however, it is important that the symplectic property of $\phi_{t_f t_i}$ be maintained, which is done by requiring that the approximation for each $\phi_{t_{k+1}t_k}$ be symplectic. Most numerical integration procedures will not guarantee this. We shall briefly discuss a few that do.

Kick Maps To begin with, the kick map produced by using an impulse approximation on a Hamiltonian system is symplectic. The idea behind the impulse approximation is an instantaneous change in the momentum coordinates. Consider more generally the map

$$\phi : \begin{pmatrix}\underline{x}\\ \underline{p}\end{pmatrix} \mapsto \begin{pmatrix}\underline{x}\\ \underline{p}\end{pmatrix} + \begin{pmatrix}\underline{g}(\underline{x},\underline{p})\\ \underline{f}(\underline{x},\underline{p})\end{pmatrix}.$$

Using the notation of Eqs. (3.48), we write the submatrices,

$$\underline{\underline{M}}_{11} = \underline{\underline{1}} + \partial \underline{g}/\partial \underline{x}, \qquad \underline{\underline{M}}_{12} = \partial \underline{g}/\partial \underline{p},$$
$$\underline{\underline{M}}_{22} = \underline{\underline{1}} + \partial \underline{f}/\partial \underline{p}, \qquad \underline{\underline{M}}_{21} = \partial \underline{f}/\partial \underline{x}.$$

The symplecticity criteria written in Eq. (3.50) require that the matrices,

$$\underline{\underline{M}}_{12}\underline{\underline{M}}_{11}^{T} = \partial \underline{g}/\partial \underline{p} + (\partial \underline{g}/\partial \underline{p}) \cdot (\partial \underline{g}/\partial \underline{x})^{T},$$
$$\underline{\underline{M}}_{21}\underline{\underline{M}}_{22}^{T} = \partial \underline{f}/\partial \underline{x} + (\partial \underline{f}/\partial \underline{x}) \cdot (\partial \underline{f}/\partial \underline{p})^{T},$$

be symmetric and that

$$\begin{aligned}\underline{\underline{M}}_{11}\underline{\underline{M}}_{22}{}^{T} - \underline{\underline{M}}_{12}\underline{\underline{M}}_{21}{}^{T} &= \underline{\underline{1}} + \partial \underline{g}/\partial \underline{x} + (\partial \underline{f}/\partial \underline{p})^{T} + (\partial \underline{g}/\partial \underline{x}) \cdot (\partial \underline{f}/\partial \underline{p})^{T} \\ &\quad - (\partial \underline{g}/\partial \underline{p}) \cdot (\partial \underline{f}/\partial \underline{x})^{T} \\ &= \underline{\underline{1}}.\end{aligned}$$

The trivial way of assuring that these conditions are satisfied is to set

$$\partial \underline{f}/\partial \underline{p} = \underline{\underline{0}} \quad \textbf{and} \quad \partial \underline{g}/\partial \underline{x} = \underline{\underline{0}} \quad \textbf{and} \quad \left(\partial \underline{g}/\partial \underline{p} = \underline{\underline{0}} \quad \textbf{or} \quad \partial \underline{f}/\partial \underline{x} = \underline{\underline{0}}\right),$$

while simultaneously requiring the matrices

$$\partial \underline{f}/\partial \underline{x} = (\partial \underline{f}/\partial \underline{x})^{T} \quad \text{and} \quad \partial \underline{g}/\partial \underline{p} = (\partial \underline{g}/\partial \underline{p})^{T}$$

be symmetric. The impulsive "kick" approximation corresponds to setting $\underline{g} = \underline{0}$, so that only the momentum changes. The symmetry condition is satisfied in the context of a Hamiltonian dynamic by finding a scalar "integrated kick" function, $F(\underline{x})$, such that $\underline{f} = -\partial F/\partial \underline{x}$. The interpretation given is that, in the impulse approximation, the position coordinates do not change appreciably while the impulse is being imparted.

For example, as written in Eq. (2.52), the symplectic Hénon map is obviously a catenation of a kick map with a rotation. In accelerator applications, the quadratic kick is an impulse approximation to the integrated effect of passage through a sextupole, the (inessential) parameter λ being proportional to an integrated field strength. (See also Eqs. (3.115) and (3.116).)

There is another possibility that generally is not used but that we propose for your consideration. Begin by specializing Eq. (2.71) to the case of a Hamiltonian vector field.

$$\underline{z}(t) = \underline{z}_0 + \int_{t_0}^{t} d\tau \underline{\underline{J}} \cdot \underline{\partial} H(\underline{z}(\tau), \tau).$$

Now we assume the time-dependence of the Hamiltonian to be a delta function impulse, say, $H(\underline{z}, \tau) = \delta(\tau - t_K)h(\underline{z})$, so that, for $t_0 < t_K < t$,

$$\underline{z}(t) = \underline{z}_0 + \underline{\underline{J}} \cdot \partial h(\underline{z}(t_K)).$$

This does not seem to help much, as we do not know $\underline{z}(t_K)$. However, time reversal requires that $\underline{z}(t_K)$ be some symmetric function of $\underline{z}(t)$ and $\underline{z}_0$. The most natural ansatz is to take the average: $\underline{z}(t_K) = (\underline{z}_0 + \underline{z}(t))/2$. This provides us with a map, $\phi : \underline{z}_i \longmapsto \underline{z}_f$, which models the impulsive kick:

$$\underline{z}_f = \underline{z}_i + \underline{\underline{J}} \cdot \partial h(\tfrac{1}{2}(\underline{z}_i + \underline{z}_f)). \tag{3.73}$$

Whether these formal manipulations mean anything or not, it is amusing that the map *implicitly* defined by Eq. (3.73) is, in fact, symplectic. To see this, we write the differential forms.

$$d\underline{z}_f = d\underline{z}_i + \tfrac{1}{2}\underline{\underline{J}} \cdot \underline{\underline{\partial\partial h}} \cdot (d\underline{z}_f + d\underline{z}_i).$$

We have suppressed writing the argument of h. Rearranging terms, we have

$$(\underline{\underline{1}} + \tfrac{1}{2}\underline{\underline{J}} \cdot \underline{\underline{\partial\partial h}}) \cdot d\underline{z}_f = (\underline{\underline{1}} - \tfrac{1}{2}\underline{\underline{J}} \cdot \underline{\underline{\partial\partial h}}) \cdot d\underline{z}_i,$$

which provides the Jacobian matrix,

$$\partial\underline{z}_f/\partial\underline{z}_i = (\underline{\underline{1}} + \tfrac{1}{2}\underline{\underline{J}} \cdot \underline{\underline{\partial\partial h}})^{-1} \cdot (\underline{\underline{1}} - \tfrac{1}{2}\underline{\underline{J}} \cdot \underline{\underline{\partial\partial h}}).$$

That this is symplectic is a straightforward application of the following.

Lemma 3.3 Let $\underline{\underline{S}}$ be a symmetric matrix, and let

$$\underline{\underline{M}} \equiv \left(\underline{\underline{1}} + \underline{\underline{J}} \cdot \underline{\underline{S}}\right)^{-1} \cdot \left(\underline{\underline{1}} - \underline{\underline{J}} \cdot \underline{\underline{S}}\right).$$

Then $\underline{\underline{M}}$ is a symplectic matrix.

The proof of this lemma is left to the reader.

Generating Functions It was demonstrated in the previous section that transformations produced by the technique of generating functions are indeed symplectic. We now apply that observation to the problem of approximating $\phi_{t+\Delta t,t} : \underline{z} = (x, p) \longmapsto \underline{z}' = (x', p')$. The method was suggested by Ruth [81] and has been developed by Channell and Scovel [20] and others. A similar method, though different in form, was independently worked out by Feng Kang [55, 53, 54].

Consider a canonical transformation with a Type 3 generating function $S(x', p, \Delta t)$, and expand the *explicit* time-dependence of S. Because $\underline{z}' = \underline{z}$ when $\Delta t = 0$, we can set the leading term to $-px'$.

$$S(x', p, \Delta t) = -px' + \Delta t\, S^{(1)}(x', p) + \tfrac{1}{2}\Delta t^2\, S^{(2)}(x', p) + O(\Delta t^3).$$

Substituting this into Eq. (3.67) provides the corresponding expansion of the transformed momentum coordinate.

$$p' = -S_{,x'}(x', p, \Delta t) = p - \Delta t\, S^{(1)}_{,x'}(x', p) - \tfrac{1}{2}\Delta t^2\, S^{(2)}_{,x'}(x', p) + O(\Delta t^3). \qquad (3.74)$$

We now require that p' be precisely the transformed momentum coordinate obtained from the equations of motion. This is done by differentiating the right-hand side (actually, a Lie derivative) and equating the result to the momentum component of the vector field.

$$\begin{aligned}\dot{p}' &= -S^{(1)}_{,x'}(x', p) - \Delta t\, S^{(1)}_{,x'x'}(x', p)\dot{x}' - \Delta t\, S^{(2)}_{,x'}(x', p) + O(\Delta t^2)\\ &= -S^{(1)}_{,x'}(x', p) - \Delta t\, S^{(1)}_{,x'x'}(x', p) H_{,p'}(x', p') - \Delta t\, S^{(2)}_{,x'}(x', p) + O(\Delta t^2)\\ &= -H_{,x'}(x', p').\end{aligned}$$

Substitute for p' from Eq. (3.74) and collect terms in powers of Δt.

$$\begin{aligned}&-S^{(1)}_{,x'}(x', p) - \Delta t\,(S^{(1)}_{,x'x'}(x', p) H_p(x', p) + S^{(2)}_{,x'}(x', p)) + O(\Delta t^2) =\\ &-H_{,x'}(x', p) - \Delta t\, H_{,x'p}(x', p) S^{(1)}_{,x'}(x', p) + O(\Delta t^2).\end{aligned}$$

Appropriate solutions for $S^{(1)}$ and $S^{(2)}$ are now obtained by matching the coefficients of like powers of Δt.

$$\begin{aligned}&S^{(1)}_{,x'}(x', p) = H_{,x'}(x', p)\\ \Rightarrow\ &S^{(1)}(x', p) = H(x', p),\\ &S^{(2)}_{,x'}(x', p) = -S^{(1)}_{,x'x'}(x', p) H_p(x', p) - S^{(1)}_{,x'}(x', p) H_{,x'p}(x', p)\\ \Rightarrow\ &S^{(2)}(x', p) = -S^{(1)}_{,x'}(x', p) H_p(x', p).\end{aligned}$$

If this calculation were continued to the third order, we would obtain

$$S^{(3)} = -S^{(2)}_{,x'} H_p + \left(S^{(1)}_{,x'}\right)^2 H_{,pp} \qquad (3.75)$$

as well.

We illustrate this by applying it to the harmonic oscillator, with normalized Hamiltonian $H(x,p) = (p^2 + x^2)/2$.

$$
\begin{aligned}
S^{(1)}(x',p) = H(x',p) &= \tfrac{1}{2}p^2 + \tfrac{1}{2}x'^2, \\
S^{(2)}(x',p) &= -px', \\
S^{(3)}(x',p) &= p^2 + x'^2.
\end{aligned}
$$

The generating function, written in full, is

$$
S(x',p,\Delta t) = -(1 + \Delta t^2/2)px' + (\Delta t + \Delta t^3/3)(p^2/2 + x'^2/2) + O(\Delta t^4).
$$

The "reverse" transformation equation, given by Eqs. (3.67),

$$
x = -S_p(x',p,\Delta t) = (1 + \Delta t^2/2)x' - (\Delta t + \Delta t^3/3)p,
$$

is used to solve for x' in terms of x and p. This provides the symplectic integration algorithm.

$$
\begin{aligned}
x' &= \frac{x + (\Delta t + \Delta t^3/3)p}{1 + \Delta t^2/2}, \\
p' = -S_{,x'}(x',p,\Delta t) &= (1 + \Delta t^2/2)p - (\Delta t + \Delta t^3/3)x'.
\end{aligned}
$$

Even though higher powers of Δt appear, this solution is accurate only up to first order (in Δt). The desirable effect of the extra terms is to make the time-evolution map exactly symplectic. Although we have developed this only to first order, such procedures can be extended to higher orders.

Exponential Maps The time-evolution map of autonomous vector fields was formally written in Eq. (2.61). Employing the notation introduced in Eq. (3.9), we rewrite this as follows:

$$
\phi_{t+\Delta t,t} = e^{\Delta t X_H} = e^{-\Delta t :H:},
$$

for a Hamiltonian vector field, with Hamiltonian function H. We shall use Dragt's notation below for Hamiltonian vector fields rather than the more traditional mathematical notation; it is awkward to have subscripted quantities in exponents, and we shall also be using subscripts to indicate derivative functions. Recall that $:H: \equiv -X_H$. It is tempting simply to expand the exponential and thereby obtain an integration procedure of arbitrarily high order. However, while the full power series in Δt would be symplectic, its truncation is not. Thus, other approaches can be tried; we shall examine one suggested by Yoshida [93].

To keep matters simple, assume that a Hamiltonian dynamical system is separable.

$$H(x,p) = K(p) + V(x).$$

By straightforward expansion, it is trivial to show that

$$e^{-\Delta t:H:} = e^{-(\Delta t/2):K:}e^{-\Delta t:V:}e^{-(\Delta t/2):K:} + O(\Delta t^3).$$

Now consider that

$$\begin{aligned} :K:x &= \{K(p),x\} = -K_{,p}(p), \\ :K:^2 x &= 0, \end{aligned}$$

which enables us to write $e^{-\Delta t:K:}$ exactly in finite form.

$$e^{-\Delta t:K:}x = x + \Delta t\, K_{,p}(p).$$

Similarly, consider the action of $:V:$ on the coordinate p.

$$\begin{aligned} :V:p &= \{V(x),p\} = V_{,x}(x), \\ :V:^2 p &= 0, \\ e^{-\Delta t:V:}p &= p - \Delta t\, V_{,x}(x). \end{aligned}$$

We put these pieces together to evaluate,

$$\begin{aligned} e^{-\Delta t:H:}x(t) &= e^{-(\Delta t/2):K:}e^{-\Delta t:V:}e^{-(\Delta t/2):K:}x(t) + O(\Delta t^3) \\ e^{-(\Delta t/2):K:}x(t) &= x(t) + \frac{\Delta t}{2}K_{,p}(p(t)), \\ e^{-\Delta t:V:}e^{-(\Delta t/2):K:}x(t) &= x(t) + \frac{\Delta t}{2}K_{,p}(p(t) - \Delta t V_{,x}(x(t))), \\ e^{-(\Delta t/2):K:}e^{-\Delta t:V:}e^{-(\Delta t/2):K:}x(t) &= x(t) + \frac{\Delta t}{2}K_{,p}(p(t)) + \frac{\Delta t}{2}K_{,p}(p(t) \\ &\quad -\Delta t V_{,x}(x(t) + \frac{\Delta t}{2}K_{,p}(p(t))). \end{aligned}$$

Normally this is done with an intermediate step, which makes the algorithm look like a sequence of kick maps.

$$\begin{aligned} x_{n+1/2} &= x_n + \frac{\Delta t}{2}K_{,p}(p_n), \\ p_{n+1} &= p_n - \Delta t V_{,x}(x_{n+1/2}), \\ x_{n+1} &= x_{n+1/2} + \frac{\Delta t}{2}K_{,p}(p_{n+1}). \end{aligned} \tag{3.76}$$

Returning to the harmonic oscillator for an example, $K_{,p}(p) = p$, $V_{,x}(x) = x$, and Eq. (3.76) becomes

$$\begin{aligned} x_{n+1/2} &= x_n + \frac{\Delta t}{2} p_n, \\ p_{n+1} &= p_n - \Delta t\, x_{n+1/2} \\ &= (1 - \Delta t^2/2) p_n - \Delta t\, x_n, \\ x_{n+1} &= x_{n+1/2} + \frac{\Delta t}{2} p_{n+1} \\ &= (1 - \Delta t^2/2) x_n + (\Delta t - \Delta t^3/4) p_n. \end{aligned}$$

3.3 TORI: ACTION-ANGLE COORDINATES

Transformations are of more than academic interest. Their greatest motivation lies in the quest for a good set of coordinates, a chart on which the vector field is represented in the *simplest possible manner.* Obviously, one of the best things that can be done is to find coordinates that are dynamical invariants. A Hamiltonian dynamical system on a $2n$-dimensional phase space is called "integrable" if there exists an atlas possessing n invariant coordinates. On such charts the Hamiltonian is represented in a form that is a straightforward generalization of the harmonic oscillator. This chapter deals with that programme. The world's most benign nonlinear dynamical systems are the shearing Hamiltonians, which produce integrable "twist maps." We shall describe this in terms of "action-angle" coordinates, $(\underline{\varphi}, \underline{I})$, originally introduced in connection with the harmonic oscillator. To begin with, however, this is only a formal use of symbols to represent the position and momentum sectors of phase space.

Liouville's Torus Theorem Assume, then, that H is a function of only the momentum coordinates, $\underline{I}$. We separate out the part linear in $\underline{I}$ and write

$$\begin{aligned} H &= H(\underline{I}) \equiv \frac{2\pi}{T} \underline{\nu} \cdot \underline{I} + H_s(\underline{I}), \\ \dot{\underline{I}} &= -\partial H / \partial \underline{\varphi} = 0, \\ \dot{\underline{\varphi}} &= \partial H / \partial \underline{I} \equiv \underline{\omega}(\underline{I}) = \frac{2\pi}{T} \underline{\nu} + \partial H_s / \partial \underline{I}. \end{aligned} \tag{3.77}$$

Here, T is a "time" scale or period, as in Eq. (2.34). The maps associated with this vector field are the most regular of what mathematicians call "twist maps"; accelerator physicists refer to H_s as the "detuning" or "shear" terms. These descriptive terms arise from viewing punctured 2D phase space as a cylinder (see Figure 3.9). If we think of the orbits as slices of the cylinder that rotate at different rates, the effect of the flow is to twist the cylinder as the orbits "shear" past each other, like laminar fluid flow.

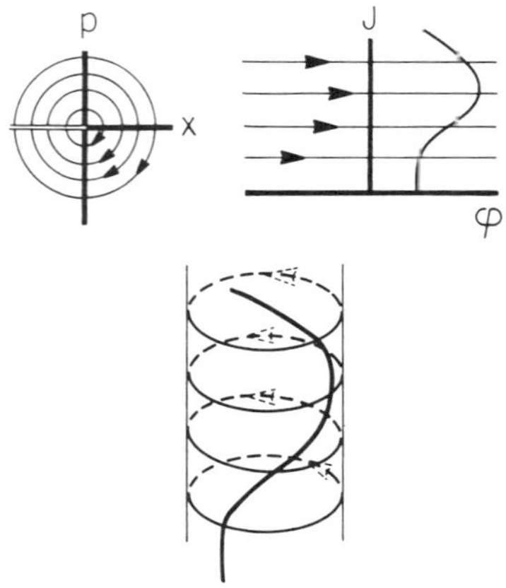

FIGURE 3.9 Twist maps on a punctured phase space.

Such a Hamiltonian is also said to be "in normal form," and much of Chapter 5 deals with attempts to find charts on which Hamiltonians assume this form.

On a single chart, the orbits of Eq. (3.77) are written trivially:

$$\underline{I}(t) = \underline{I}(0), \quad \underline{\varphi}(t) = \underline{\varphi}(0) + \omega(\underline{I})t.$$

A rather boring possibility is that one chart is sufficient to represent the entire phase space, in which case the position coordinates, $\underline{\varphi}$, simply grow without bound linearly in time at a rate that depends on the momentum coordinates, $\underline{I}$. For example, this is exactly what happens in free-particle motion. The more interesting possibility is that the $\underline{\varphi}$ coordinates really represent angle coordinates, as with the harmonic oscillator. In this case, the orbits lie on tori, but unlike the harmonic oscillator, in which all tori possessed the same $\underline{\omega}$, now $\underline{\omega}$ depends explicitly on $\underline{I}$.

The n coordinates of the momentum sector, $I_1, I_2, \ldots, I_n$, viewed as observables, have three important properties:

(a) They are dynamical invariants; $dI_k/dt = 0$, for all k. Since time does not appear explicitly in the definition of I_k, this is equivalent to saying that $[I_k, H] = 0$, so that each I_k "commutes" with H.

(b) They are "in involution," which means that they all "commute":

$$\forall i, j : [I_i, I_j] = 0.$$

This is the classical analog of the quantum mechanical concept of a "complete" set of commuting observables.

(c) They are independent. Formally, the differential forms $\{dI_k, k = 1, \ldots, n\}$ are linearly independent everywhere within the region of interest. In particular, the level sets of these observables are nowhere tangent: they intersect transversely.

It turns out that these three conditions are all that are required to assure the existence of a local twist map. This is incorporated into the following theorem, sometimes associated with Arnold, although Arnold himself attributes it to Liouville [5].

Theorem 3.7 (Liouville-Arnold tori) Let a Hamiltonian system defined over a $2n$ dimensional phase space, $\mathcal{P}$, possess n independent, continuously differentiable, invariants in involution, $\underline{G} = (G_1, G_2, \ldots, G_n)$. From dynamical invariance, orbits must lie in the level sets of $\underline{G}$, which we symbolize as

$$M_{\underline{g}} \equiv \{\mathbf{p} \in \mathcal{P} \mid \underline{g} = \underline{G}(\mathbf{p})\} \subset \mathcal{P}.$$

Each $M_{\underline{g}}$ is a smooth, invariant manifold. Furthermore, if $M_{\underline{g}}$ is compact and connected, then it is diffeomorphic to the n-torus, $\mathcal{T}^n$. Orbits on $M_{\underline{g}}$ are "conditionally periodic." That is, it is possible to introduce a chart of angle coördinates, $\underline{\varphi}$, on the torus so that, for some n-tuple of real numbers, $\underline{\omega}_{\underline{g}}$, which depend only on the torus,

$$\dot{\underline{\varphi}} = \underline{\omega}_{\underline{g}}.$$

There is no such thing as a "fundamental theorem of Hamiltonian dynamics," but if there were, this would be it. By virtue of this theorem, a Hamiltonian over a $2n$ dimensional phase space possessing n independent, invariants in involution is called "integrable." The theorem states that tori are the universal type of manifold describing the behavior of bounded orbits for *all* integrable Hamiltonian systems, not just the harmonic oscillator.

Rather than writing the theorem's proof, we shall be content with sketching a few of its key features. First, notice that the n gradient vector fields, $\underline{\partial} G_k$, are everywhere linearly independent, by hypothesis, and orthogonal to the surface $M_{\underline{g}}$. They therefore span the n-dimensional subspace orthogonal to $TM_{\underline{g}}$ at any point of evaluation. Now, consider the n vector fields $X^{(k)} \equiv X_{G_k} = \underline{\underline{J}} \cdot \underline{\partial} G_k$, constructed by using the n observables as Hamiltonian functions. These also must be linearly independent. Furthermore, each $X^{(k)}$ is orthogonal to every $\underline{\partial} G_j$, by the hypothesis of involution and the defining Eq. (3.7). Therefore each $X^{(k)}$ is tangent to $M_{\underline{g}}$, and the collection $\{X^{(k)} \mid k = 1, \ldots, n\}$ spans the tangent space at the point of evaluation. They can be viewed as a "complete" collection of linearly independent vector fields *on the manifold* $M_{\underline{g}}$ *itself*. Finally, the vector fields are themselves pairwise commutative under Lie brackets, because of the (anti-)morphism property expressed in Eq. (3.9).

These few observations provide an enormous topological constraint on the manifold, $M_{\underline{g}}$. The *only* compact manifolds on which one can construct such families of vector fields — complete, pairwise commutative, and linearly independent everywhere — are tori. This topological result is established by constructing a coordinate chart on $M_{\underline{g}}$ using the vector fields to lay out the grid. Let $\phi_\tau^{(k)}$ be the one-parameter time-evolution map of the vector field $X^{(k)}$. Because the vector fields commute, their maps commute also; there are no "loop defects" (see Section 2.3.2.) One then can define a chart, with an arbitrary point, $\mathbf{p}$, at the origin, by interpreting a coordinate,

$\underline{\tau} = (\tau_1, \tau_2, \ldots, \tau_n) \in R^n$, as an instruction to follow the vector field $X^{(1)}$ for "time" τ_1, $X^{(2)}$ for "time" τ_2, and so forth. Formally, we define the n-parameter family of maps, $\phi_{\underline{\tau}} = \prod_k \phi^{(k)}_{\tau_k} : M_{\underline{g}} \rightarrow M_{\underline{g}}$, which when applied to a "point of origin," $\mathbf{p}$, induces the chart, $C : R^n \rightarrow M_{\underline{g}}, \underline{\tau} \longmapsto \phi_{\underline{\tau}}(\mathbf{p})$.

As with most charts, we must restrict the domain of C. The hypothesis states that $M_{\underline{g}}$ is compact, whereas R^n is not. Therefore $\phi_{\underline{\tau}}$ cannot be a one-to-one, differential map over all of R^n; the two spaces are not topologically equivalent. Now, because of the Chapman-Kolomogorov property expressed in Eq. (1.6), and because all the individual factors in $\phi_{\underline{\tau}}$ commute, the set of maps, $\{\phi_{\underline{\tau}} \mid \underline{\tau} \in R^n\}$, inherits the Abelian group structure of R^n under component-wise addition. In particular,

$$\phi_{\underline{\tau}} \circ \phi_{\underline{\tau}'} = \phi_{\underline{\tau}'} \circ \phi_{\underline{\tau}} = \phi_{\underline{\tau}+\underline{\tau}'}. \tag{3.78}$$

The subset that corresponds to those maps for which $\mathbf{p}$ is a fixed point is a discrete subgroup of this group, called the "stationary group" of $\mathbf{p}$ and symbolized Γ. In fact, it is fairly easy to see that Γ is independent of $\mathbf{p}$. Let $\underline{\sigma}$ be coordinates of another point, say, $\mathbf{p}'$, using $\mathbf{p}$ as origin, and let $\phi_{\underline{\tau}}$ be in the stationary group of $\mathbf{p}$. Then

$$\phi_{\underline{\tau}}(\mathbf{p}') = \phi_{\underline{\tau}}(\phi_{\underline{\sigma}}(\mathbf{p})) = \phi_{\underline{\sigma}}(\phi_{\underline{\tau}}(\mathbf{p})) = \phi_{\underline{\sigma}}(\mathbf{p}) = \mathbf{p}',$$

so that $\phi_{\underline{\tau}}$ is also in the stationary group of $\mathbf{p}'$. The stationary group is nontrivial — that is, it contains more than the identity element — since it is the same group for all the points of $M_{\underline{g}}$, and we've already noted that C is not one-to-one.

Invoking the morphism of Eq. (3.78), Γ is imaged by a discrete subgroup of R^n. The only possible discrete subgroups are sets of lattice points generated by base sets of elements of R^n, $\{\underline{e}_j \mid j = 1, \ldots, k, k \leqslant n\}$; that is, by sets of points in R^n which can be written as superpositions of the $\underline{e}_j$ with integer coefficients. Now it turns out that because $M_{\underline{g}}$ is compact, no indices can be left over: we must have $k = n$. The generators, $\underline{e}_j$, are then boundaries of a basic cell, which repeated, covers all of R^n. In short, we are describing a universal covering space for a torus, like that in Chapter 1 (following Eq. (2.12)), and $M_{\underline{g}}$ is therefore topologically equivalent to a torus.

We repeat: none of this proves anything; we have only indicated the directions that legitimate proofs generally follow. Nonetheless, one can begin to understand how these problems from classical mechanics spurred the development of modern topological concepts.

Coordinates from Generating Functions Suppose that we are given an autonomous, integrable, Hamiltonian system, complete with n invariant, continuously differentiable, independent observables, $\underline{G}$ — one of which is typically the Hamiltonian itself — and that we are sufficiently clever to construct the tori, $M_{\underline{g}}$, and their "angle" coordinate charts, provided by the universal covering of the preceding section. A state can be determined by (a) n components of $\underline{g}$, which identify the torus it is in, and (b) n angle coordinates which determine where it is on that torus. We can therefore think of these $2n$ numbers as providing a coordinate chart over a local region of phase space foliated by the tori provided that we are certain to connect the angle coordinates smoothly between them. However, given one complete set of invariant observables, it is easy to construct any number of others by combining them in various ways. The "action" coordinates are a preferred set that label the tori with n observables conjugate to the angle coordinates. They are given that name both because, on dimensional grounds, they must have the units of an action integral and because, as we shall see, their calculation proceeds along very similar lines. Action coordinates have the advantage of leaving Hamilton's equations in canonical form, as in Eq. (3.4). Since the tori are invariant, the Hamiltonian must be a function of the action coordinates only, and we have the exact analytic representation of Eq. (3.77).

A canonical transformation to conjugate action-angle coordinates, $(\underline{x},\underline{p}) \rightsquigarrow (\underline{\varphi},\underline{I})$, can be accomplished with a Type 2 generating function, using Eqs. (3.65) and (3.66).

$$\begin{aligned} dS &= \underline{p}\cdot d\underline{x} + \underline{\varphi}\cdot d\underline{I} + (K-H)\,dt, \\ \underline{p} &= \partial S/\partial \underline{x}, \\ \underline{\varphi} &= \partial S/\partial \underline{I}. \end{aligned} \tag{3.79}$$

S itself is obtained by integrating the differential form dS along paths on the torus.

$$S(\underline{x},\underline{I}) \equiv \int_{\underline{x}_0}^{\underline{x}} dS = \int_{\underline{x}_0}^{\underline{x}} \underline{p}\cdot d\underline{x}. \tag{3.80}$$

The integral is taken along a path within $M_{\underline{g}}$ connecting points with configuration coordinates $\underline{x}_0$ and $\underline{x}$; of course $dt = 0$, and we are demanding $d\underline{I} = \underline{0}$ along γ, as we want $\underline{I}$ to label the torus. The point $\underline{x}_0$ is arbitrary; changing it merely provides an additive constant to S, which does not alter the transformation. Although we have not proved it, *the value of this integral is the same for any two paths on the torus that are homotopic,* that is, which can be deformed (continuously) into each other without leaving the torus. However, dS is not actually the differential of a single-valued function on the torus; it is not an "exact" differential form. If γ be a closed path that winds around the torus, then $\oint_\gamma dS$ does not vanish. This is to be expected, because at least one angular coordinate $\underline{\varphi}$, which is related to S via Eq. (3.79), must increase by 2π when going around the torus. Indeed, let γ_k be the closed path corresponding to a generator of the stationary group,

$$\gamma_k = \gamma(\underline{e}_k) = \{\phi_{u\underline{e}_k}\mathbf{p} \mid u \in [0,1]\}, \tag{3.81}$$

and let $\Delta_k q$ be the change in quantity q when traversing γ_k. Then we want $\Delta_k\varphi_j = 2\pi\delta_{kj}$ in order that φ_k be the angle coordinate associated with generator $\underline{e}_k$, and with path γ_k. But using Eq. (3.79) we have

$$2\pi\delta_{kj} = \Delta_k\varphi_j = \Delta_k(\partial S/\partial I_j) = \frac{\partial}{\partial I_j}\Delta_k S = \frac{\partial}{\partial I_j}\oint_{\gamma_k} dS.$$

This provides the definition of the action coordinate I_k.

$$I_k = \frac{1}{2\pi} \oint_{\gamma_k} dS = \frac{1}{2\pi} \oint_{\gamma_k} \underline{p} \cdot d\underline{x} \tag{3.82}$$

The array, $\underline{I} \equiv (I_1, I_2, \ldots, I_n)$ — corresponding to the collection of generators $\{\underline{e}_1, \underline{e}_2, \ldots, \underline{e}_n\}$, or the nonhomotopic, closed curves $\{\gamma_1, \gamma_2, \ldots, \gamma_n\}$ that wind around the torus in different directions — comprise the n action coordinates we seek to identify $M_{\underline{g}}$.

This heuristic romp has left many unanswered questions. For example, we stated that the integral in Eq. (3.80) takes the same value on two paths that are homotopic. This means that it vanishes taken over any closed curve obtained by joining two such paths (reversing the direction of one of them). But it does *not* vanish over the closed paths γ_k. What is the distinction between these two? Furthermore, if a path γ_k' homotopic to γ_k is substituted for γ_k in Eq. (3.82), the value of the integral remains the same. This property is needed in order that the action coordinates be associated with the torus itself. But why does it work? The appropriate framework for pursuing these questions and others like them is within the theories of differential forms and homology.[13] The answer to the first is that one closed path is a boundary; the other is not. The symplectic two-form alluded to in Section 3.2.1, in connection with Darboux's theorem, vanishes when applied to vector fields on the torus, $M_{\underline{g}}$, and a "Stokes' theorem" allows us to convert line integrals of $\underline{p} \cdot d\underline{x}$ (a one-form) over boundaries into volume integrals of this vanishing two-form over regions of the torus. This same theorem answers the second question as well. It is not a great exaggeration to reverse our previous statement and say that the theories of differential forms and homology were originally created in order to establish the correct framework for discussing action-angle coordinates. As such, a little knowledge of this problem does much to motivate their key ideas. Regrettably, we shall not pursue these subjects here: compromises must be made, and we must move on to other matters. Excellent introductions have been written by Flanders [35] and Nash and Sen [74], although few connections are made with Hamiltonian mechanics; slightly more formidable are the treatments in Abraham and Marsden [1] or Lang [58]; most formidable of all is the classic of Hilton and Wylie [47].

It should also be noted that we have not yet extended the generating function S beyond a single torus. This is done smoothly by assigning a curve cutting through the nested[14] tori to be the set of points for which $S = 0$.

Other Action-Angle Coordinates The "quadrature" prescription that we have outlined for creating action-angle coordinates does not determine a unique set of such coordinates. Action coordinates were written as integrals over curves corresponding

[13] Arnold stated the matter without compromise when he wrote: 'Hamiltonian mechanics cannot be understood without differential forms" [5]. This is what he was referring to: the foundations of symplectic structure and action-angle coordinates.

[14] And how do we know that the tori actually foliate a region of phase space?

to generators of the stationary group, $\Gamma \subset R^n$, that is, a set of basis vectors for Γ. Now there is more than one possible set of generators, so that the action coordinates are not unique. We shall examine how their definition changes as we alter the basis.

Using the original basis, we write Γ as the set of all superpositions with integer coefficients.

$$\Gamma = \{\underline{v} = \sum m_k \underline{e}_k \mid m_k \in Z\}.$$

The coefficients, m_k, are the winding numbers of the closed path $\gamma(\underline{v})$ (see Eq. (3.81)) on $M_{\underline{g}}$ *relative to* the basis. To change basis, we choose a complete, linearly independent set of elements from Γ, say, $\underline{u}_i, i = 1, \ldots, n$. Let $u_k^{(i)}$ be the components of $\underline{u}_i$ in the $\underline{e}$ basis.

$$\underline{u}_i = \sum u_k^{(i)} \underline{e}_k \in \Gamma.$$

That the $\underline{u}_i$ form a basis means that any $\underline{v} \in \Gamma$ can be written as a superposition of them with integer coefficients.

$$\underline{v} = \sum m_k \underline{e}_k = \sum \xi_i \underline{u}_i = \sum \xi_i u_k^{(i)} \underline{e}_k.$$

Let $\underline{\underline{U}}$ be the matrix with elements $u_k^{(i)}$. Since the $\underline{e}_k$ are linearly independent, we have the matrix equations,

$$\begin{aligned} \underline{m} &= \underline{\xi} \cdot \underline{\underline{U}}, \\ \underline{\xi} &= \underline{m} \cdot \underline{\underline{U}}^{-1} \\ &= \frac{1}{\det(\underline{\underline{U}})} \left(\underline{m} \cdot \operatorname{cof}(\underline{\underline{U}})\right), \end{aligned} \tag{3.83}$$

where $\operatorname{cof}(\underline{\underline{U}})$, whose elements are integers, is the matrix of cofactors of $\underline{\underline{U}}$. It is required that $\underline{\xi}$ be an array of integers, which means that $\det(\underline{\underline{U}})$ must divide every component of $\underline{m} \cdot \operatorname{cof}(\underline{\underline{U}})$ for any integer array $\underline{m}$. Therefore it must be that $\underline{\underline{U}}$ is unimodular: $\det(\underline{\underline{U}}) = 1$. Figure 3.10 illustrates two simple examples. We have here the multidimensional version of the linear automorphism of the torus that was introduced in Eq. (2.77), what is sometimes called the "cat map" [7]; it is also a generalization of the "egg carton" transformation shown in Figure 2.14.

As $\underline{\underline{U}}$ is unimodular we can use the parallelepiped, $\mathcal{U} = \{\sum x_i \underline{u}_i \mid 0 \leqslant x_i \leqslant 1\}$, formed from the $\underline{u}_i$ rather than that formed from the $\underline{e}_k$, $\mathcal{E} = \{\sum x_k \underline{e}_k \mid 0 \leqslant x_k \leqslant 1\}$, as a set of representatives of the factor group, $R^n/\Gamma \simeq \mathcal{T}^n \simeq M_{\underline{g}}$, which is the torus. Geometrically (see Figure 3.10), think of tiling R^n with copies of $\mathcal{U}$, rather than copies of $\mathcal{E}$, and identifying a point in any cell with the corresponding point in the base cell, $\mathcal{U}$. Algebraically, Γ sets up an equivalence relation in R^n: $\underline{v} \sim \underline{w}$ if and only

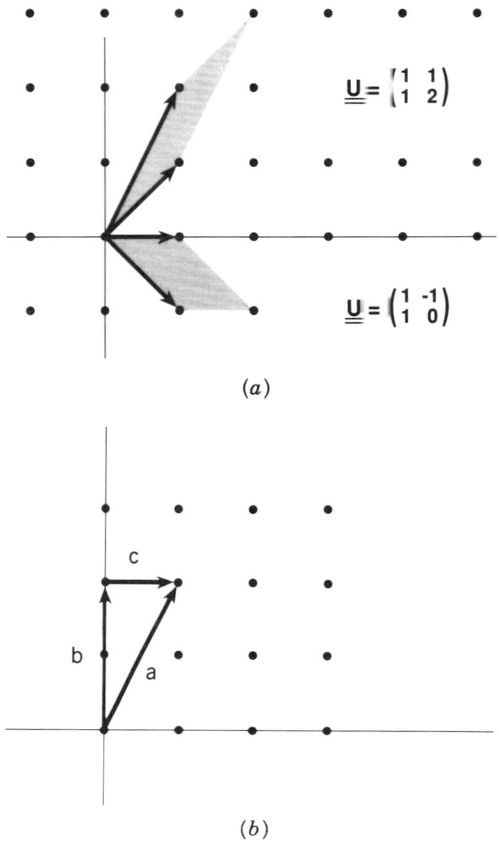

FIGURE 3.10 Changing the generators of the stationary group, Γ, alters the definition of action-angle coordinates: (*a*) matrices associated with two new $\underline{u}$ bases, relative to the original $\underline{e}$ basis vectors; (*b*) the integral of $\underline{p} \cdot d\underline{x}$ along path a is the sum of the integrals along paths b and c.

if $\underline{v} - \underline{w} \in \Gamma$. Then any $\underline{v}$ is equivalent to some $\underline{w} = \sum w_k \underline{e}_k$, where all components satisfy $0 \leqslant w_k < 1$. That is, $\mathcal{E}$ is a set of representatives of equivalence classes in R^n. The change in basis merely changes these representatives to elements of $\mathcal{U}$, which induces a linear transformation (mod 2π) among the angle coordinates.

The new action coordinates are given as follows:

$$J_i = \frac{1}{2\pi} \oint_{\gamma(\underline{u}_i)} \underline{p} \cdot d\underline{x}. \tag{3.84}$$

This can be evaluated in R^n, the universal covering space, by integrating along the path $\{t\underline{u}_i \mid 0 \leqslant t \leqslant 1\}$, as shown in Figure 3.10(b). If we (at least provisionally) accept the propositions that (a) any Jordan curve in R^n maps to a boundary on $M_{\underline{g}}$ and (b) the integral of $\underline{p} \cdot d\underline{x}$ around any boundary vanishes — which result from the symplectic two-form vanishing on $M_{\underline{g}}$ — then the integral can equally well be evaluated along any path homotopic to $\gamma(\underline{u}_i)$, in particular, along a path that follows the grid laid out by the $\underline{e}_k$ basis. We then see immediately that the action J_i must be a superposition of the old actions I_k.

$$
\begin{aligned}
J_i &= \frac{1}{2\pi}\oint_{\gamma(\sum u_k^{(i)}\underline{e}_k)} \underline{p}\cdot d\underline{x} \\
&= \sum \frac{1}{2\pi}\oint_{\gamma(u_k^{(i)}\underline{e}_k)} \underline{p}\cdot d\underline{x} \\
&= \sum u_k^{(i)} \frac{1}{2\pi}\oint_{\gamma(\underline{e}_k)} \underline{p}\cdot d\underline{x} = \sum u_k^{(i)} I_k.
\end{aligned}
$$

Written as a matrix relation,

$$\underline{J} = \underline{\underline{U}} \cdot \underline{I}. \tag{3.85}$$

Taken together, the symplectic transformation of Eqs.(3.85) and (3.83) are exceedingly useful in the study of resonance normal form Hamiltonians, which we shall take up in the next chapter. For whatever it is worth, the manipulations we have been doing fall within the branches of mathematics called homology and cohomology theory, your first exposure to which probably came when you studied Cauchy's theorem in complex analysis.

3.4 ACCELERATOR MODELS

Ignoring spin degrees of freedom, the phase space of a particle in an accelerator is a six-dimensional manifold. It frequently happens in large synchrotrons that the time scale for motion in the two-dimensional "longitudinal" subspace is much larger than that for oscillations in the four-dimensional "transverse" space. The fortunate consequence of this is that we may frequently (but not always) model transverse and longitudinal motion quasi-independently: the longitudinal effects of transverse motion tend to average out over a synchrotron period, while longitudinal oscillations serve to parametrically modulate the transverse dynamics. We shall, at any rate, adopt this picture in considering the two most frequently used models in accelerator theory: the "bucket" Hamiltonian of longitudinal motion, and the "transverse" Hamiltonian of a storage ring.

3.4.1 Longitudinal Model

There is an autonomous Hamiltonian model of an accelerator's longitudinal dynamics that can be motivated heuristically from the map expressed in Eqs. (2.46) and (2.50) by changing these finite difference equations to ordinary differential equations. It is useful to introduce a new variable,

$$W \equiv \frac{2\pi}{\omega_s} y.$$

The differential equations appropriate to discrete sampling equations are written by approximating the time derivative of a quantity as its discrete difference divided by the cycling period, $2\pi/\omega$. Applying this to Eqs. (2.46) and (2.50), we obtain

$$\begin{aligned} \frac{dW}{dt} &= \frac{2\pi}{\omega_s(t)} \frac{\omega(E,t)}{2\pi} eV(\sin\varphi - \sin\varphi_s) \\ &\approx eV(\sin\varphi - \sin\varphi_s), \\ \frac{d\varphi}{dt} &\approx -\frac{h}{2\pi}\left(\frac{\omega^2 E}{p^2 \underline{c}^2}\eta\right)_s W, \end{aligned} \tag{3.86}$$

where it is formally noted that (a) ω_s depends on t and (b) ω depends on both E and t.

One could try to justify this passage to a continuous model for orbits near the elliptic fixed point on the grounds that a large number of iterates are required to change the state appreciably. It could also be derived, starting from scratch, as the dominant harmonic in a traveling wave representation of the accelerating field—a rather strained attempt, one of several in the field, to use Fourier transforms where they are least appropriate. However, we are not going to dwell on the apologetics of this bizarre substitution but move on to characterizing its flow.

As with the discrete model, the fixed points of Eqs. (3.86) are at $W = 0$ with $\varphi = \varphi_s$ or $\pi - \varphi_s$. Also as before, we classify these fixed points via linearization. For example, expanding the equations to first order in $\varphi - \varphi_s$,

$$\begin{aligned} \frac{d}{dt}\begin{pmatrix} \varphi - \varphi_s \\ W \end{pmatrix} &= \begin{pmatrix} 0 & -\Omega^2/eV\cos\varphi_s \\ eV\cos\varphi_s & 0 \end{pmatrix} \cdot \begin{pmatrix} \varphi - \varphi_s \\ W \end{pmatrix} \\ &\equiv \underline{\underline{A}} \cdot \begin{pmatrix} \varphi - \varphi_s \\ W \end{pmatrix}. \end{aligned}$$

Here we have defined

$$\Omega^2 \equiv \frac{h}{2\pi}\left(\frac{\omega^2 E}{p^2 c^2}\eta\right)_s eV\cos\varphi_s \tag{3.87}$$

and have tacitly used the assumption that $d\varphi_s/dt = 0$. The characteristic polynomial of $\underline{\underline{A}}$ is $P(\lambda) = \lambda^2 + \Omega^2$. Provided that $\Omega^2 > 0$ the eigenvalues are pure imaginary, $\lambda = \pm i\Omega$, and the fixed point is elliptic; nearby orbits will wind around it at frequency Ω, which is called the "synchrotron oscillation frequency." This is, happily, the same conclusion we reached using the map equations (see the discussion surrounding

Eq. (2.49)). Written in terms of Ω, the φ component of the dynamic is

$$\frac{d\varphi}{dt} = -(\Omega^2(t)/eV\cos\varphi_s)W, \tag{3.88}$$

and we have formally written Ω as a function of t. (If $\varphi_s \neq 0$ the synchronous energy is not constant.)

The first order differential equations given in Eqs. (3.86) and/or Eq. (3.88) define a Hamiltonian system; they are derivable, in canonical form, from a "Hamiltonian" in which W and φ play the roles of conjugate coordinates:

$$H = -\frac{\Omega^2(t)}{2eV\cos\varphi_s}W^2 + eV(\cos\varphi + \varphi\sin\varphi_s). \tag{3.89}$$

Under acceleration, the Hamiltonian is *not* an invariant, because E_s — and, through it, Ω — depends explicitly on time. Ignore that for the moment: provisionally accept the nontrivial assertion that if E_s changes little in time $1/\Omega$, that is,

$$\frac{1}{2\pi}\frac{\omega_s}{\Omega}\frac{eV\sin\varphi_s}{E_s} \ll 1.$$

Then we can treat the dynamics as though H were exactly constant and later superpose the effects of its adiabatic variation. Formally, the problem is then equivalent to that of a particle on a line moving in the potential sketched in Figure 3.11, for $\pi/2 < \varphi_s < \pi$; also shown is a mechanical analog: the biased pendulum. For a given value of H, φ is confined to definite regions, one of which is unbounded. If the conditions are such that the orbit is stably confined, to oscillate about $\varphi = \varphi_s$, then it is said that the particle is trapped in a "bucket." While there are multiple buckets separated by 2π in phase, the physical situations described by a particle with Hamiltonian H trapped in a bucket at φ_s and that of one with Hamiltonian $H + 2\pi eV\sin\varphi_s$ in a bucket at $\varphi_s + 2\pi$ are identical.

Level sets of H in (φ, W) phase space can be visualized easily in two steps. First, use Eq. (3.89) to solve for W^2 and plot these curves for various values of H. Only $W^2 > 0$ is valid, so only the upper half-plane is important. To get the orbit paths, simply plot $W = \pm\sqrt{W^2}$, resulting in the curves in Figure 3.12; these orbits should be compared to those in Figure 2.11. In a triumph of mixed metaphors this particular sketch is sometimes called the "fish diagram," and the stable region bounded by the separatrix is called a "bucket." In the "unstable region" outside the buckets, W decreases without limit at an average rate, according to Eq. (3.86), of

$$\left\langle\frac{dW}{dt}\right\rangle = -eV\sin\varphi_s.$$

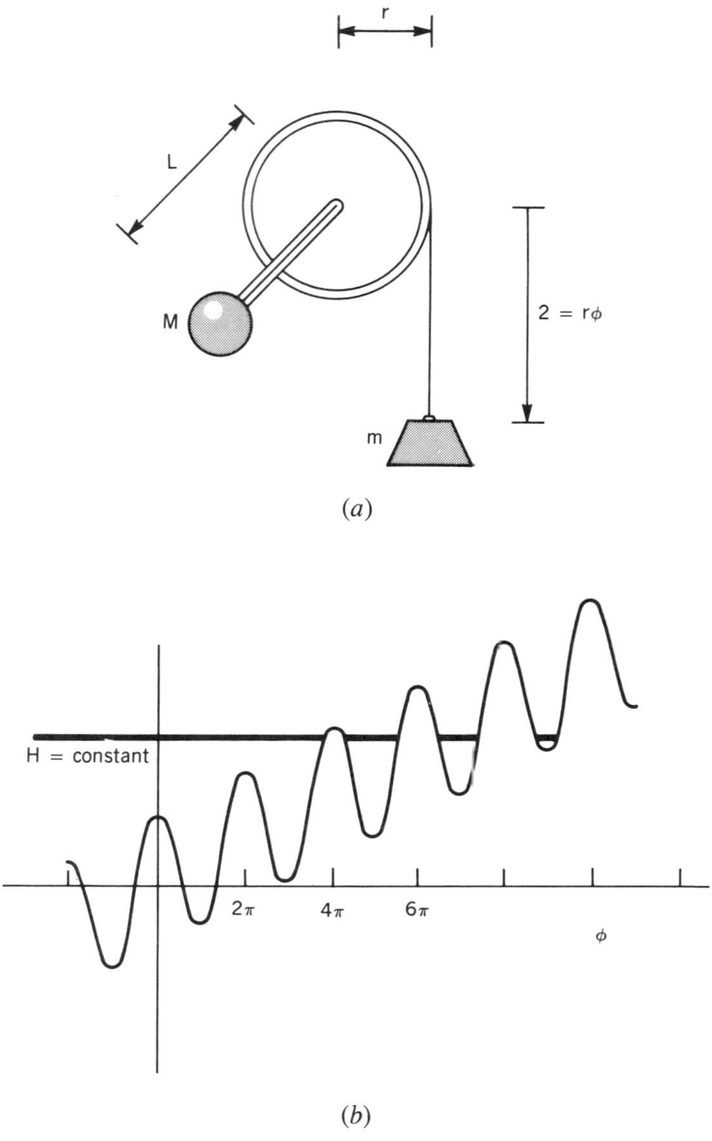

FIGURE 3.11 Equivalent potential fo- RF problem.

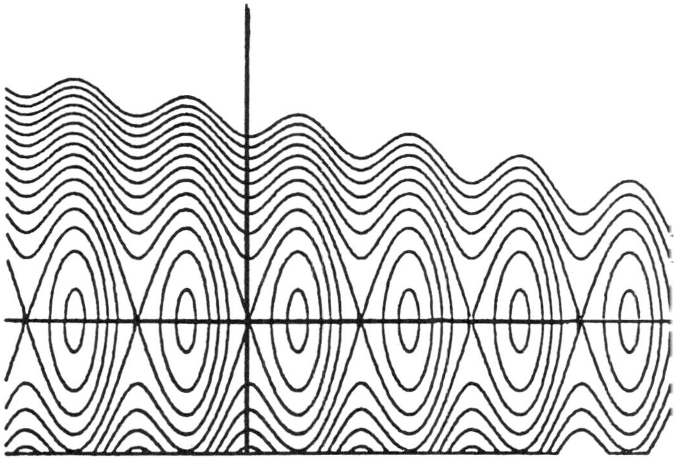

FIGURE 3.12 Fish diagram.

This is precisely what might have been expected. It means that on the average the particle energy is not increasing:

$$\left\langle \frac{dE}{dt} \right\rangle = 0.$$

3.4.2 Transverse Model

Particle trajectories in modern, large synchrotrons are on the scale of kilometers, while the transverse deviations that can be tolerated are typically no more than a few millimeters. It simply does not make sense from a computational standpoint to represent the dynamics using a global coordinate patch that covers the entire orbit, because the oscillations we need to study would be buried in the seventh or eighth significant figure. It is obviously prudent to use an atlas of coordinate charts with origins in the neighborhood of the particles. This can be done in more ways than one. The traditional method is to identify a "design orbit" for the synchrotron, a closed orbit that a particle would follow if it could be injected with infinite precision into a perfectly constructed machine. This strategy produces an atlas, which we shall call the "design atlas," that represents phase space in some neighborhood of the design orbit, around its entire circumference. However, working within this atlas at all locations in the ring may make it more difficult to integrate the equations of motion, in which case we have complete freedom to use other local charts, whose coordinates may simplify the calculations.[15] Whatever atlas is used, when a chart's coordinates are referred to a local curve segment as their "fiducial"— either to pieces of the design

[15]This point was made earlier, when we discussed the Chapman-Kolmogorov equation in the form Eq. (1.5). Its importance has been stressed by Forest [37] and Forest et al. [38].

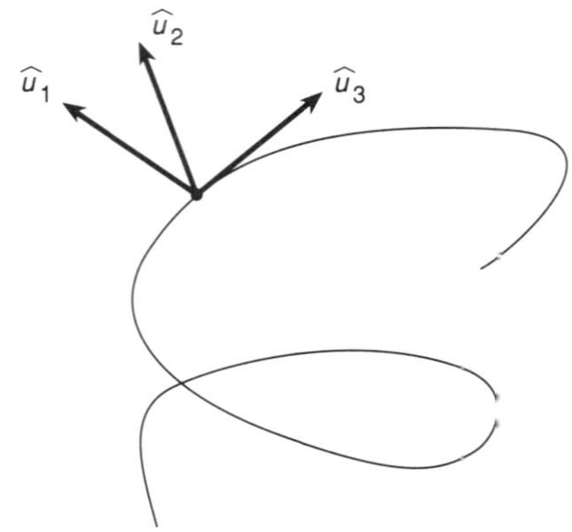

FIGURE 3.13 Fiducial coordinates are defined relative to the fiducial curve using a traveling frame.

orbit or to convenient curves motivated by the geometry of nearby magnets, kickers, and so forth — we shall call them the local "fiducial coordinates."

Fiducial Coordinate Charts A procedure for finding fiducial coordinates of an arbitrary curve was laid out for accelerator physicists by Courant and Snyder [25] using the classic method of traveling frames [26]. We first represent the local fiducial curve, $\vec{r}_o(s)$, parametrically in terms of an arc length along the curve. Because s is arc length, we have $d\vec{r}_o \cdot d\vec{r}_o = (ds)^2$, so that $d\vec{r}_o/ds$ must be a unit vector. We take this as the "third" member of a local triad of orthonormal unit vectors which comprise our traveling frame: $\{\hat{u}_1(s), \hat{u}_2(s), \hat{u}_3(s)\}$.

$$\begin{aligned} \hat{u}_3(s) &= d\vec{r}_o(s)/ds, \\ \hat{u}_1(s) &= -\rho(s)\, d\hat{u}_3/ds, \\ \hat{u}_2(s) &= \hat{u}_3(s) \times \hat{u}_1(s). \end{aligned} \tag{3.90}$$

The second equation serves to define the function $\rho(s)$, which is interpreted as the local radius of curvature. For future reference, our conventions will be that (a) $\hat{u}_3$ is along the beam *current* and (b) $\rho \geqslant 0$, so that $\hat{u}_1$ points "outward." This is sketched in Figure 3.13. Differentiation leads to the celebrated formulas of Serret and Frenet, which will prove useful later [26].

$$\begin{aligned} \frac{d\hat{u}_1}{ds} &= \frac{1}{\rho(s)}\hat{u}_3(s) + \tau(s)\hat{u}_2(s), \\ \frac{d\hat{u}_2}{ds} &= -\tau(s)\hat{u}_1(s), \\ \frac{d\hat{u}_3}{ds} &= -\frac{1}{\rho(s)}\hat{u}_1(s). \end{aligned}$$

These are the most general expressions that can be written. The last does nothing more than repeat the definition of $\rho(s)$ in Eq. (3.90). Since $\hat{u}_1(s) \cdot \hat{u}_1(s) = 1$, $d\hat{u}_1/ds$

must be orthogonal to $\hat{u}_1$. Furthermore, because $\hat{u}_1 \cdot \hat{u}_2 = \hat{u}_1 \cdot \hat{u}_3 = 0$, we must have $\hat{u}_3 \cdot d\hat{u}_1/ds = 1/\rho$ and $\hat{u}_2 \cdot d\hat{u}_1/ds = -\hat{u}_1 \cdot d\hat{u}_2/ds$. For the latter we introduce the local "torsion" of the curve, $\tau(s)$. These functions are called "connections." They provide instructions for relating the basis vectors of a frame as one moves along the one-dimensional manifold that is the fiducial curve.

Along segments that bend, ρ is finite, and the torsion, τ, is completely determined. However, along straight segments, $1/\rho$ vanishes, and τ is *completely arbitrary.* Normally one would expect to keep things simple by setting $\tau = 0$ along straight segments. However, if one uses a design atlas based on a closed fiducial orbit that is not planar, the torsion on straight segments is arranged so that (a) $\tau(s)$ is a continuous function and (b) $\oint d\hat{u}_k = \vec{0}$, for all k. In this way the traveling frame returns to itself after one circuit of the ring.[16]

We are finally ready to describe the local fiducial chart. An arbitrary spatial point is represented by a triplet, (x_1, x_2, s), defined according to the relation

$$\vec{r} = \vec{r}_o\,(s) + x_1\hat{u}_1(s) + x_2\hat{u}_2(s).$$

We shall refer to x_1 and x_2 as the "transverse" coordinates. These are to be fitted into a *canonical* chart, $(x_1, x_2, s; p_1, p_2, p_s; t)$, which will become the (extended) phase space of this system. This is accomplished, starting from canonical, global Cartesian coordinates, say, $(\vec{r}, \vec{q}; t)$, by applying a generating function of Type 3.

$$S\,(\vec{q}; x_1, x_2, s) \equiv -\vec{q} \cdot \vec{r}(x_1, x_2, s) = -\vec{q} \cdot [\vec{r}_o\,(s) + x_1\hat{u}_1(s) + x_2\hat{u}_2(s)].$$

Using Eq. (3.67), we obtain the momenta conjugate to (x_1, x_2, s).

$$\begin{aligned} p_1 &= \vec{q} \cdot \hat{u}_1(s), \\ p_2 &= \vec{q} \cdot \hat{u}_2(s), \\ p_s &= (1 + x_1/\rho\,(s))\vec{q} \cdot \hat{u}_3(s) + \tau(s)(x_1p_2 - x_2p_1). \end{aligned} \tag{3.91}$$

The last term drops out in the case of planar fiducial curves, or in general along segments of the fiducial curves on which the torsion vanishes.

A Hamiltonian describing a charged particle influenced by electromagnetic fields was written in Eq. (3.29).

$$E = \sqrt{(\underline{q} - e\underline{A})^2c^2 + m^2c^4} + e\phi.$$

(We use the symbol E here rather than H, which is reserved for later.) Cartesian coordinates are used so that the dot product is the usual quadratic form with unit matrix. To express E on the fiducial chart, we substitute from Eqs. (3.91). A little rearrangement gives us

$$(E - e\phi)^2 = (\underline{p} - e\underline{A})^2_\perp c^2 + m^2c^4 + \left(\frac{p_s - \tau(x_1p_2 - x_2p_1)}{1 + x_1/\rho} - eA_3\right)^2 c^2,$$

[16] This is assumed to be advantageous.

where the subscript $\perp$ has been used to indicate transverse coordinates, and we have suppressed writing the explicit dependence on s. Note that $\underline{p}_\perp = \underline{q}_\perp$.

In accelerator applications we are not so much interested in studying the orbit as a function of time but as a function of location in the accelerator. Looking ahead to this application, it is normal to change the phase space orbit parameter from time to arc length, which is clearly allowed, as $ds/dt > 0$ always. The procedure for doing this was described above Eq. (3.45): if we want to treat time as a dynamical variable and arc length as the orbit-history parameter, then the variable conjugate to t becomes $-E$, and the new Hamiltonian is $-p_s$.

$$
\begin{aligned}
&(x_1, x_2, s; p_1, p_2, p_s; t) \Rightarrow (x_1, x_2, t; p_1, p_2, -E; s)\\
&H \equiv -p_s = -\tau(x_1 p_2 - x_2 p_1) \qquad (3.92)\\
&\qquad - (1 + x_1/\rho)\left(eA_3 + \frac{1}{c}\sqrt{(E - e\phi)^2 - (\underline{p} - e\underline{A})_\perp^2 c^2 - m^2c^4}\right).
\end{aligned}
$$

Hill's Equation for a Storage Ring Equation (3.92) has been written very generally; we now introduce approximations and restrictions. Our immediate goal is to establish the connection between Eq. (3.92) and a linear Hill's equation commonly used to analyze the smallest deviations from the fiducial curve: the Kerst-Serber [56], Courant-Snyder [25] formalism. There are important situations in which the approximations made will not be valid. The linearization process that connects to Hill's equation, with its use of the design atlas, itself becomes awkward when dealing with certain magnet types, such as rectangular bends or wigglers. Other approximations we shall make will be most suitable for describing orbits in large storage rings with no RF cavities and no ramps, so that t does not appear explicitly in either $\underline{A}$ or ϕ. These difficulties can be overcome by not using the design atlas. Nonetheless, in this section we shall proceed traditionally and derive some standard expressions.

If (a) there are static magnetic fields and no RF cavities, (b) synchrotron radiation is negligible, and (c) in general, we ignore all mechanisms of energy loss or gain, then E is an invariant and can be treated as a number. Within that context, it is traditional to introduce the "magnetic rigidity."

$$(e|B\rho|)^2c^2 \equiv \mathbf{p}_3^2c^2 \equiv (E - e\phi)^2 - m^2c^4.$$

The symbol ρ that appears in this expression corresponds to the radius of curvature of a charged particle with momentum $\mathbf{p}_3$ undergoing planar, circular motion in a constant magnetic field of strength B. However, it should not be confused with the *function* $\rho(s)$ appearing everywhere else up to this point. To keep this clear the rigidity will always appear as $|B\rho|$. Similarly, the symbol $\mathbf{p}_3$ has been written in bold font to emphasize that it is *not* the true, longitudinal, kinematic momentum of the projectile, which is $q_3 - eA_3$. These numbers are simply different ways of expressing the energy (ignoring synchrotron radiation) of a particle in a storage ring.

$$H = -\tau(x_1p_2 - x_2p_1) - (1 + x_1/\rho)\left[eA_3 + \sqrt{\mathbf{p}_3^2 - (\underline{p} - e\underline{A})_\perp^2}\right]. \qquad (3.93)$$

To obtain the linearized equations of motion, we expand this Hamiltonian to second order in the transverse dynamical variables, $\underline{z} = (x_1, x_2; p_1, p_2)$. For the purpose of this expansion, we assume that $e\underline{A}_\perp \lessdot \underline{p}_\perp$; that is, the vector potential is dominated by A_3, and its transverse components are not significantly larger than the transverse momentum coordinates. Formally, we write

$$H = H^{(0)} + H^{(1)} + H^{(2)} + H^{(hot)}, \tag{3.94}$$

where $H^{(n)}$ is an nth degree polynomial in $\underline{z}$, and $H^{(hot)}$ contains all the higher order terms: $H^{(hot)} = O(z^3)$. Expanding the square root in powers of $\underline{p}_\perp^2/\mathbf{p}_3^2$ and collecting terms of like degree provide us with the following:

$$\begin{aligned} H^{(0)} &= -\mathbf{p}_3, \\ H^{(1)} &= -\mathbf{p}_3 x_1/\rho - eA_3^{(1)}, \\ H^{(2)} &= -\tau(x_1 p_2 - x_2 p_1) + \frac{1}{2\mathbf{p}_3}(\underline{p} - e\underline{A}^{(1)})_\perp^2 - \frac{x_1}{\rho} eA_3^{(1)} - eA_3^{(2)}. \end{aligned}$$

The next approximation we make is to set both ϕ and $\underline{A}_\perp$ to zero while keeping A_3. Physically this means that *we are only going to consider the effects of transverse magnetic fields on the orbit*. The expressions we shall use for $A_3^{(1)}$ and $A_3^{(2)}$ are the "dipole" and "quadrupole" polynomials, which we lift in advance from Eq. (3.112), where they will be discussed further:

$$\begin{aligned} -A_3^{(1)}/B_o &= b_0(s)x_1 - a_0(s)x_2, \\ -A_3^{(2)}/B_o &= \tfrac{1}{2}(b_1(s) - b_0(s)/\rho(s))x_1^2 - \tfrac{1}{2}b_1(s)x_2^2 - a_1(s)x_1x_2. \end{aligned}$$

Briefly, B_o serves only to scale the vector potential; typically, B_o is the value of the dipole field in a dipole magnet. The "multipole coefficients" b_0, a_0, b_1, and a_1 represent, respectively, the "normal" and "skew" dipole and "normal" and "skew" quadrupole components of the field. The normal dipole field is responsible for bending the orbit. If B_o is taken to be the dipole field, then $b_0 = 1$ within the dipole. The skew dipole field is typically an error or correction term. We shall usually ignore both it and the skew quadrupole field. The normal quadrupole field provides a quadratic focusing term in the Hamiltonian, much like the quadratic potential of a harmonic oscillator. That the term b_0/ρ appears to modify with the normal quadrupole, b_1, in $A_3^{(2)}$ comes from using a fiducial curve, which bends, at least inside the dipoles, and provides extra focusing from the dipoles.

Substituting these fields back into the expressions for $H^{(n)}$ and once again collecting terms provide the following:

$$\begin{aligned} H^{(1)} &= (eB_o b_0(s) - \mathbf{p}_3/\rho(s))x_1 - eB_o a_0(s)x_2, \\ H^{(2)} &= -\tau(x_1 p_2 - x_2 p_1) + \frac{1}{2\mathbf{p}_3}(p_1^2 + p_2^2) \\ &\quad + \tfrac{1}{2}eB_o(b_1(s) + b_0(s)/\rho(s))x_1^2 - \tfrac{1}{2}eB_o b_1(s)x_2^2 \\ &\quad - eB_o(a_1(s) + a_0(s)/\rho(s))x_1x_2. \end{aligned} \tag{3.95}$$

Ideally, $H^{(1)}$ vanishes. Using $\mathbf{p}_3 = e|B\rho|$, we see that the coefficient of x_1 is zero provided the fiducial curve through a dipole is taken to lie along the design orbit. The coefficient of x_2 comes from misaligned rotated dipoles or the correction elements used to adjust the orbit. The x_1x_2 term in $H^{(2)}$ produces linear coupling between the horizontal (1) and vertical (2) degrees of freedom. Applying Hamilton's equations to $H^{(1)} + H^{(2)}$ gives us the following dynamic:

$$\frac{d}{ds}\begin{pmatrix} x_1 \\ x_2 \\ p_1 \\ p_2 \end{pmatrix} = \begin{pmatrix} 0 \\ 0 \\ -(eB_o b_0(s) - \mathbf{p}_3/\rho(s)) \\ eB_o a_0(s) \end{pmatrix} \tag{3.96}$$

$$+ \begin{pmatrix} 0 & \tau(s) & 1/\mathbf{p}_3 & 0 \\ -\tau(s) & 0 & 0 & 1/\mathbf{p}_3 \\ -eB_o(b_1(s) + b_0(s)/\rho(s)) & eB_o(a_1(s) + a_0(s)/\rho(s)) & 0 & \tau(s) \\ eB_o(a_1(s) + a_0(s)/\rho(s)) & eB_o b_1(s) & -\tau(s) & 0 \end{pmatrix} \cdot \begin{pmatrix} x_1 \\ x_2 \\ p_1 \\ p_2 \end{pmatrix}$$

The only role that torsion, τ, plays in all of this is to mix the "horizontal" and "vertical" sectors via a rotation. The "skew" fields, a_0 and a_1, do the same but dynamically, rather than geometrically.

Edwards-Teng Representation Solutions of Eq. (3.96) are constructed by superposing a particular solution, including the inhomogeneous driving term, with general solutions, omitting it. The particular solution, with periodic boundary conditions imposed, is proportional to what is called the "dispersion function" of the ring. The general solutions obey a Hill's equation, obtained from zeroing the inhomogeneous term. As was done in Section 2.1.2, Floquet's analysis proceeds by constructing the single-turn, period-advance map, which is done by propagating the unit matrix once around the ring, as in Eqs. (2.15) and (2.18). That is, we integrate to construct solutions of Eq. (3.96) possessing as initial conditions the columns of the unit matrix. A commonly used substitution is the "bang-bang" model[17] in which a_0, a_1, b_0, b_1, and τ are approximated as step functions. This allows us to build the matrix $\underline{\underline{Z}}_s(2\pi)$ as a catenation of maps corresponding to locally autonomous linear dynamics, equivalent to the "time-ordered product" implied by the Chapman-Kolmogorov equation, Eq. (1.4) or Eq. (1.5).

Recall from Section 2.1.2 that in two-dimensional phase space we expressed the maps $\underline{\underline{M_{2\pi}}}^{[\theta]}$ in the form

$$\underline{\underline{M_{2\pi}}}^{[\theta]} = \exp\left[\mu \underline{\underline{J}}^{[\theta]}\right],$$

where we have defined $\mu \equiv 2\pi\nu$, and that the matrix $\underline{\underline{J}}$ was written as

$$\underline{\underline{J}} = \begin{pmatrix} \alpha & \beta \\ -\gamma & -\alpha \end{pmatrix},$$

[17] A term borrowed from control theory.

with $\beta\gamma - \alpha^2 = 1$. Viewed as free parameters, α, β, and μ, can serve as the three coordinates needed to specify a 2×2 symplectic matrix; that is, they serve as a chart on the manifold S_1, the symplectic group of order 1. As we follow θ around the ring, μ remains constant while α and β are periodic "lattice" functions of θ or, equivalently, of s. α and β determine the invariant torus in the two-dimensional phase space — the "Courant-Snyder ellipse" — while $\mu/2\pi = \nu$ is the fractional part of the winding number on the torus corresponding to a single period advance of the orbit. The complete set of period-advance matrices, $\{\underline{\underline{M_{2\pi}}}^{[\theta]} \mid \theta \in [0, 2\pi)\}$, is an imbedding of the unit circle into the three-dimensional manifold S_1.

Return now to the four-dimensional transverse phase space of Eq. (3.96). $\underline{\underline{M_{2\pi}}}$ is now an element of the ten-dimensional manifold S_2, the symplectic group of order 2. Eight of these coordinate "lattice functions" will determine the shape and orientation of the two-dimensional invariant torus of the period-advance map; the other two, corresponding to the eigenvalues of $\underline{\underline{M_{2\pi}}}$, specify the winding numbers on this torus.

There are, of course, a variety of charts on S_2 that can be used to represent $\underline{\underline{M_{2\pi}}}$. The one we shall consider was suggested originally by Edwards and Teng [32] and has become more or less a standard representation among accelerator physicists. Its description is made easier by rearranging the phase space coordinates: rather than using "position" and "momentum" sectors as we have been, we shall group the coordinates into "horizontal" and "vertical" sectors, that is, $(x_1, p_1; x_2, p_2)$ instead of $(x_1, x_2; p_1, p_2)$. In this representation, if there were no coupling ($\tau = a_0 = a_1 = 0$), $\underline{\underline{M_{2\pi}}}$ would be a block diagonal matrix,

$$\underline{\underline{M_{2\pi}}} = \begin{pmatrix} \underline{\underline{H}} & \underline{\underline{0}} \\ \underline{\underline{0}} & \underline{\underline{V}} \end{pmatrix},$$

where the 2×2 submatrices $\underline{\underline{H}}$ and $\underline{\underline{V}}$ are both of the form described above. If the eigenvalues of $\underline{\underline{M_{2\pi}}}$ are $e^{\pm i\mu_1}$ and $e^{\pm i\mu_2}$, we could identify

$$\underline{\underline{H}} = \exp[\mu_1 \underline{\underline{J}}_1] \quad \text{and} \quad \underline{\underline{V}} = \exp[\mu_2 \underline{\underline{J}}_2].$$

This gives us six of the ten coordinates we need; coupling provides the other four. Without further ado we write the Edwards-Teng representation,

$$\underline{\underline{M_{2\pi}}} = \begin{pmatrix} \cos\phi \underline{\underline{1}} & \sin\phi \underline{\underline{D}}^{-1} \\ -\sin\phi \underline{\underline{D}} & \cos\phi \underline{\underline{1}} \end{pmatrix} \cdot \begin{pmatrix} \underline{\underline{H}} & \underline{\underline{0}} \\ \underline{\underline{0}} & \underline{\underline{V}} \end{pmatrix} \cdot \begin{pmatrix} \cos\phi \underline{\underline{1}} & -\sin\phi \underline{\underline{D}}^{-1} \\ \sin\phi \underline{\underline{D}} & \cos\phi \underline{\underline{1}} \end{pmatrix},$$

where $\underline{\underline{D}}$ is a 2×2 unimodular matrix. Each matrix on the right side is symplectic, so $\underline{\underline{M_{2\pi}}}$ is guaranteed to be symplectic for arbitrary ϕ and symplectic $\underline{\underline{D}}$. In addition, this provides a similarity transformation relating $\underline{\underline{M_{2\pi}}}$ to its block-diagonal form.

Preparatory to calculating the 2×2 matrices, $\underline{\underline{D}}, \underline{\underline{H}}, \underline{\underline{V}}$, and the coupling angle ϕ, we obtain the tunes by finding the eigenvalues of $\underline{\underline{M_{2\pi}}}$. It is also convenient to write $\underline{\underline{M_{2\pi}}}$ in the block matrix form of Eq. (3.48). Assuming that $|\phi| \leqslant \pi/4$, in order to avoid confusing $\underline{\underline{H}}$ with $\underline{\underline{V}}$, the submatrices are then calculated as follows:

$$\begin{aligned}
\cos 2\phi &= \mathrm{Tr}\left(\underline{\underline{M}}_{11} - \underline{\underline{M}}_{22}\right)/2(\cos\mu_1 - \cos\mu_2), \\
\underline{\underline{D}} &= -\frac{1}{\sin 2\phi(\cos\mu_1 - \cos\mu_2)}\left(\underline{\underline{M}}_{21} - \underline{\underline{J}}\,\underline{\underline{M}}_{12}{}^{T}\underline{\underline{J}}\right), \\
\underline{\underline{H}} &= \underline{\underline{M}}_{11} - \underline{\underline{D}}^{-1}\cdot\underline{\underline{M}}_{21}\tan\phi, \\
\underline{\underline{V}} &= \underline{\underline{M}}_{22} + \underline{\underline{D}}\cdot\underline{\underline{M}}_{12}\tan\phi.
\end{aligned} \tag{3.97}$$

Remember: $\underline{\underline{J}}$ here has a representation different from that in Eq. (3.2).

Action-Angle Coordinates In what follows we shall assume that (a) the inhomogeneous driving term in Eq. (3.96) is zero, and (b) there is no torsion, $\tau(s) = 0$. Even in the presence of solenoid detectors, the latter is generally a reasonable assumption when the fiducial curve is confined to a plane. The former is basically indefensible: it cannot hold simultaneously for all particles in a bunch, as they will possess a distribution of values of $\mathbf{p}_3$. The "dispersion" function, which is the particular, closed orbit solution corresponding to the inhomogeneous term, takes this into account.

Two transformations are sometimes performed at this point, both employing variations of Eq. (3.95). The first is *not* a symplectic transformation, a feature connected with the distinction between "emittance" and "invariant emittance." Rather than using the transverse momenta, $\underline{p}_\perp$, as coordinates, many accelerator physicists prefer to use the tangents of the transverse angles. That is, they normalize $\underline{p}_\perp$ with $\mathbf{p}_3$: $\underline{p}_\perp \to \underline{p}_\perp/\mathbf{p}_3$. The new momentum is thus the tangent of an angle. This is done to enhance the analogy with rays in geometric optics (e.g., see Guillemin and Sternberg [44]); two particles with the same tangent are traveling in the same direction. We compensate for this transformation by dividing the Hamiltonian by the same factor: $H \to H/\mathbf{p}_3$, so that—upon overloading and abusing the notation somewhat and dropping the skew terms a_0 and a_1 for brevity—Eq. (3.95) becomes

$$H^{(2)}(x_1, x_2; p_1, p_2; s) = \frac{1}{2}(\underline{p}_\perp)^2 + \frac{1}{2}\frac{B_o}{|B\rho|}[(b_1 + b_0/\rho)x_1^2 - b_1 x_2^2]. \tag{3.98}$$

This is acceptable provided that $\mathbf{p}_3 = e|B\rho|$ is a *legitimate constant of motion*; it should be treated with remarkable suspicion otherwise.

The second variation is more trivial: merely changing the orbit parameter from arc length, s, to angle, θ, the relationship between the two being $ds = R d\theta$. On the design atlas, $2\pi R$ is clearly the total path length of the design orbit, and we can use that definition on other charts as well: R merely carries the dimensions of length. The orbit parameter θ is introduced for later convenience. To compensate correctly, the Hamiltonian must be multiplied by R, so that Eq. (3.98) becomes

$$H^{(2)}(x_1, x_2; p_1, p_2; \theta) = \frac{R}{2}(\underline{p}_\perp)^2 + \frac{1}{2}\frac{B_o R}{|B\rho|}[(b_1 + b_0/\rho)x_1^2 - b_1 x_2^2]. \tag{3.99}$$

Again, this assumes $e|B\rho|$ is indeed constant; if not, the expression must be multiplied by it; it is not legitimate to absorb it into the Hamiltonian.

With a view toward establishing an analytic description of transverse resonances, we shall find a set of action-angle coordinates for the Hamiltonian, Eq. (3.93). The two transverse degrees of freedom are independent: the four-dimensional vector field is a cross-product of two two-dimensional vector fields, each governed by a Hamiltonian of the form[18]

$$H(x,p;s) = \tfrac{1}{2}(p^2 + K(s)x^2), \quad \text{where} \quad K = \frac{B_o}{|B\rho|}\begin{cases} b_1 + b_0/\rho, & \text{horizontal} \\ -b_1, & \text{vertical} \end{cases}.$$

The lattice functions, $\alpha_k, \beta_k, \psi_k, k = 1,2,$ associated with the Floquet theory for this Hamiltonian, and introduced in Eqs.(2.28) and (2.32) can be used to write its action-angle coordinates. It will be convenient to reintroduce the "phase deviation" functions,

$$\tilde{\psi}_k \equiv \psi_k - \nu_k\theta = \psi_k - \nu_k s/R, \quad k = 1,2,$$

which, unlike the ψ_k themselves, possess the advantage of being periodic. That property makes possible a periodic canonical transformation to the action-angle coordinates, $(\underline{\varphi}; \underline{I})$,

$$\begin{aligned} x_k &= \sqrt{2I_k\beta_k}\sin(\tilde{\psi}_k + \varphi_k), \\ q_k &= \sqrt{2I_k\beta_k}\cos(\tilde{\psi}_k + \varphi_k) \\ &= \alpha_k x_k + \beta_k p_k, \\ \alpha_k &= -\frac{1}{2}\frac{d\beta_k}{ds} = -\frac{1}{2}\beta_k', \end{aligned} \tag{3.100}$$

which can be accomplished via the following generating functions of Type 1 [86]:

$$\begin{aligned} S_k(x_k, \varphi_k; s) &= (x_k^2/2\beta_k)(-\alpha_k + \cot[\tilde{\psi}_k + \varphi_k]) \\ &= (x_k^2/2\beta_k)(\tfrac{1}{2}\beta_k' + \cot F_k), \quad \text{where} \\ F_k \equiv \psi_k - \nu_k\theta + \varphi_k &= \psi_k(s_0) + \int_{s_0}^{s}\frac{ds}{\beta_k} - \nu_k\frac{s}{R} + \varphi_k. \end{aligned}$$

This transformation should be compared to the harmonic oscillator Eqs. (3.52) and (3.69). By inspection, the correct linear orbits are reproduced if (a) $\varphi_k = \nu_k\theta$ and (b) I_k = constant, which is exactly the sort of dynamic we expect for action-angle coordinates.

[18] I have inadvertently slipped back to s as the orbit parameter rather than θ. Originally, this was done to make a connection with Eq. (2.26).

To establish our result, that S_k generates the desired transformation, we employ Eqs. (3.64), while temporarily suppressing the subscripts.

$$p = \frac{\partial S}{\partial x} = \frac{x}{\beta}\left(\frac{1}{2}\beta' + \cot F\right),$$
$$I = -\frac{\partial S}{\partial \varphi} = \frac{x^2}{2\beta}\frac{1}{\sin^2 F}.$$

A little rearrangement then reproduces Eqs. (3.100). The transformed representation of the Hamiltonian is

$$H_{\text{new}} = \tfrac{1}{2}(p^2 + K(s)x^2) + \partial S/\partial s. \tag{3.101}$$

For the first two terms we substitute from Eqs. (3.100),

$$x^2 = 2\beta I \sin^2 F,$$
$$p^2 = \frac{2I}{\beta}\left(\frac{1}{2}\beta' \sin F + \cos F\right)^2,$$

while the third is

$$\partial S/\partial s = I\left(\frac{\nu}{R} - \frac{1}{\beta}\left(1 - \frac{1}{2}\left((\beta\beta'' - \beta'^2)\sin^2 F - 2\beta' \sin F \cos F\right)\right)\right).$$

Upon substituting these into Eq. (3.101) and using the result of Problem 2.9,

$$\tfrac{1}{2}\beta\beta'' - \tfrac{1}{4}\beta'^2 + K\beta^2 = 1,$$

to simplify, we obtain the compact expression

$$H_{\text{new}} = \nu I/R. \tag{3.102}$$

This is equivalent to Eq. (3.53), as R merely scales the orbit parameter, s. We can formally eliminate it by rescaling, using θ, with $d/d\theta = R d/ds$. Abusing notation slightly, we have

$$H(\varphi, I; \theta) = H(\varphi, I; s) \times R = \nu I.$$

Identical arguments hold for both transverse degrees of freedom. Again abusing the notation, we write the quadratic part of the transverse Hamiltonian, Eq. (3.99), as follows:

$$H^{(2)}(\underline{\varphi}, \underline{I}; \theta) = \nu_1 I_1 + \nu_2 I_2 = \underline{\nu} \cdot \underline{I}, \tag{3.103}$$

which confirms the identification of I_1 and I_2 as the action coordinates appropriate to this system. This describes, of course, a system equivalent to two uncoupled harmonic oscillators: I_1 and I_2 are dynamical invariants, and φ_1 and φ_2 increase linearly with θ at rates ν_1 and ν_2.

A parallel analysis holds for coupled dynamics, but (a) the Floquet solutions can no longer be identified with "horizontal" and "vertical" motion, (b) the transformations are more complicated, because they now depend on ten lattice functions: eight envelope functions and two phases, and (c) it is not worth the effort; map-based algorithms which do not depend on explicitly writing lattice functions are a better way of dealing with the added degrees of freedom (see Section 5.3).

3.4.3 Multipolar Field Model

We turn now to the nonlinear terms contained in $H^{(hot)}$ of Eqs. (3.93) and (3.94). In principle these should include all the higher order terms neglected in going from the correct expression in Eq. (3.93) to the simple quadratic Hamiltonian of Eq. (3.95) or Eq. (3.99). However, our purpose in this chapter is to connect the linear model of the previous section with what may be called "traditional" transverse resonance theory. That "traditional" theory comes from assuming that $H^{(hot)}$ is dominated by the longitudinal component of the vector potential, A_3. It amounts to the ansatz,

$$H^{(hot)} \approx -eA_3^{(hot)},$$

in the normalization of Eq. (3.95), or

$$H^{(hot)} \approx -RA_3^{(hot)}/|B\rho|, \tag{3.104}$$

in that of Eq. (3.99), which, in fact, will be used throughout the rest of this chapter. Although it seems – and *is* – drastic, this approximation has been used for many years to model the physics of very large, very high energy synchrotrons, by assuming that (a) edge effects are unimportant and (b) terms of higher order in $\underline{x}/\rho$ or $\underline{p}_\perp/\mathbf{p}_3$ can be neglected. These assumptions would not be valid for small storage rings, for example. Nonetheless, this "large ρ, high constant energy" approximation historically produced a good deal of intuition concerning real resonances excited by magnetic field nonlinearities or nonlinear elements purposely introduced into machines (e.g., to perform slow extraction).

The question of how best to represent a ring's nonlinear magnetic fields is an important issue, one to which we, regrettably, will not do justice. Here we will simply introduce the "multipolar model," which has been used (perhaps overused) in studying the behavior of beams in "large" accelerators. The physical interpretation is that nonlinear fields arise from a series of magnetic multipoles — quadrupoles, sextupoles, octupoles, and so forth. This model has been invoked frequently within the context of large accelerators, especially when it is believed that (a) edge effects are negligible and (b) nonlinear fields can be treated in an impulse approximation. It probably owes its historic popularity to a confluence of three circumstances:

(a) Error fields in dipoles are frequently measured by rotating coils. The signals obtained from these coils are generally interpreted in terms of multipolar magnetic fields via harmonic analysis.

(b) Multipolar fields are in a form that easily lends itself to isolated resonance models (see Chapter 4). It was therefore natural for accelerator physicists working with these most simple models of particle instability to think in terms of multipoles.

(c) Partly because of (b), many of the correction and extraction elements in accelerators are designed and constructed to be magnetic multipoles. It was easy to predict the effects of isolated resonances and to design correction and extraction systems using the multipolar model.

None of these imply that multipoles are the best, or even a correct, representation under all circumstances: any model has its limits as well as its advantages.

Harmonic Polynomials Assuming that the model *is* appropriate, the simplest multipolar representation is predicated on translational symmetry of time-invariant sources (and fields) along the direction of the fiducial current, such as a proton moving in the $+\hat{u}_3$ or an antiproton in the $-\hat{u}_3$ direction on the fiducial curve. The magnetic field is thus postulated to be transverse, $B_3 = 0$, which — in a source-free region through a straight magnet, with the fiducial curve also straight — allows us to connect it with an analytic function of the complex transverse coordinate, $\zeta = x_1 + ix_2$. To see how this comes about, we shall quickly review the Cauchy-Riemann equations associated with analytic functions. Let $\Gamma(\zeta) \equiv u(x_1, x_2) + iv(x_1, x_2)$ be a complex-valued function, and write the differential form,

$$\begin{aligned} d\Gamma &= du + i dv \\ &= (\partial_1 u\, dx_1 + \partial_2 u\, dx_2) + i(\partial_1 v\, dx_1 + \partial_2 v\, dx_2) \\ &= \tfrac{1}{2}[(\partial_1 u + \partial_2 v) + i(\partial_1 v - \partial_2 u)]\, d\zeta \\ &\quad + \tfrac{1}{2}[(\partial_1 u - \partial_2 v) + i(\partial_1 v + \partial_2 u)]\, d\zeta^*. \end{aligned}$$

If the coefficient of $d\zeta^*$ vanishes, the function Γ is called "analytic." This requires that

$$\partial_1 u = \partial_2 v \quad \text{and} \quad \partial_2 u = -\partial_1 v,$$

which are, of course, the Cauchy-Riemann equations. One obvious consequence is the directional independence of $d\Gamma/d\zeta$, so that, for example,

$$\frac{d\Gamma}{d\zeta} = \partial_1 u + i\partial_1 v = \frac{\partial\Gamma}{\partial x_1} = \partial_2 v - i\partial_2 u = \frac{\partial\Gamma}{\partial(ix_2)},$$

and so forth. A more useful form for our purposes, however, is the mixed expression,

$$\frac{d\Gamma}{d\zeta} = \partial_1 u - i\partial_2 u. \tag{3.105}$$

The connection with transverse magnetic fields is made by observing that the condition $B_3 = 0$ allows us to choose a vector potential that satisfies the longitudinal gauge, $A_1 = A_2 = 0$. From $\vec{B} = \nabla \times \vec{A}$ we then write

$$\begin{aligned} B_2 + iB_1 &= -\partial_1 A_3 + i\partial_2 A_3 \\ &= -(\partial_1 A_3 - i\partial_2 A_3). \end{aligned}$$

Comparing this to Eq. (3.105) suggests that we interpret A_3 as the real part of an analytic function, $\Gamma(\zeta)$, for which $B_2 + iB_1 = -d\Gamma/d\zeta$. However, *this is only possible in source-free regions,* as A_3 would then necessarily satisfy Laplace's equation.

The "multipoles" appear as the terms in a power series representation of Γ. Unfortunately, there are several notational conventions for writing this; there has even been disagreement on this totally insignificant point among people working together on the same problem at the same institution. We shall adopt one that seems to have wide acceptance.

$$B_2 + iB_1 = B_o \sum_{n=0}^{\infty} k_n \zeta^n = B_o \sum_{n=0}^{\infty} (b_n + ia_n)\zeta^n, \tag{3.106}$$

$$\Gamma(\zeta) = -B_o \sum_{n=1}^{\infty} \frac{1}{n}(b_{n-1} + ia_{n-1})\zeta^n, \tag{3.107}$$

$$A_3 = \operatorname{Re}\Gamma(\zeta). \tag{3.108}$$

Here, B_o is an arbitrary number that establishes the scale of the magnetic field and carries its dimensions. Typically, its value is that of a representative dipole field in the ring, especially when the higher order terms model the "error" fields of the dipole. The b_n's and a_n's are, respectively, the "normal" and "skew" multipole coefficients. The principal property of the former is that a purely horizontal (vertical) displacement will produce a horizontal (vertical) force; this is not true of the latter. (It also does not mean that normal multipoles do not couple the horizontal and vertical motion when both are present.) In order to understand the association with actual, physical multipolar magnets there is no substitute for investing 15 minutes, or so, in sketching a few fields for different values of n. Figure 3.14 illustrates one such sketch for $n = 2$, which in our convention index sextupolar fields. The key point to note is that if one traverses a circle about the origin, the symmetry point of the field, then the field of a $2(n + 1)$ pole will wind n times, in accordance with the behavior of ζ^n. The field orientation is also established to be in agreement with Eq. (3.106) for the normal

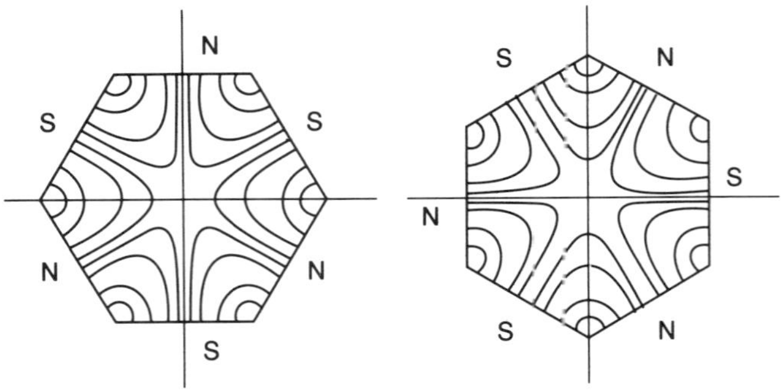

FIGURE 3.14 Sketches of normal and skew sextupole fields.

($a_n = 0$) and skew ($b_n = 0$) cases. Equation (3.108) establishes A_3 as a polynomial in ζ_1 and ζ_2.

$$A_3 = -B_o \sum_{m,n=0}^{\infty} c_{mn} \zeta_1^m \zeta_2^n,$$

$$c_{mn} = \frac{1}{m+n} \binom{m+n}{n} \begin{cases} (-1)^{n/2} b_{m+n-1}, & \text{even } n \\ (-1)^{(n+1)/2} a_{m+n-1}, & \text{odd } n \end{cases}. \tag{3.109}$$

Identification of normal and skew coefficients organizes the terms into a series of harmonic polynomials. The first of these are

$$\begin{aligned} A_3/(-B_o) = {} & b_0\zeta_1 - a_0\zeta_2 \\ & + (b_1/2)(\zeta_1^2 - \zeta_2^2) - a_1\zeta_1\zeta_2 \\ & + (b_2/3)(\zeta_1^3 - 3\zeta_1\zeta_2^2) \\ & - (a_2/3)(3\zeta_1^2\zeta_2 - \zeta_2^3) \\ & + (b_3/4)(\zeta_1^4 - 6\zeta_1^2\zeta_2^2 + \zeta_2^4) \\ & - (a_3/4)(4\zeta_1^3\zeta_2 - 4\zeta_1\zeta_2^3) + \cdots . \end{aligned} \tag{3.110}$$

The pattern should be evident.

Curvature The assumption of translation symmetry necessary to support this model is only approximately correct within the body of a large magnet and completely breaks down near its edges. If these edge effects are important, as they are known to be in small rings, then the edge fields must be handled more carefully. In addition, the magnet itself need not be straight; for example, large dipoles may be (and have been) bent so that particles remain near the center of the beam pipe throughout the length of the magnet. At least in theory this means that $\vec{B} \cdot \hat{u}_3(s) = 0$ along a *bent* fiducial curve following the path of the design orbit within the dipole. This modifies our original condition of transverse fields and gives rise to another representation. We choose a vector potential satisfying

$$\vec{A} = A_3(x_1, x_2)\hat{u}_3(s).$$

$\vec{A}$ still has zero divergence and, in a source-free region, still obeys Laplace's equation: $\nabla \times \vec{B} = -\nabla^2 \vec{A} = \vec{0}$. If ρ is locally constant, the coordinates are locally cylindrical, and this is written

$$(\rho + x_1)^2(\partial^2/\partial x_1^2 + \partial^2/\partial x_2^2)A_3 + (\rho + x_1)\partial A_3/\partial x_1 = A_3. \tag{3.111}$$

Suppose we expand A_3 in a power series,

$$A_3 = -B_o \sum_{m,n=0}^{\infty} (c_{mn} + \varepsilon_{mn}) x_1^m x_2^n,$$

where the c_{mn} are as given in Eq. (3.109) and satisfy the condition

$$(\partial^2/\partial x_1^2 + \partial^2/\partial x_2^2) \sum c_{mn} x_1^m x_2^n = 0,$$

while the ε_{mn} are "corrections" necessary to satisfy Eq. (3.111). Applying Eq. (3.111) provides a recursion for determining these numbers,

$$\begin{aligned}
&(m + 2)(m + 1)\rho^2 \varepsilon_{m+2n} + (n + 2)(n + 1)\rho^2 \varepsilon_{mn+2} \\
&+ (m + 1)(2m + 1)\rho\varepsilon_{m+1n} + 2(n + 2)(n + 1)\rho\varepsilon_{m-1n+2} \\
&+ (m + 1)(m - 1)\varepsilon_{mn} + (n + 2)(n + 1)\varepsilon_{m-2n+2} \\
&= -(m + 1)\rho c_{m+1n} - (m - 1)c_{mn},
\end{aligned}$$

the seeds of which are

$$\forall n : \varepsilon_{1n} = \varepsilon_{0n} = 0.$$

Specifically working out the first few of these, we get the following expression for A_3:

$$\begin{aligned}
A_3/(-B_o) = {} & b_0 x_1 - a_0 x_2 \\
& + \tfrac{1}{2}(b_1 - b_0/\rho)x_1^2 - \tfrac{1}{2} b_1 x_2^2 - a_1 x_1 x_2 \\
& + (b_2/3 - b_1/6\rho + b_0/2\rho^2)x_1^3 - b_2 x_1 x_2^2 \\
& - (a_2 - a_1/2\rho + a_0/2\rho^2)x_1^2 x_2 + \tfrac{1}{3} a_2 x_2^3 + \cdots .
\end{aligned} \tag{3.112}$$

Fortunately, it frequently happens that $b_{k+1}\rho \ll b_k$ and $a_{k+1}\rho \ll a_k$ for $k \geqslant 1$, so that all the extra terms can be ignored.

This is one of many possible representations,[19] which should be treated with at least as much skepticism as the rectilinear model. Both are founded on differing assumptions: in the first case the magnetic field is rectilinearly transverse; in the second it is cylindrically transverse. The coefficients in either case must be related to data obtained, typically, from rotating coils. Harmonic analysis of those signals produces numbers *that are interpreted* as normal and skew multipole coefficients. Complications develop when one starts thinking carefully of the geometry of these measurements. In general, the nontrivial tasks of obtaining good field data and representing it appropriately depend on the problem at hand.

Impulse Approximation: Kick Maps As mentioned earlier, multipolar models are frequently used in conjunction with an impulse approximation. Upon substitution from Eqs. (3.108), Eq. (3.109), or Eq. (3.110) into Eq. (3.104),

$$H^{(hot)} = \frac{B_o R}{|B\rho|} \operatorname{Re}\left(\sum_{n=1}^{\infty} \frac{1}{n} k_{n-1}(\theta) \zeta^n \right),$$

we obtain the multipolar contribution to the transverse vector field.

$$\begin{aligned}
\frac{dp_1}{d\theta} &= -\frac{\partial H}{\partial x_1} = -\frac{B_o R}{|B\rho|} \operatorname{Re}\left(\frac{d}{d\zeta} \sum_{n=1}^{\infty} \frac{1}{n} k_{n-1}(\theta)\zeta^n \right) \\
&= -\frac{B_o R}{|B\rho|} \operatorname{Re}\left(\sum_{n=0}^{\infty} k_n(\theta)\zeta^n \right), \\
\frac{dp_2}{d\theta} &= -\frac{\partial H}{\partial x_2} = -\frac{B_o R}{|B\rho|} \operatorname{Re}\left(i\frac{d}{d\zeta} \sum_{n=1}^{\infty} \frac{1}{n} k_{n-1}(\theta)\zeta^n \right) \\
&= \frac{B_o R}{|B\rho|} \operatorname{Im}\left(\sum_{n=0}^{\infty} k_n(\theta)\zeta^n \right).
\end{aligned}$$

The impulse — erstwhile "thin lens" — approximation consists of ignoring the variation in $\underline{x}$ as the particle traverses the multipolar field.[20] The total impulse imparted by the field is then related to the integrated multipole coefficients. In the accelerator literature these are written in a number of equivalent ways:

$$K_n L \equiv \int R d\theta \, B_o k_n = \int ds \, B_o k_n = \int ds \, B^{[n]}/n! \, .$$

[19] For a review of some of these, see Mane [67]. One should also study carefully the computational methods described by Berz [14].

[20] The reader can worry about consistency with ignoring edge effects at the same time.

Another way of thinking of this is that the impulse approximation amounts to replacing the distribution $k_n(\theta)$ with a periodic delta distribution.

$$k_n(\theta) = \frac{K_n L}{B_o R}\delta_{2\pi}(\theta - \theta_m). \tag{3.113}$$

Here, θ_m symbolizes the location of the multipolar field. The kick map, $\mathcal{M} : (x_1, x_2; p_1, p_2) \longmapsto (x_1, x_2; p_1 + \Delta p_1, p_2 + \Delta p_2)$, which this produces is given by the following equations:

$$\begin{aligned} \Delta p_1 &= -\frac{1}{|B\rho|}\operatorname{Re}\left(\sum_{n=0}^{\infty} K_n L \zeta^n\right), \\ \Delta p_2 &= \frac{1}{|B\rho|}\operatorname{Im}\left(\sum_{n=0}^{\infty} K_n L \zeta^n\right). \end{aligned} \tag{3.114}$$

As an important example, consider the kick map associated with a thin sextupole. From Eq. (3.110), the perturbation arising from a sextupolar field is

$$H_1 = \frac{1}{3}\frac{B_o R}{|B\rho|} b_2(\theta)(x_1^3 - 3x_1 x_2^2),$$

which contributes to the equation of motion for p_1,

$$\frac{dp_1}{d\theta} = -\frac{\partial H_1}{\partial x_1} = -\frac{B_o R}{|B\rho|} b_2(\theta)(x_1^2 - x_2^2).$$

In the impulse approximation, this gives rise to a "kick" applied to p_1 proportional to the integrated second derivative of the field,

$$\Delta p_1 = \int d\theta \frac{dp_1}{d\theta} \approx -\frac{1}{|B\rho|} K_2 L (x_1^2 - x_2^2), \tag{3.115}$$

where, in this example, $K_2 L = \int ds B_o b_2$.

We now establish a correspondence between this and the Hénon map. First, we confine the motion to the horizontal plane by setting $x_2 = 0$. Second, consider that this thin sextupole is the only nonlinear element installed within a storage ring of tune ν. The map representing this configuration is more easily written using the (x_1, q_1) coordinates from Eq. (3.100), because the matrix for its linear part is simply a rotation through $2\pi\nu$. Recognizing that within the impulse approximation, $\Delta q_1 = \Delta(\alpha_1 x_1 + \beta_1 p_1) = \beta_1 \Delta p_1$, we write the complete map as the symplectic transformation,

$$\mathcal{M} : \begin{pmatrix} x_1 \\ q_1 \end{pmatrix} \longmapsto \begin{pmatrix} \cos 2\pi\nu & \sin 2\pi\nu \\ -\sin 2\pi\nu & \cos 2\pi\nu \end{pmatrix} \begin{pmatrix} x_1 \\ q_1 - \lambda x_1^2 \end{pmatrix}, \tag{3.116}$$

which is precisely the Hénon map of Eq. (2.52). The inessential scale parameter, λ, has been reintroduced with value:

$$\lambda = \frac{\beta_1(K_2L)}{|B\rho|}. \tag{3.117}$$

3.5 ABSTRACT LIE ALGEBRAS

In the preceding sections we have met several concrete Lie algebras: commutators of vector fields, Lie brackets of their representative arrays, and Poisson brackets of observables. We shall consider here some of the more formal theory of this category of objects, culminating in a discussion of the Baker-Campbell-Hausdorff formula, which connects Lie algebras back to Lie groups via exponentiation.

Lie Algebras Define an abstract Lie algebra as an object, $\mathcal{A} \equiv \langle A, \circ \rangle$, where A is a vector space over a field, $\mathcal{F}$, and $\circ$ is an extra binary operation over A, called the "Lie product." The criteria that make this a Lie algebra are that the Lie product must (a) be skew, (b) be bilinear, and (c) obey the Jacobi identity. That is, $\circ$ must satisfy, for all $a, b, c \in A$, and all $x, y \in \mathcal{F}$

$$\begin{aligned} \textbf{skew:} \quad & a \circ b = -b \circ a; \\ \textbf{bilinear:} \quad & a \circ (xb + yc) = x(a \circ b) + y(a \circ c), \\ & (xb + yc) \circ a = x(b \circ a) + y(c \circ a); \\ \textbf{Jacobi identity:} \quad & a \circ (b \circ c) + b \circ (c \circ a) + c \circ (a \circ b) = 0. \end{aligned}$$

For now, we shall leave $\mathcal{F}$ unspecified, but whenever necessary we shall assume $\mathcal{F} = R$. We shall call the operation $\circ$ a "Lie product" over A.

Lie algebras are nonassociative. The Jacobi identity is a generalization of the associativity condition obeyed by such algebras as real or complex numbers, matrices, and quaternions. If $\circ$ were an associative operation, we would be able to write

$$a \circ (b \circ c) = (a \circ b) \circ c.$$

Using skew-symmetry we can write the above equation as follows:

$$a \circ (b \circ c) + c \circ (a \circ b) = 0.$$

The Jacobi identity looks like this, except that an extra term, $b \circ (c \circ a)$, is added. In this sense it generalizes associativity.

Adjoint Algebra Let $L(A)$ be the set of all linear operators on A, typically represented by matrices, given some basis for A. Using bilinearity, a subset of these can be associated with A itself. We define the function $\phi : A \longrightarrow L(A)$ as follows:

$$\phi(a) : A \longrightarrow A; \quad \phi(a) : b \longmapsto a \circ b.$$

Since $\circ$ is a bilinear operation, $\phi(a)$ is indeed a linear operator and thus a member of $L(A)$. It is called the "adjoint operator" associated with a. We have seen this before, of course: in the context of the Poisson bracket Lie algebra, Eq. (3.10) means that the adjoint operator of an observable is the Hamiltonian vector field that it generates.

The notation $\phi(a)$ is a bit awkward for the manipulations that follow. Dragt's notation [30], $:a:$, is better, but it is associated with Hamiltonian dynamics and Poisson brackets. We shall use $\bar{a}$ to signify the adjoint operator associated with a. Overloading this notation, we shall signify the set of all adjoint operators as

$$\bar{A} = \{\bar{a} \mid a \in A\} \subset L(A).$$

As an example, consider the Lie algebra studied first by most physicists: the vector cross product defined on the space of three-dimensional vectors, an algebra usually introduced in high school. To show that $a \circ b = \vec{a} \times \vec{b}$ obeys the three defining conditions is no more than a trivial exercise. Returning to the three-vector cross product, the adjoint algebra comprises a set of 3×3 matrices, a set that must be topologically equivalent to R^3. To find a representation, consider that the adjoint operators associated with an orthonormal set of basis vectors must satisfy

$$\bar{\vec{e}_i}\, \vec{e}_j = \vec{e}_i \times \vec{e}_j = \sum_k \epsilon_{ijk} \vec{e}_k,$$

so that the matrix representation of $\bar{\vec{e}_1}$, for example, would be given by

$$\left(\bar{\vec{e}_1}\right) \longrightarrow \begin{pmatrix} 0 & 0 & 0 \\ 0 & 0 & -1 \\ 0 & 1 & 0 \end{pmatrix},$$

with similar results for $\bar{\vec{e}_2}$ and $\bar{\vec{e}_3}$. We then use linearity to write the adjoint operator associated with a general three-vector, $\vec{v}$.

$$\left(\bar{\vec{v}}\right) = \sum_k v_k \bar{\vec{e}_k} = \begin{pmatrix} 0 & -v_3 & v_2 \\ v_3 & 0 & -v_1 \\ -v_2 & v_1 & 0 \end{pmatrix}.$$

Morphism The overwhelming importance of the adjoint operators follows from the fact that their commutator algebra is a (homo)morphism of the original Lie algebra. This follows immediately from the Jacobi identity, which we rearrange below.

$$a \circ (b \circ c) - b \circ (a \circ c) = (a \circ b) \circ c.$$

Using the definition of the adjoint operators, this becomes

$$\bar{a}\bar{b}c - \bar{b}\bar{a}c = [\bar{a},\bar{b}]c = \overline{a \circ b}c. \tag{3.118}$$

Since c is completely arbitrary, this becomes an operator identity.

$$[\bar{a},\bar{b}] = \overline{a \circ b}. \tag{3.119}$$

This equation indicates a morphism between the commutator algebra of adjoint operators — or, as it is called, the "adjoint algebra" — and the original Lie algebra from which we started. That this is so is not trivial. Indeed, $\bar{A}$ is not even closed under ordinary concatenation, or multiplication, of operations. That is, if $\bar{a}$ and $\bar{b}$ are members of $\bar{A}$, *it does not follow that their product, $\bar{a}\bar{b}$, belongs to $\bar{A}$*. It is therefore a remarkable fact, contingent upon the Jacobi identity, that their commutator always does.

Notice that we did not have to group operations in parentheses on the left-hand side of Eq. (3.118): operator algebras are associative. In this way a morphism is established between nonassociative Lie algebra and a corresponding associative algebra. This is what makes classification of Lie algebras possible.

Equation (3.119) can be written in another form,

$$\bar{\bar{a}}\bar{b} = \overline{\bar{a}b},$$

which is suggestive of a recursion. Indeed, using this as the seed, we can demonstrate the identity

$$\bar{\bar{a}}^n\bar{b} = \overline{\bar{a}^n b},$$

for assuming that it is true for $n - 1$ we have

$$\begin{aligned}
\bar{\bar{a}}^n\bar{b} &= \bar{\bar{a}}(\bar{\bar{a}}^{n-1}\bar{b}) \\
&= \bar{\bar{a}}\left(\overline{\bar{a}^{n-1}b}\right) \\
&= \overline{\bar{a}\left(\bar{a}^{n-1}b\right)} \\
&= \overline{\bar{a}^n b}.
\end{aligned}$$

Now, let Λ be any univariate, polynomial expression or convergent infinite series with coefficients c_n.[21] Because of linearity, we can multiply each nth instance of this identity by the corresponding c_n and sum to get the result

$$\Lambda(\bar{\bar{a}})\bar{b} = \overline{\Lambda(\bar{a})b}. \tag{3.120}$$

[21] We are dealing a little too loosely with the issue of convergence and shall continue to do so.

Again, this is just a consequence of the morphism that exists between the original Lie bracket algebra and the commutator algebra of its adjoint operators.

Lie Group Absorption Once again employing skew-symmetry of $\circ$ we write the Jacobi identity in yet another way:

$$a \circ (b \circ c) = (a \circ b) \circ c + b \circ (a \circ c).$$

Or, upon introducing the adjoint operator, $\bar{a}$,

$$\bar{a}(b \circ c) = (\bar{a}b) \circ c + b \circ (\bar{a}c).$$

Any adjoint operator thus acts as a derivation with respect to the Lie product. It is a simple exercise, left to the reader, to develop the analog of Liebniz's formula from differential calculus.

$$\bar{a}^n(b \circ c) = \sum_{k=0}^{n} \binom{n}{k} (\bar{a}^{n-k}b) \circ (\bar{a}^k c). \tag{3.121}$$

Using this as a lemma, we shall now obtain the absorption (distributive) formula,

$$e^{\bar{a}}(b \circ c) = (e^{\bar{a}}b) \circ (e^{\bar{a}}c).$$

This is little more than straightforward calculation. From Eq. (3.121) we have

$$\begin{aligned} e^{\bar{a}}(b \circ c) &= \sum_{n=0}^{\infty} \frac{1}{n!} \bar{a}^n(b \circ c) \\ &= \sum_{n=0}^{\infty} \frac{1}{n!} \sum_{k=0}^{n} \frac{n!}{k!(n-k)!} (\bar{a}^{n-k}b) \circ (\bar{a}^k c) \\ &= \sum_{n=0}^{\infty} \sum_{k=0}^{n} \sum_{m=0}^{n} \delta_{m+k,n} \left(\frac{1}{m!} \bar{a}^m b\right) \circ \left(\frac{1}{k!} \bar{a}^k c\right). \end{aligned}$$

With the delta function, the triple sum simply ranges over all points in the quadrant, Z^2, each point being counted once (see Figure 3.15). Therefore we can replace the triple sum with an independent double sum

$$\sum_{n=0}^{\infty} \sum_{k=0}^{n} \sum_{m=0}^{n} \delta_{m+k,n} \cdots = \sum_{k=0}^{\infty} \sum_{m=0}^{\infty} \cdots,$$

to obtain the final result.

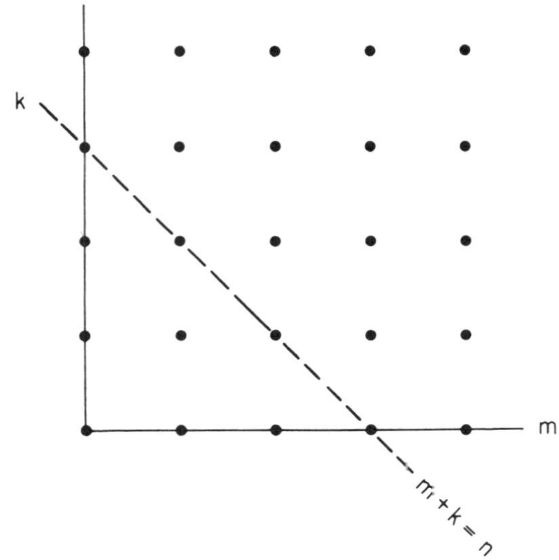

FIGURE 3.15 The triple sum ranges over all points on the lattice Z^2.

$$\begin{aligned}
e^{\bar{a}}(b \circ c) &= \sum_{k=0}^{\infty} \sum_{m=0}^{\infty} \left(\frac{1}{m!} \bar{a}^m b \right) \circ \left(\frac{1}{k!} \bar{a}^k c \right) \\
&= \left(\sum_{m=0}^{\infty} \frac{1}{m!} \bar{a}^m b \right) \circ \left(\sum_{k=0}^{\infty} \frac{1}{k!} \bar{a}^k c \right) \\
&= (e^{\bar{a}} b) \circ (e^{\bar{a}} c).
\end{aligned}$$

The exponential map, $e^{\bar{a}}$, therefore produces an automorphism of the Lie algebra. Taking this at face value and applying it to observables and Hamiltonians, we have a weaker version of Theorem 3.4 for autonomous Hamiltonians.

$$e^{-t:H:}[f, g] = [e^{-t:H:}f, e^{-t:H:}g].$$

Lie algebras arose originally in the study of continuous groups. Now we shall invert the process and indicate how to construct a group from a formal Lie algebra. The procedure for doing this is exponentiation: let $a \in A$ be an element of a Lie algebra and $\bar{a}$ the associated adjoint operator; then the group element corresponding to a is $e^{\bar{a}}$. The tricky part in showing that the set of all such objects forms a group is to prove that they are algebraically closed. That this is so is the content of the Baker-Campbell-Hausdorff (BCH) theorem, which we shall state and prove below.

Some Terminology Before embarking on this demonstration, we shall collect a few results from previous sections and define some useful terminology. A denotes our original Lie algebra,[22] and $L(A)$ was the vector space of all linear maps of A into

[22]More literally, A was the vector space on which we built the Lie algebra by appending a Lie operator.

itself (i.e., automorphisms of A as a vector space). The subspace of adjoint operators we denoted $\bar{A}$.

$$\bar{A} = \{\bar{a} \mid a \in A\} \subset L(A).$$

We now define the space of exponential maps,

$$expL(A) \equiv \{e^{\alpha} \mid \alpha \in L(A)\},$$

and the subset associated with $\bar{A}$,

$$exp\bar{A} \equiv \{e^{\bar{a}} \mid \bar{a} \in \bar{A}\} \subset expL(A).$$

$L(A)$ is automatically a commutator Lie algebra. From the previous section we know already that

(i) $\bar{A}$ is an invariant subalgebra,
(ii) furthermore, it is a homomorphism of the original Lie algebra, and
(iii) any element of $exp\bar{A}$ induces an automorphism on $\mathcal{A}$.

The goal of the BCH theorem is to establish that $exp\bar{A}$ is closed under composition. A crucial step is the observation that $L(A)$ is itself a Lie algebra, and therefore we can associate adjoint operators with each of *its* elements. Accordingly, we let $L(L(A))$ be the space of all linear maps, $L(A) \longrightarrow L(A)$. With each $\alpha \in L(A)$ we associate the adjoint, $\bar{\alpha} \in L(L(A))$, according to the same rules as before.

$$\bar{\alpha} : L(A) \longrightarrow L(A); \bar{\alpha} : \beta \longmapsto [\alpha, \beta], \quad \forall \beta \in L(A).$$

The space of all such maps we denote, in natural fashion, as $\overline{L(A)}$.

$$\overline{L(A)} \equiv \{\bar{\alpha} \mid \alpha \in L(A)\}.$$

Included among all such possible $\bar{\alpha}$'s are the elements of $\bar{A}$; that is, some α's are $\bar{a}$'s, and therefore some $\bar{\alpha}$'s are $\bar{\bar{a}}$'s. We single out this subset,

$$\bar{\bar{A}} \equiv \{\bar{\bar{a}} \mid \bar{a} \in \bar{A}\} \subset \overline{L(A)}.$$

It is important to note that, because $\bar{A}$ is a subalgebra, closed under commutation, $\bar{\bar{A}}$ is an automorphism of $\bar{A}$, $\bar{\bar{A}} : \bar{A} \longrightarrow \bar{A}$.

This is a lot of terminology to absorb at one sitting. If you are confused, go through the definitions a few times until they make sense before continuing.

Lemmas Development of the BCH formula will go a lot more smoothly with the help of a few lemmas, which we present here. Rather than repeatedly writing expressions like "$\forall \alpha \in L(A)$," just remember that in what follows all simple Greek symbols represent elements of $L(A)$.

Lemma 3.4 $e^{\bar{\alpha}}\xi = e^{\alpha}\xi e^{-\alpha}$.

Proof: Define a function $f : R \to L(A)$ as follows:

$$f(x) = e^{x\alpha}\xi e^{-x\alpha} \in L(A)$$

Differentiation leads immediately to the following development:

$$\begin{aligned} f'(x) &= (e^{x\alpha}\alpha)\xi e^{-x\alpha} + e^{x\alpha}\xi(-\alpha e^{-x\alpha}) \\ &= e^{x\alpha}[\alpha, \xi]e^{-x\alpha} \\ &= [\alpha, e^{x\alpha}\xi e^{-x\alpha}] \\ &= [\alpha, f(x)] \\ &= \bar{\alpha} f(x). \end{aligned}$$

This is just a linear equation for $f(x)$, whose solution is

$$f(x) = e^{x\bar{\alpha}}f(0) = e^{x\bar{\alpha}}\xi.$$

Setting $x = 1$ then produces the statement of our lemma. □

Having done that, it is now easy to prove our second lemma, which is a statement of homomorphism between exponential maps.

Lemma 3.5 If $e^{\alpha}e^{\beta} = e^{\gamma}$, then $e^{\bar{\alpha}}e^{\bar{\beta}} = e^{\bar{\gamma}}$.

Proof: This is merely a matter of calculation, using the preceding lemma thrice in succession. For any ξ,

$$\begin{aligned} e^{\bar{\alpha}}(e^{\bar{\beta}}\xi) &= e^{\bar{\alpha}}(e^{\beta}\xi e^{-\beta}) \\ &= e^{\alpha}(e^{\beta}\xi e^{-\beta})e^{-\alpha} \\ &= (e^{\alpha}e^{\beta})\xi(e^{\alpha}e^{\beta})^{-1} \\ &= e^{\gamma}\xi e^{-\gamma} \\ &= e^{\bar{\gamma}}\xi. \end{aligned}$$

Since ξ is completely arbitrary, this becomes an operator identity,

$$e^{\bar{\alpha}}e^{\bar{\beta}} = e^{\bar{\gamma}},$$

which is the required result. □

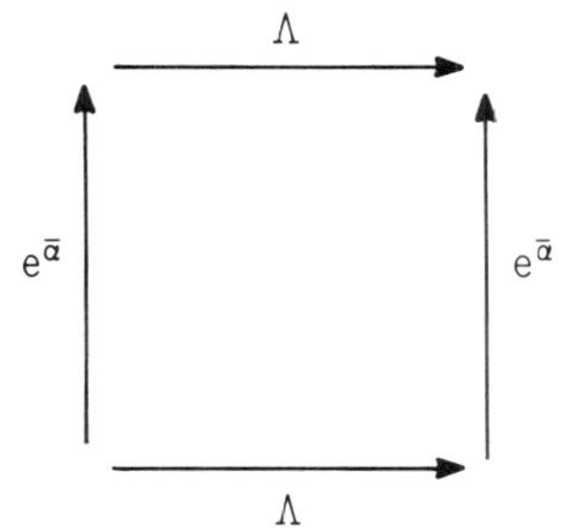

FIGURE 3.16 Commutative diagram illustrating a Lie algebraic identity.

A Few Parenthetic Remarks and Further Consequences The converse of Lemma 3.5 is not valid, but from the manipulations leading to the proof of Lemma 3.5, we have that

$$\begin{aligned}\forall \alpha, \beta, \gamma, \xi \in L(A): \quad & (e^{\alpha}e^{\beta})\xi(e^{\alpha}e^{\beta})^{-1} = e^{\gamma}\xi e^{-\gamma}, \\ & \xi = (e^{-\gamma}e^{\alpha}e^{\beta})\xi(e^{-\gamma}e^{\alpha}e^{\beta})^{-1}, \\ & [e^{-\gamma}e^{\alpha}e^{\beta}, \xi] = 0.\end{aligned}$$

This holds for all $\xi \in L(A)$, and in particular for those invertible elements that form a group. That group has an irreducible representation of maximum dimension, so by Schur's lemma we have that $e^{-\gamma}e^{\alpha}e^{\beta}$ must be proportional to the identity, or that

$$e^{\alpha}e^{\beta} = ce^{\gamma}.$$

Lemma 3.4 has a remarkable consequence, which extends it considerably. If $\xi \in L(A)$ then $\xi^n \in L(A)$, $\forall n \in Z^+$, and so $e^{\alpha}\xi^n e^{-\alpha} = e^{\bar{\alpha}}\xi^n$. On the other hand, $(e^{\bar{\alpha}}\xi)^n = (e^{\alpha}\xi e^{-\alpha})^n = e^{\alpha}\xi^n e^{-\alpha}$. We therefore have the triple identity,

$$(e^{\bar{\alpha}}\xi)^n = e^{\alpha}\xi^n e^{-\alpha} = e^{\bar{\alpha}}\xi^n. \tag{3.122}$$

Now, let Λ be a univariate, polynomial expression with coefficients, c_n. Because the exponential operators are linear, we can multiply each nth instance of Eq. (3.122) by the corresponding c_n and sum to get the result,

$$\Lambda(e^{\bar{\alpha}}\xi) = e^{\alpha}\Lambda(\xi)e^{-\alpha} = e^{\bar{\alpha}}\Lambda(\xi).$$

Treating both Λ and $e^{\bar{\alpha}}$ as mappings of $L(A)$ into itself, algebraists would draw a commutative diagram like that in Figure 3.16.

If the issue of convergence can be overcome, Λ could be any expression representable as an infinite series. If ξ is invertible, then we can extend the condition $n \in Z^+$ to $n \in Z$, so that Λ could be a Laurent expression as well (I think).

If we let A be the vector space of observables over a phase space and take the Poisson bracket as the Lie product, we can use Lemma 3.4 to prove an identity, which we write using Dragt's "colon" notation for adjoint operators:

$$e^{-:f:}e^{:F:}e^{:f:} = \exp[:e^{-:f:}F:] \tag{3.123}$$

This can be established trivially provided we demonstrate

$$e^{-:f:}:F:e^{:f:} = :e^{-:f:}F:. \tag{3.124}$$

Indeed, Eq. (3.124) is much stronger than Eq. (3.123). It implies that, if Λ is any polynomial expression in a single variable, then

$$e^{-:f:}\Lambda(:F:)e^{:f:} = \Lambda(:e^{-:f:}F:).$$

By extension, this would apply to Λ's expressing infinite series, provided convergence is not an issue, and Eq. (3.123) is then merely a special case.

To establish Eq. (3.124), and therefore Eq. (3.123), as identities we revert temporarily to the notation of this section. Let $\alpha = -\bar{f}$ and $\xi = \bar{F}$, for $f, F \in A$. Lemma 3.4 followed by the identity Eq. (3.120), with $\Lambda = \exp$, yields the following:

$$\begin{aligned} e^{-\bar{f}}\bar{F}e^{\bar{f}} &= e^{-\bar{\bar{f}}}\bar{F} \\ &= \overline{e^{-\bar{f}}F}. \end{aligned}$$

Converting this back to Dragt's notation produces the desired Eq. (3.124).

Development (Proof) and Statement of BCH Formula Returning to the BCH formula, recall that our goal is to prove the operator algebra $exp\bar{A} \equiv \{e^{\bar{a}} \mid a \in A\}$ to be closed under multiplication. This means we must demonstrate that

$$\forall a, b \in A\ \exists c \in A : e^{\bar{a}}e^{\bar{b}} = e^{\bar{c}}.$$

The key to accomplishing this is construction of a function, $\sigma : R \longrightarrow L(A)$, according to

$$\sigma(x) = \ln[e^{\bar{a}}e^{x\bar{b}}], \quad \text{that is,} \quad e^{\sigma(x)} = e^{\bar{a}}e^{x\bar{b}}. \tag{3.125}$$

That we can do this follows from the fact that $expL(A)$ is closed under composition, which assures us of the existence of $\sigma(x)$. We need to show that $\sigma(1) \equiv \bar{c} \in \bar{A}$. This will be done by writing $\sigma(1)$ in the form

$$\sigma(1) = \bar{a} + \bar{b} + \Lambda(\bar{\bar{a}}, \bar{\bar{b}})\bar{b}, \tag{3.126}$$

where Λ is an expression which we shall develop as a polynomial. Since both $\bar{\bar{a}}$ and $\bar{\bar{b}}$ map $\bar{A}$ into itself, then so does any such expression. Therefore, if we can demonstrate this form, it must be that $\sigma(1) \in \bar{A}$, the desired result.

We proceed, of course, by differentiating both sides of Eq. (3.125). The right-hand side is trivial,

$$\frac{d}{dx}(e^{\bar{a}}e^{x\bar{b}}) = e^{\bar{a}}e^{x\bar{b}}\bar{b} = e^{\sigma(x)}\bar{b}.$$

Regrettably, the left-hand side is more difficult. We cannot write $(e^{\sigma})' = e^{\sigma}\sigma'$ because σ and σ' do not necessarily commute. We must proceed more carefully and calculate the derivative as a limit, $(e^{\sigma})'(x) = \lim_{\epsilon\to 0}(e^{\sigma(x+\epsilon)} - e^{\sigma(x)})/\epsilon$, taking this noncommutativity into account.

$$\begin{aligned} e^{\sigma(x+\epsilon)} &= e^{\sigma(x)+\epsilon\sigma'(x)+O(\epsilon^2)} \\ &= \sum_{n=0}^{\infty}\frac{1}{n!}(\sigma(x)+\epsilon\sigma'(x)+O(\epsilon^2))^n \\ &= \sum_{n=0}^{\infty}\frac{1}{n!}\sigma^n(x) + \epsilon\sum_{n=1}^{\infty}\sum_{m=0}^{n-1}\frac{1}{n!}\sigma^m(x)\sigma'(x)\sigma^{n-1-m}(x) \\ &\quad + O(\epsilon^2). \end{aligned}$$

The first sum is just $e^{\sigma}(x)$, which gives us the result

$$(e^{\sigma})'(x) = \sum_{n=1}^{\infty}\sum_{m=0}^{n-1}\frac{1}{n!}\sigma^m(x)\sigma'(x)\sigma^{n-1-m}(x). \tag{3.127}$$

Now reverse the order of summation.

$$\sum_{n=1}^{\infty}\sum_{m=0}^{n-1}\cdots = \sum_{m=0}^{\infty}\sum_{n=m+1}^{\infty}\cdots = \sum_{m=0}^{\infty}\sum_{\substack{k=0\\ n\leftarrow m+k+1}}^{\infty}\cdots.$$

Equation (3.127) gets rewritten as follows:

$$(e^{\sigma})'(x) = \sum_{m=0}^{\infty}\sum_{k=0}^{\infty}\frac{1}{(m+k+1)!}\sigma^m\sigma'\sigma^k.$$

Now comes an elegant trick: we introduce the integral form of the beta function,

$$\int_0^1 du\,(1-u)^m u^k = \frac{\Gamma(m+1)\Gamma(k+1)}{\Gamma(m+k+2)} = \frac{m!k!}{(m+k+1)!},$$

which is the reciprocal of a binomial coefficient when m and k are integers. By introducing this we decouple the summations, thereby making them independent.

$$
\begin{aligned}
(e^{\sigma})'(x) &= \sum_{m=0}^{\infty}\sum_{k=0}^{\infty}\frac{1}{m!k!}\int_0^1 du\,(1-u)^m u^k\,\sigma^n\sigma'\sigma^k \\
&= \int_0^1 du\left[\sum_{m=0}^{\infty}\frac{1}{m!}((1-u)\sigma)^m\right]\cdot\sigma'\cdot\left[\sum_{k=0}^{\infty}\frac{1}{k!}(u\sigma)^k\right] \\
&= \int_0^1 du\,e^{(1-u)\sigma}\,\sigma'\,e^{u\sigma} \\
&= e^{\sigma}\int_0^1 du\,e^{-u\sigma}\sigma' e^{u\sigma}.
\end{aligned}
$$

The integral is evaluated with the help of Lemma 3.4 above.

$$
\begin{aligned}
(e^{\sigma})' &= e^{\sigma}\left(\int_0^1 du\,e^{-u\bar{\sigma}}\right)\sigma' \\
&= e^{\sigma}\left(\frac{1-e^{-\bar{\sigma}}}{\bar{\sigma}}\right)\sigma'.
\end{aligned}
$$

That last expression in parentheses is formal. It should be interpreted as representing a power series, as $\bar{\sigma}$ does *not* possess an operator inverse.

We have written the derivatives of both the right-hand and left-hand sides of Eq. (3.125). The time has come to equate these. Since e^{σ} certainly possesses an inverse, $e^{-\sigma}$, we can use this equality to solve for σ'.

$$
\begin{aligned}
e^{\sigma}\bar{b} &= e^{\sigma}\left(\frac{1-e^{-\bar{\sigma}}}{\bar{\sigma}}\right)\sigma' \\
\sigma' &= \left(\frac{\bar{\sigma}}{1-e^{-\bar{\sigma}}}\right)\bar{b} \\
&= \left(\frac{e^{\bar{\sigma}}\bar{\sigma}}{e^{\bar{\sigma}}-1}\right)\bar{b} \\
\sigma(x) &= \sigma(0) + \int_0^x dt\left(\frac{e^{\bar{\sigma}(t)}\bar{\sigma}(t)}{e^{\bar{\sigma}(t)}-1}\right)\bar{b}.
\end{aligned}
$$

(Recall that $b \in A$, but $\sigma \in L(A)$.) For convenience in what follows we define the auxiliary expression,

$$
\lambda(z) = \frac{z\ln z}{z-1} - 1.
$$

Again, we shall interpret this formally as specifying a power series in its argument. Using $\sigma(0) = \bar{a}$, from the definition of σ, we then have

$$\sigma(1) = \bar{a} + \bar{b} + \left[\int_0^1 dt\, \lambda(e^{\bar{\sigma}(t)})\right]\bar{b}.$$

It does not seem as though we have made progress through all these manipulations. On the contrary, we are on the verge of the final result. All that remains is to apply Lemma 3.4 and the definition, Eq. (3.125), to the integrand.

$$\sigma(1) = \bar{a} + \bar{b} + \left[\int_0^1 dt\, \lambda(e^{\bar{\bar{a}}} e^{t\bar{\bar{b}}})\right]\bar{b}.$$

Comparing with Eq. (3.126) we see that we have found our Λ,

$$\Lambda(\bar{\bar{a}}, \bar{\bar{b}}) = \int_0^1 dt\, \lambda(e^{\bar{\bar{a}}} e^{t\bar{\bar{b}}}),$$

and the construction is complete. □

The preceding manipulations may have seemed more formal than satisfactory. The importance of this proof for us is that it is constructive. That is, it provides a procedure for constructing the Lie algebraic element, $\bar{c}$, from $\bar{a}$ and $\bar{b}$.

PROBLEMS

3.1 Show that the property of being a gradient field is not geometric, but being Hamiltonian is.

3.2 Meditate on the following specious argument, remaining ever alert for plausible sounding nonsense, and resolve its implied paradox. Given any two orbits $\underline{z}_1(t)$ and $\underline{z}_2(t)$ of a Hamiltonian system, construct the superposition, $\underline{z}(t) \equiv \underline{z}_1(t) + \underline{z}_2(t)$, and develop its equation of motion. Since the Poisson bracket is bilinear, we have

$$\begin{aligned}\dot{\underline{z}}(t) = \frac{d}{dt}(\underline{z}_1(t) + \underline{z}_2(t)) &= \dot{\underline{z}}_1(t) + \dot{\underline{z}}_2(t) \\ &= \{\underline{z}_1(t), H\} + \{\underline{z}_2(t), H\} \\ &= \{(\underline{z}_1(t) + \underline{z}_2(t)), H\} \\ &= \{\underline{z}, H\}.\end{aligned}$$

Therefore $\underline{z}$ is also an orbit; it obeys the same Hamilton's equations of motion as $\underline{z}_1$ and $\underline{z}_2$. But this is precisely the characteristic of linearity: the superposition of two arbitrary orbits is yet another orbit. We appear to have proved that *all* Hamiltonian systems are linear! Why, in fact, is this *not* a proof?

3.3 Let $\mathcal{P} = R^2$, $\underline{z} = (x,p) \in \mathcal{P}$, $H(\underline{z}) = x^m p^n$, and let X_H be the vector field associated with H. Find an algebraic expression for the time-evolution map $\underline{\phi}$ defined by $\exp tX_H : \underline{z} \longmapsto \underline{\phi}(\underline{z})$. Demonstrate explicitly that it is symplectic. Describe the difference between $m + n$ being even or odd. In either case, does the collection of maps $\exp tX_H, t \in R$, satisfy the Chapman-Kolmogorov equation?

3.4 Verify that the Poisson bracket generates a Lie algebra over the linear vector space of observables by showing that it satisfies the Jacobi identity.

3.5 Show that quadratic polynomials span an invariant subalgebra of the Poisson bracket Lie algebra.

3.6 Show that linear, Hamiltonian vector fields form a subalgebra of the Lie bracket algebra; that is, show that the Lie bracket of any two of them is another. Show explicitly that the set of all their matrices is closed under commutation but that the product of any two belongs to the set if and only if they commute.

3.7 Given a vector space, V, a "two-form" on V is a function $\gamma : V \times V \longrightarrow R$ that is bilinear, nondegenerate, and antisymmetric. Consider the complex plane, $C \simeq R^2$, as a vector space over the reals. Show that $\gamma(a,b) \equiv \mathrm{Im}[ab^*]$ satisfies these conditions and has $O(2) \simeq SU(1)$ as its symmetry group.

3.8 Complete the proof of Van Hove's theorem by

(a) showing that Dirac quantization

$$[\bar{f}, \bar{g}] = i\overline{\{f,g\}}$$

requires that

$$\overline{x^2} = \bar{x}^2, \quad \overline{p^2} = \bar{p}^2, \quad \text{and} \quad \overline{xp} = \tfrac{1}{2}(\bar{x}\bar{p} + \bar{p}\bar{x}),$$

and (b) deriving the identity we left unfinished,

$$\{x^3, p^3\} = 3\{px^2, p^2x\}.$$

3.9 Let $(x(\lambda), 0)$ be a critical point of the autonomous Hamiltonian,

$$H(x,p;\lambda) = p^2/2m + V(x;\lambda),$$

with control parameter λ. Find a vector field, $v(x;\lambda)$, over R such that $x(\lambda)$ is an orbit of v. What is the interpretation of the other orbits? (Numerically solving this first order differential equation leads to the "homotopic methods" for calculating critical points.)

3.10 Why is the Lie derivative of $\underline{A}$ *not* used in Eq. (3.36)?

3.11 In the brachistochrone problem, establish the correspondence between the *orbits* of a bead on minimal time paths and those of markers on cycloids. In particular, show that if the reference circle is rolled according to $\dot{\phi} \equiv \omega(R) =$ constant, then the orbit of the marker indeed corresponds to that of the bead. Find the functional form of $\omega(R)$.

3.12 What *is* the minimal amount of time required for a bead to travel from one point to another in a constant gravity field?

3.13 In the bead on a rotating ring problem, show explicitly that if one starts from

$$T = \tfrac{1}{2}m(\dot{r}^2 + r^2\dot{\theta}^2 + r^2\sin^2\theta\dot{\phi}^2),$$

which is a quadratic form in generalized velocity coordinates, and uses $H = T + V$ *without* imposing the conditions $\dot{r} = 0$ and $\dot{\phi} = \omega$ until *after* writing Hamilton's equations, then the equations of motion are the same as before.

3.14 In the same problem, derive the two stable, secondary fixed points from the simple observation that the bead's acceleration must be caused by the centripetal force provided by the ring.

3.15 Reduce the "bead on a rotating ring" problem to quadrature in terms of elliptic functions. Assume the energy is sufficiently small so that the motion is bounded. You may want to consider the following milestones: (a) Using the scaled variables, $h = H/m\omega^2a^2$, $\kappa = g/\omega^2 a$, and $\eta = \cos\theta$, establish that

$$\omega\, dt = \pm \frac{d\eta}{\sqrt{(\eta^2 - 1)(\eta^2 + 2\kappa\eta - 2h - 1)}}.$$

(b) Factorize the radical as

$$(\eta^2 - 1)(\eta^2 + 2\kappa\eta - 2h - 1) = (\eta + 1)(\eta - 1)(\eta - \eta_-)(\eta - \eta_+)$$

$$\eta_\pm \equiv -\kappa \pm \sqrt{\kappa^2 + 2h + 1}.$$

Establish that $\eta_\pm$ are real. (c) Define

$$m^2 \equiv \frac{2(\eta_+ - \eta_-}{(1 - \eta_-)(1 + \eta_+)},$$

$$z^2 \equiv \frac{1}{2}(1 - \eta_-)\frac{\eta + 1}{\eta - \eta_-}.$$

Estab[illegible] 1 and $0 \leqslant z^2$. (d) Show that

$$[illegible] = \mp \frac{2}{\sqrt{(1 - \eta_-)(1 + \eta_+)}} \frac{dz}{\sqrt{(1 - z^2)(1 - m^2z^2)}}.$$

[illegible]at $(1 - z^2)(1 - m^2z^2) \geqslant 0$.

(e) Now make use of the definition for elliptic functions of the first kind:

$$F(z,m) = \int_0^z dz\,[(1-z^2)(1-m^2z^2)]^{-1/2}.$$

3.16 Show that the set of all $2n \times 2n$ symplectic matrices form a group.

3.17 Verify that the Jacobian of the transformation from polar coordinates to Cartesian, given in Eq. (3.52), is symplectic.

3.18 Can the electromagnetic tensor $F^{\mu\nu} = \partial^\mu A^\nu - \partial^\nu A^\mu$ serve as the matrix of a symplectic two-form over four-dimensional space-time?

3.19 Let $\mathbf{M}$ be a $2n \times 2n$ symplectic matrix with characteristic polynomial $P(\lambda) \equiv \det[\mathbf{M} - \lambda \mathbf{1}]$. Show that $P(\lambda)$ is palindromic. That is, show that if

$$P(\lambda) = \sum_{k=0}^{2n} a_k \lambda^k,$$

then $a_k = a_{2n-k}$.

3.20 Use the previous problem to show that eigenvalues of a real symplectic matrix can be associated in (a) complex conjugate pairs and (b) reciprocal pairs.

3.21 Sketch (a) all possible "generic" configurations for the eigenvalues of real 2×2 and 4×4 symplectic matrices and (b) the bifurcations that are the transitions from one configuration to another.

3.22 Let $\underline{v}(\underline{z}) = \underline{\underline{A}} \cdot \underline{z}$ be a linear, autonomous, Hamiltonian vector field. Show *explicitly* that the family of matrices $\exp(t\underline{\underline{A}})$, $t \equiv -\infty \cdots \infty$, which are Jacobians of the time-evolution maps of $\underline{v}$, comprise a one-parameter subgroup of a symplectic group.

3.23 Confirm that the generating functions given in Eqs. (3.68) and (3.69) produce the transformations of Eqs. (3.48) and (3.52), respectively.

3.24 Prove Lemma 3.3: if $\underline{\underline{S}}$ is a symmetric matrix and $\underline{\underline{M}} \equiv \left(\underline{\underline{1}} - \underline{\underline{J}} \cdot \underline{\underline{S}}\right)^{-1} \cdot \left(\underline{\underline{1}} + \underline{\underline{J}} \cdot \underline{\underline{S}}\right)$, then $\underline{\underline{M}}$ is symplectic.

3.25 Compare the behavior of two symplectic numerical integrators against Runge-Kutta and against the exact solution of a harmonic oscillator. Discuss their relative advantages and disadvantages.

3.26 Extend the generating function symplectic numerical integrator to third order: derive Eq. (3.75).

3.27 Let $dS_1 = \underline{p} \cdot d\underline{x} - \underline{I} \cdot d\underline{\varphi}$ be the differential form of a Type 1 generating function for a transformation to action-angle coordinates. Show that $\oint_{\gamma_k} dS_1 = 0$ for all γ_k given in Eq. (3.81). Is this sufficient to ensure that $\oint_\gamma dS_1 = 0$ for *all* closed curves γ on the torus?

3.28 Is the value of the integral defining an action coordinate,

$$I_k = \frac{1}{2\pi} \oint_{\gamma_k} \underline{p} \cdot d\underline{x},$$

invariant under a canonical transformation, $(\underline{x}, \underline{p}) \rightsquigarrow (\underline{x}', \underline{p}')$?

3.29 Perform the eigenvalue analysis of the synchrotron dynamic given in Eqs. (3.86) and Eq. (3.88) at the fixed point $W = 0$, $\varphi = \pi - \varphi_s$. Calculate the rate at which nearby orbits diverge.

3.30 Argue that the synchrotron stability criterion is $\eta_s \cos \varphi_s > 0$ by using simple pictorial reasoning on what happens to the phase of a particle when it is on either side of synchrony.

3.31 Confirm the procedure given in Eqs. (3.97) for finding the Edwards-Teng lattice functions of a coupled ring.

3.32 Using the notation of Eqs. (3.97), show that

$$\cos \mu_1 - \cos \mu_2 = u\sqrt{1 + v/u^2},$$

where $2u = \mathrm{Tr}(\underline{\underline{M}}_{11} - \underline{\underline{M}}_{22})$ and $v = 2\det(\underline{\underline{M}}_{21}) + \mathrm{Tr}(\underline{\underline{M}}_{12} \cdot \underline{\underline{M}}_{21})$.

3.33 Show that when the fiducial orbit is the arc of a circle, then the local $\rho(s)$ is indeed the radius of the circle and that $\hat{u}_1$ indeed points outward, as promised.

3.34 Consider a planar section of a fiducial orbit, that is, a section that is confined to a plane. Show that if the section is *not* a straight line, the torsion, $\tau(s)$, vanishes on it.

3.35 A particle undergoes planar, circular orbits in a constant magnetic field. Show that the magnetic rigidity is $eB\rho$, where ρ is the radius of the circle.

3.36 Write the two uncoupled second order Hill's equations implied by the quadratic terms of the transverse Hamiltonian, Eq. (3.99).

3.37 Fill in the missing details necessary to verify Eq. (3.102).

3.38 Show that the polar coordinates of Eq. (3.100) are legitimate action-angle coordinates of the quadratic Hamiltonian.

3.39 Reconsider the one-dimensional Hill's problem. Define $u = x/\sqrt{\beta}$ and $\phi = \psi/\nu$. Treating ϕ as the orbit parameter, show that $d^2u/d\phi^2 + \nu^2 u = 0$. Why can this *not* be extended to two degrees of freedom, as in Eq. (3.103)?

3.40 Derive a multipolar kick map by simply considering the force on a charged particle passing through a magnetic field.

3.41 Calculate the transfer map through a solenoid. Choose a local chart associated with the fiducial curve passing straight through the solenoid, along the field lines.

3.42 For integer m and k, show that

$$\int_0^1 du(1-u)^m u^k = \frac{\Gamma(m+1)\Gamma(k+1)}{\Gamma(m+k+2)} = \frac{m!k!}{(m+k+1)!}$$

by using induction.

3.43 Prove Leibniz's formula for adjoint operators,

$$\bar{a}^n(b \circ c) = \sum_{k=0}^{n} \binom{n}{k} (\bar{a}^{n-k} b) \circ (\bar{a}^k c)$$

3.44 Show that nilpotence and bilinearity implies skew-symmetry.

3.45 In three-dimensional Euclidean space, consider a rotation through an angle ϕ about the axis $\underline{u} = (u_1, u_2, u_3)$. Let $\underline{\underline{R}}$ be the matrix representing this rotation, so that $\underline{x} \longmapsto \underline{\underline{R}} \cdot \underline{x}$, for all $\underline{x} \in R^3$. Show that

$$\underline{\underline{R}} = \underline{\underline{1}} + \underline{\underline{W}} \sin\phi + \underline{\underline{W}}^2(1 - \cos\phi),$$

where

$$\underline{\underline{W}} = \begin{pmatrix} 0 & -u_3 & u_2 \\ u_3 & 0 & -u_1 \\ -u_2 & u_1 & 0 \end{pmatrix}.$$

Repeat this calculation using the 2×2 Pauli spin matrices as a basis.

3.46 Evaluate the Baker-Campbell-Hausdorff formula through terms of fourth order.

$$\begin{aligned} e^{\alpha} e^{\beta} &= e^{\sigma} \Rightarrow \\ \sigma &= \alpha + \beta + \tfrac{1}{2}[\alpha, \beta] \\ &\quad + \tfrac{1}{12}[\alpha, [\alpha, \beta]] - \tfrac{1}{12}[\beta, [\alpha, \beta]] \\ &\quad + \tfrac{1}{12}[\alpha, [\alpha, [\alpha, \beta]]] + \tfrac{1}{24}[\alpha, [\beta, [\alpha, \beta]]] + \tfrac{1}{36}[\beta, [\beta, [\alpha, \beta]]] \end{aligned}$$

Some hints on how to proceed: First write $\sigma = \alpha + \beta + F \cdot \beta$ where F is the linear operator

$$F = \int_0^1 dt\, f(e^{\bar{\alpha}} e^{t\bar{\beta}})$$

$$f(z) = \frac{z \ln z}{z-1} - 1.$$

Then expand f as follows

$$f(z) = \sum_{n=1}^{\infty}(-1)^{n+1}\frac{w^n}{n(n+1)}$$
$$w \equiv z - 1$$

Let us understand ξ_k to mean "the terms of order k in expression ξ." Then we can write

$$\sigma = \alpha + \beta + \sum_{k=1}^{\infty} F_k \cdot \beta$$
$$F_k = \sum_{n=1}^{k}(-1)^{n+1}\frac{1}{n(n+1)}\int_0^1 dt(w^n)_k.$$

The primitives $(w^n)_k$ are obtained recursively.

$$\text{for } k \geqslant 1: \quad (w)_k = \sum_{m=0}^{k}\frac{t^m}{m!(k-m)!}\overline{\alpha}^{k-m}\overline{\beta}^m$$
$$\text{for } k \geqslant n: \quad (w^n)_k = \sum_{l=n-1}^{k-1}(w)_{k-l}(w^{n-1})_l$$

3.47 Let $< A, \circ >$ be a Lie algebra, and let $\{e_k | k = 1 \ldots n\}$ be a basis for A. The "structure constants" of the algebra, f_{ij}^k, are defined

$$e_i \circ e_j = \sum_k f_{ij}^k e_k.$$

Show that matrices $\underline{\underline{M}}_i, i = 1 \ldots n$ defined by

$$(\underline{\underline{M}}_i)_{jk} = -f_{ij}^k$$

satisfy

$$[\underline{\underline{M}}_i, \underline{\underline{M}}_j] = \sum_k f_{ij}^k \underline{\underline{M}}_k.$$

(This is called the algebra's "regular representation.")

3.48 Show that $\underline{\underline{M}}^T \underline{\underline{J}}\,\underline{\underline{M}} = \underline{\underline{J}}$ for any symplectic matrix $\underline{\underline{M}}$.

CHAPTER 4

Integrable Resonance Models

Harry Pena . . . a trap fisherman . . . scowled at a picture of a French girl in a bikini. Fred, understanding that he seemed a bleak, sexless person to Harry, tried to prove that Harry had him wrong. He nudged Harry, man-to-man. "Like that, Harry?" he asked.

"Like what?"

"The girl there."

"That's not a girl. That's a piece of paper."

"Looks like a girl to me." Fred Rosewater leered.

"Then you're easily fooled," said Harry. "It's done with ink on paper . . . If this were a real girl, all I'd have to do for a living would be to stay home and cut out pictures of big fish."

— Kurt Vonnegut, Jr.
God Bless You, Mr. Rosewater

The most definitive phase space structure associated with dissipative dynamics is the attractor; for conservative or, more strongly, symplectic dynamics it is the separatrix, sometimes called a "vague attractor."[1] It is regrettable that the concept "resonance" is frequently introduced as a disease of perturbation theory. A commonly presented scenario is this: an attempt is made to construct a transformation that will put a Hamiltonian into "normal form," as in Eq. (3.77). This attempt eventually flounders: some terms in the perturbation series may become arbitrarily large because of division by "small denominators." This problem is associated with the existence of a resonance condition, and the connection is complete. The overall effect of such presentations is to suggest that a resonance has more to do with the way things are calculated than with real, physical phenomena — the sort of feeling one originally gets about

[1] Abraham and Marsden [1] credit Thom [89] with coining this descriptive phrase. We are extending it somewhat here; Thom evidently applied it to periodic orbits.

renormalization in quantum field theory. They ignore what should be the central geometric features of a resonance: (a) the dimensional collapse of the invariant torus, which we have already seen in the first chapter, and (b) the existence of a *separatrix*, the utility of which is its ability to organize phase space via partitioning. A separatrix is a topological obstacle: no continuous transformation can deform phase space so as to make it disappear. "Small denominators" are not the real stumbling block but only *its manifestation within the context of a perturbation theory.*

In this chapter we shall study integrable Hamiltonian dynamics possessing separatrices, the "isolated resonance models," with emphasis on (a) "detuning," (b) the topological structure of separatrices, and (c) applications to accelerator physics. Resonance phenomena are important enough that their geometry should be understood before becoming enmeshed in the logistics of perturbative expansions, which are deferred to the next chapter. In this chapter we will depend on "averaging techniques" to provide the heuristic arguments necessary to move the discussion along, leaving it to the next to justify more fully the results obtained. Our objective is to gain a semi-intuitive, geometric picture of separatrix structure as quickly as possible. We begin with a quick overview of averaging and its extension to resonances. This is followed by a discussion of a methodology for characterizing the separatrices associated with multidimensional isolated resonances, after which we move on to specific accelerator models. The basic accelerator orbit theory required is quickly reviewed preparatory to a study of the "principal" resonances that arise. As our primary goal is to understand the phase space structure of resonance models themselves, the reader who is less interested in their connections with accelerators may skip to the final sections on first reading. In the same spirit, it may be worthwhile to browse the concrete examples beginning with Section 4.3.2 before reading Section 4.2.

4.1 AVERAGING

We introduce a "method of averaging," of which there are actually several. If the physical, heuristic arguments presented here seem (justifiably) unsatisfactory, please keep in mind that (a) this indicates you are discerning and not easily swayed by hand-waving arguments, (b) the expressions are at best valid only to first order in the strength of the nonlinearities, and (c) a more complete treatment will be laid out in the next chapter, especially in Section 5.2 for flows and in Section 5.3 for maps. In either case, flows or maps, the objective for symplectic dynamics is a resonance normal form, like the one that we shall write in Eq. (4.6). The purpose of this chapter is to examine the physical consequences of this form, not to derive it rigorously.

The heuristic motivation we will follow here was suggested originally by an alleged separation of dynamical coordinates into two categories: "fast" and "slow." The fundamental physics assumption is that the rates of change of the "slow" coordinates are much smaller than those of the "fast." The goal is to replace a given vector field with one that is separable in these coordinates — that is, one that can be written as a superposition of terms, which are products of functions of "fast" coordinates with functions of "slow" coordinates. Invariably, within the context of Hamiltonian dynamics, angle coordinates belong to the former, and action coordinates to the latter. In addition, the angle coordinates appear in the equations of motion for the action coordinates in the arguments of oscillatory functions characterized by a "small time scale," inducing the central thesis that *the effect of their rapid oscillations on the slowly varying action coordinates will tend to cancel out*. This extends to non-Hamiltonian dynamics as well when the system is expressed in polar-like coordinates, with the amplitude-like coordinate taking the place of the action. The net result of these arguments is to replace the original dynamical system with one obtained by averaging over all angle coordinates, with the hope that the new dynamic represents the mean behavior of the original when averaged over the time scale of the rapid fluctuations.

Averaging Van der Pol's Equation Before burying ourselves once again in Hamiltonian formalism, if only to demonstrate that averaging is not restricted to Hamiltonian systems, we shall apply it to the Van der Pol oscillator, one representation of which was given in Eq. (2.38).

$$\begin{aligned} \dot{x} &= p, \\ \dot{p} &= -x + \gamma(\mu - 3x^2)p. \end{aligned}$$

Averaging is done most transparently using the complex eigencoordinates of the harmonic oscillator, $\gamma = 0$ limit, which were introduced in Eq. (2.8) and Eq. (2.9).

$$a = \sqrt{\tfrac{1}{2}}\,(x + ip) = i\sqrt{I}e^{-i\varphi}.$$

Using a, the Van der Pol dynamic is written

$$\begin{aligned} \dot{a} &= (p - ix)/\sqrt{2} + i\gamma\mu p/\sqrt{2} - i\gamma 3x^2 p/\sqrt{2} \\ &= -ia + \tfrac{1}{2}\gamma\mu(a - a^*) - \tfrac{3}{4}(a^2 + a^{*2} + 2a^*a)(a - a^*). \end{aligned} \tag{4.1}$$

In the spirit of averaging, we concentrate on the behavior of the amplitude, $r^2 = x^2 + p^2 = 2I$.

$$\begin{aligned} \frac{d}{dt}(\tfrac{1}{2}r^2) = r\dot{r} = \frac{d}{dt}(a^*a) &= 2\,\mathrm{Re}(a^*\dot{a}) \\ &= 2\,\mathrm{Re}\Big[-ia^*a + \tfrac{1}{2}\gamma\mu(a^*a - a^{*2}) \\ &\quad -\tfrac{3}{4}(a^2 + a^{*2} + 2a^*a)(a - a^*)a^*\Big]. \end{aligned}$$

The averaging rule declares that we neglect all the oscillatory terms containing φ; that is, we shall throw out all terms with differing powers of a and a^*. What remains

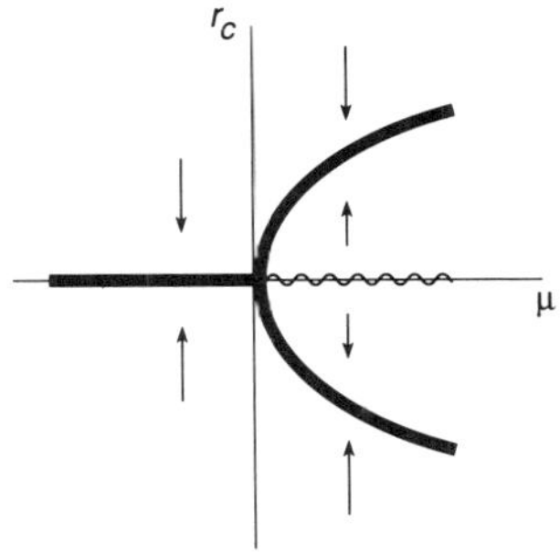

FIGURE 4.1 Diagram of a pitchfork bifurcation at $\mu = 0$.

is the following:

$$\begin{aligned} r\dot{r} &\approx \gamma\mu a^* a - 3\gamma(a^* a)^2 + \tfrac{3}{2}\gamma(a^* a)^2 \\ &= \tfrac{1}{2}\gamma\mu r^2 - \tfrac{3}{8}\gamma r^4, \\ \dot{r} &\approx \tfrac{1}{2}\gamma r(\mu - \tfrac{3}{4}r^2) = -\tfrac{3}{8}\gamma r(r^2 - r_c^2). \end{aligned} \tag{4.2}$$

We have set $r_c \equiv \sqrt{4\mu/3}$, which agrees with the value obtained in Eq. (2.41). $\dot{r}$ is negative (positive) when $r^2 > r_c^2 (r^2 < r_c^2)$, establishing once again the nature of the limit cycle as an attractor.

Considered independently, Eq. (4.2) represents a one-dimensional dynamic over R^1 possessing a "pitchfork" bifurcation, as shown in the bifurcation diagram of Figure 4.1.

For $\mu < 0$, the origin, $r = 0$, is the only fixed point. It is an attractor whose basin of attraction is the entire phase space. For $\mu > 0$, the origin is a repellor, and the two additional fixed points at $r = \pm r_c$ are attractors. The basin of one is the positive reals, of the other, the negative. The reason for the name "pitchfork bifurcation" is evident from the figure.

Hamiltonian Shear Terms It is easier to formulate averaging prescriptions for periodic Hamiltonian systems than for more general dynamics, as the former are governed by a single scalar function. Thus, having expressed a periodic Hamiltonian in polar, or action-angle, representation, $H(\underline{\varphi}, \underline{I}; \theta)$, averaging replaces it with one depending only on the amplitude coordinates:

$$H(\underline{\varphi}, \underline{I}; \theta) \rightarrow H_{\text{avg}}(\underline{I}) \equiv \langle H(\underline{\varphi}, \underline{I}; \theta)\rangle_{\text{all angles}} = \oint \frac{d\theta}{2\pi}\frac{d\underline{\varphi}}{(2\pi)^N} H(\underline{\varphi}, \underline{I}; \theta). \tag{4.3}$$

The vector fields generated by these approximate Hamiltonians are precisely those given in Eq. (3.77). They foliate phase space with invariant, KAM tori, indexed by $\underline{I}$, and each having its own characteristic frequencies (or tunes) that vary continuously (according to Eq. (3.77)) as one passes through them along an arbitrary path. Their time-evolution operators produce multidimensional twist maps.

The preconditions for averaging — all that business about "fast" and "slow" coordinates — are especially relevant to perturbed systems, when H can be written in the form

$$H(\underline{\varphi}, \underline{I}; \theta) = H_0(\underline{I}) + \epsilon H_1(\underline{\varphi}, \underline{I}; \theta, \epsilon),$$

with ϵ playing the role of some small, perturbative order parameter, and $\lim_{\epsilon \to 0} H_1 \neq 0$. From Hamilton's equation, $\dot{\underline{I}} = \epsilon\, \partial H_1 / \partial \underline{\varphi}$, it can be argued that the "time" for significant variation in $\underline{I}$ scales like $1/\epsilon$ and therefore can be made arbitrarily large, while that for $\underline{\varphi}$ is bounded by $1/|\underline{\omega}_0(\underline{I})|$.

Isolated Resonance Models However, this averaged Hamiltonian is not the exact Hamiltonian, and in going from the former to the latter, the most optimistic view is that the differences can be explained as arising from a small perturbation. Were the exact Hamiltonian autonomous, the KAM theorem would then guarantee survival of "sufficiently nonresonant" tori. By itself this would not mean that the resonant tori do not survive, but in fact that is generically the case: resonant tori are typically replaced by chaotic separatrices, and we shall devote much of what follows to a treatment of their integrable counterparts. Fortunately, there do exist simple, integrable dynamics possessing separatrices worthy of study: the "isolated resonance" models. These allow us to study "regular" (as opposed to "chaotic") separatrices analytically and thereby discover their main features. In this and the following sections we shall define a methodology for doing such analyses and illustrate it with a few examples. Within the spirit of this chapter, we will motivate isolated resonance models heuristically, once again deferring a more exact treatment to the next chapter. As described so far, averaging requires accepting that the oscillatory terms, discarded by Eq. (4.3), fluctuate sufficiently rapidly that their influence averaged to zero, or at least something negligible. We shall extend this argument a little. Consider expanding the periodic Hamiltonian, $H(\underline{\varphi}, \underline{I}; \theta)$, in harmonics.

$$H(\underline{\varphi}, \underline{I}; \theta) = \sum_{\underline{m}\, n} H_{\underline{m}\, n}(\underline{I}) e^{i(\underline{m} \cdot \underline{\varphi} + n\theta)}. \tag{4.4}$$

Within this decomposition, the vector field components are

$$\begin{aligned} \dot{\underline{\varphi}} &= \partial H / \partial \underline{I} = \sum_{\underline{m}\, n} \underline{\omega}_{\underline{m}\, n}(\underline{I}) e^{i(\underline{m} \cdot \underline{\varphi} + n\theta)}, \\ \dot{\underline{I}} &= -\partial H / \partial \underline{\varphi} = -\sum_{\underline{m}\, n} i\underline{m} H_{\underline{m}\, n}(\underline{I}) e^{i(\underline{m} \cdot \underline{\varphi} + n\theta)}. \end{aligned} \tag{4.5}$$

Here we have introduced $\underline{\omega}_{\underline{m}n} \equiv \partial H_{\underline{m}n}/\partial \underline{I}$.[2] The averaging heuristic is essentially one of *stationary phase*: the terms that dominate the integrated behavior of an orbit are those for which the argument of the complex exponential is as slowly varying as possible. Equation (4.3) corresponded to isolating the $\underline{m} = \underline{0}, n = 0$, "shear" term, but that is not the only possibility. Consider that, from the shear term alone — presumed to dominate the dynamic — the $\underline{\varphi}$ orbit history is approximately

$$\underline{\varphi} \approx \underline{\varphi}(0) + \underline{\omega}(\underline{I})\theta,$$

where we've shortened $\underline{\omega}_{\underline{0}0} \equiv \underline{\omega}$. Substituting this directly into Eq. (4.5) provides, for better or worse, an approximation:

$$\dot{\underline{I}} \approx -\sum_{\underline{m}n} \left(i\underline{m} H_{\underline{m}n}(\underline{I}) e^{i\underline{m}\cdot\underline{\varphi}(0)} \right) e^{i(\underline{m}\cdot\underline{\omega}(\underline{I})+n)\theta}.$$

This is not necessarily easier to solve than the original dynamic we started from, but it does provide a way of characterizing terms with stationary phase: they are the ones that satisfy

$$m \cdot \underline{\omega}(\underline{I}) + \frac{2\pi}{T} n = \underline{m} \cdot \underline{\omega}(\underline{I}) + n \approx 0,$$

for *some* $\underline{m}$ and n. This is, of course, a resonance condition — of the form introduced in Eq. (2.34) in connection with striated tori — and so the terms are called, not surprisingly, "resonance terms." For these terms, the heuristic of "rapid fluctuation" is not even approximately correct; their frequency is too small, their period too long, and they cannot be discarded.

An isolated resonance model, or a "resonance normal form" Hamiltonian, amounts to accepting only one such term as dominant, at least within a local phase space region of interest.

$$\begin{aligned} H_{\text{res}}(\underline{\varphi}, \underline{I}; \theta) &= H_{\text{avg}}(\underline{I}) + H_{\underline{m}n}(\underline{I}) e^{i(\underline{m}\cdot\underline{\varphi}+n\theta)} + H^*_{\underline{m}n}(\underline{I}) e^{-i(\underline{m}\cdot\underline{\varphi}+n\theta)} \\ &= \underline{\nu}\cdot\underline{I} + H_s(\underline{I}) + H_r^{(c)}(\underline{I}) \cos[\underline{m}\cdot\underline{\varphi} + n\theta] \\ &\quad + H_r^{(s)}(\underline{I}) \sin[\underline{m}\cdot\underline{\varphi} + n\theta]. \end{aligned} \tag{4.6}$$

Here, H_s symbolizes the shear terms, the ones independent of angle coordinates, while $H_r^{(c)}$ and $H_r^{(s)}$ are the new resonance terms. We can combine these two into one, at least locally, by introducing an amplitude-dependent phase function, $\phi(\underline{I})$.

$$H(\underline{\varphi}, \underline{I}; \theta) = \underline{\nu}\cdot\underline{I} + H_s(\underline{I}) + H_r(\underline{I}) \cos[\underline{m}\cdot\underline{\varphi} + n\theta + \phi(\underline{I})]. \tag{4.7}$$

[2]Here, $\dot{\varphi} = d\varphi/d\theta$, and so forth. There is a small notational difficulty. With θ as the orbit-parameter, these frequencies are actually winding numbers (or tunes); it may have been more consistent to symbolize them as "$\underline{\nu}_{\underline{m}n}$." However, in this chapter I wanted "ν" to mean the tune at zero-amplitude. In any case, since the period of θ is 2π, $\underline{\omega} = 2\pi\underline{\nu}/T = \underline{\nu}$ anyway.

In the simplest models $\phi(\underline{I})$ is simply a number; we shall encounter a few such shortly.

There are a number of flaws in the line of reasoning we've employed, not the least of which is the impression given that a resonance must arise from the Fourier expansion given in Eq. (4.4). In fact, an important resonance may not appear analytically until higher orders in a perturbation series. We repeat: what has been presented here should be accepted as, at best, a crude path that leads to the isolated resonance model of Eq. (4.6) or Eq. (4.7). We should also dampen the impression that resonances are analytic phenomena: a resonance is a *topological* attribute of a dynamical system.

A critical property of isolated resonance models is obtained trivially upon employing Hamilton's equation.

$$\dot{\underline{I}} = \underline{m}\left(H_r^{(c)}(\underline{I})\sin[\underline{m}\cdot\underline{\varphi} + n\theta] - H_r^{(s)}(\underline{I})\cos[\underline{m}\cdot\underline{\varphi} + n\theta]\right) \propto \underline{m}$$

$$\underline{I}(\theta) = \underline{I}(0) + \underline{m}f(\theta).$$

Projected onto the amplitude coordinates, $\underline{I}$, all orbits lie along parallel lines with direction $\underline{m}$. If only positive values of $\underline{I}$ are physically meaningful, as in Eq. (3.100) or Eq. (2.9), then the existence of unbounded orbits requires that all components of $\underline{m}$ have the same sign: if they do not, then orbits must intersect the boundary of the physical quadrant, $\{\underline{I} \mid \forall k : I_k > 0\}$, in two points, thereby bounding the motion. This property has been expressed in the folklore that "sum resonances are dangerous, and difference resonances are safe." Of course, merely bounding an orbit does not make it "safe"; finite excursions in an accelerator may bump against such realities as beam pipes. A better quote may be that, "sum resonances are dangerous, and some are not."

Another consequence is that, in a $2N$-dimensional phase space, we can trivially construct $N-1$ dynamical invariants in involution, which are hyperplanes in the action sector orthogonal to $\underline{m}$. That is, let $\underline{m}_k^{\perp}, k = 1,\ldots,N-1$ be linearly independent integer multiplets orthogonal to $\underline{m}$, and define the observables $F_k \equiv \underline{m}_k^{\perp}\cdot\underline{I}$. This family of observables is obviously in involution, and each is an invariant.

$$\dot{F}_k = \underline{m}_k^{\perp}\cdot\dot{\underline{I}} \propto \underline{m}_k^{\perp}\cdot\underline{m} = 0 \tag{4.8}$$

Only one more invariant is necessary to establish the model's integrability, and it will be obtained in the next section.

4.2 RESONANCE ANALYSIS, SEPARATRIX GEOMETRY

Prominent among the highly desirable features possessed by isolated resonance models is their understandability; they are integrable dynamics with analyzable separatrices. These make them only slightly more difficult to classify than the shear models, but much richer in their behaviors and more useful for modeling real, physical systems near instability. In this section we shall describe a methodology for doing global analysis on isolated resonance models of arbitrary (even) dimensionality. Our goal is to achieve a geometric understanding of their orbit spaces. To this end we shall show that each can be related to a reduced orbit space that exists on a two-dimensional phase space.

It seems logical that this section be placed here. However, the discussion will be rather dry and abstract. You may gain more from it by first spending an enjoyable hour, or so, worrying the two-dimensional examples in Section 4.3.2. This beneficial, invigorating activity will help lay a foundation of "intuition," which may smooth the rather tortured path that follows immediately.

Angle Reduction: Untwisted Tori The first step of the methodology takes advantage of the happy circumstance that angle coordinates appear in Eq. (4.6) only in the linear combination $\underline{m} \cdot \underline{\varphi}$. This suggests employing a linear, canonical transformation of the form Eq. (3.48) or Eq. (3.68) to a new set of coordinates, $(\underline{\varphi}, \underline{I}) \rightarrow (\underline{\eta}, \underline{J})$.

$$\underline{I} = \underline{\underline{M}}^T \cdot \underline{J}, \quad \underline{\eta} = \underline{\underline{M}} \cdot \underline{\varphi}. \tag{4.9}$$

The matrix $\underline{\underline{M}}$ is written in row form using a linearly independent family of integer multiplets of the kind introduced in Eq. (4.8).

$$\underline{\underline{M}} \equiv \begin{pmatrix} \underline{m} \\ \underline{m}_1^{\perp} \\ \underline{m}_2^{\perp} \\ \vdots \\ \underline{m}_{N-1}^{\perp} \end{pmatrix}.$$

In the particular case of a four-dimensional phase space, we may choose

$$\underline{\underline{M}} \equiv \begin{pmatrix} m_1 & m_2 \\ m_2 & -m_1 \end{pmatrix}, \quad \text{so that} \quad \underline{\underline{M}}^{-1} = \frac{1}{m_1^2 + m_2^2} \quad \underline{\underline{M}}.$$

On this new chart the Hamiltonian is represented in the form

$$\begin{aligned} K(\underline{\eta}, \underline{J}; \theta) &= (\underline{\nu} \cdot \underline{\underline{M}}^T) \cdot \underline{J} + H_s(\underline{\underline{M}}^T \cdot \underline{J}) \\ &\quad + H_r^{(c)}(\underline{\underline{M}}^T \cdot \underline{J}) \cos[\eta_1 + n\theta] + H_r^{(s)}(\underline{\underline{M}}^T \cdot \underline{J}) \sin[\eta_1 + n\theta] \\ &\equiv K_l(\underline{J}) + K_s(\underline{J}) \\ &\quad + K_r^{(c)}(\underline{J}) \cos[\eta_1 + n\theta] + K_r^{(s)}(\underline{J}) \sin[\eta_1 + n\theta]. \end{aligned}$$

There are a few points to be noted:

(a) Equation (4.9) is a cheat, as the angle coordinates, $\underline{\varphi}$, are not defined globally; put another way, $(\underline{\varphi}, \underline{J})$ and $(\underline{\varphi} + 2\pi \underline{n}, \underline{J})$ represent the same point in phase space. We can avoid specifying detailed charts by noting that $\underline{\eta}$ possesses similar equivalence classes to $\underline{\varphi}$. Therefore we can think of the angle part of Eq. (4.9) as mapping one quotient space into another. (It might help to review the discussion that followed Eq. (2.12).) Everything is fine as long as the vector field is periodic in these coordinates.

(b) K_l is a linear function of the coordinates, $J_1, \ldots, J_N$, with coefficients that are themselves linear functions of the tunes, $\nu_1, \ldots, \nu_N$. Its most important attribute is the coefficient of J_1: $\partial K_l / \partial J_1 = \underline{m} \cdot \underline{\nu}$.

(c) $\eta_2, \ldots, \eta_N$ are ignorable coordinates, so $J_2, \ldots, J_N$ are dynamical invariants, as was observed already in the previous section. We can treat them either as dynamical coordinates of the original Hamiltonian or effectively as control parameters of this "reduced" Hamiltonian. This provides a natural method of collapsing the problem to two dimensions: we project orbits into the (η_1, J_1) subspace and classify the flow as functions of the $N-1$ "parameters," $J_2, \ldots, J_N$.

(d) η_1 has modular range $2\pi \times \sum |m_k|$. Since K is of period 2π in η_1, the flow consists of $\sum |m_k|$ identical copies of the fundamental domain, $\eta_1 \in [0, 2\pi)$.

(e) Redefining coordinates in this way amounts to choosing a new set of homotopically inequivalent closed curves, $\{\gamma_k\}$, introduced in conjunction with definition of action given in Eqs. (3.82) and (3.84), a set that conforms more closely to the way resonant orbits actually wind around their tori. These transformations are precisely those introduced in Section 3.3 — see Eqs. (3.83) and (3.85) — which confirms that they indeed connect different sets of action coordinates of the "unperturbed" Hamiltonian. Like the cat map of Eq. (2.77), this change of angle coordinates is an instance of the "egg carton" transformation sketched in Figure 2.14. Resonant orbits are thus straightened out on this chart; the effect is to untwist the tori of Figure 2.5.

A final transformation $(\underline{\eta}, \underline{J}) \rightarrow (\underline{\xi}, \underline{J}')$ gets rid of the explicit dependence of the Hamiltonian on the orbit parameter, θ, thereby rendering it autonomous. This can be accomplished with a Type 2 canonical transformation generated by

$$S(\underline{\eta}, \underline{J}'; \theta) = \underline{\eta} \cdot \underline{J}' + n\theta J_1'.$$

This produces the following transformation equations

$$\begin{aligned} \underline{J} &= \underline{J}', \\ \xi_1 &= \eta_1 + n\theta, \\ \xi_k &= \eta_k, \quad k > 1. \end{aligned} \tag{4.10}$$

This use of the orbit parameter, θ, preserves the equivalence classes of ξ_1. The effect of this transformation is to collapse resonant orbits to fixed points; this will be discussed later.

Utilizing the fact that $\underline{J} = \underline{J}'$, the representation of our Hamiltonian changes only slightly.

$$K(\underline{\xi}, \underline{J}; \theta) = K_l(\underline{J}) + K_s(\underline{J}) + K_r^{(c)}(\underline{J}) \cos \xi_1 + K_r^{(s)}(\underline{J}) \sin \xi_1. \tag{4.11}$$

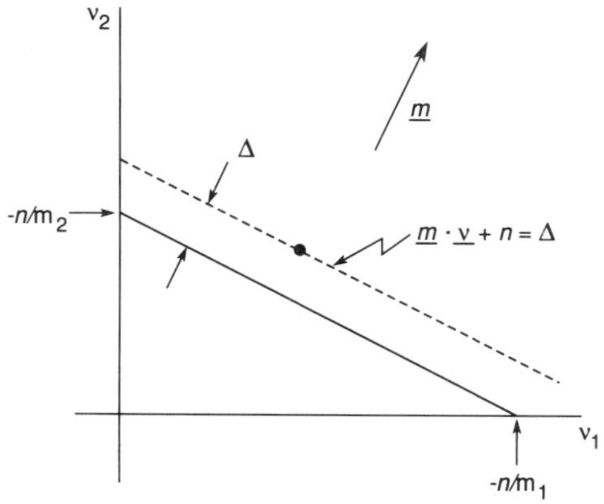

FIGURE 4.2 Isolated resonance models predict that the flow is characterized by the distance to a resonance line, $\underline{m} \cdot \underline{\nu} + n = 0$.

We have overloaded the symbol K_l here. The only difference between this K_l and the previous K_l is the coefficient of J_1: $\partial K_l/\partial J_1 = \underline{m} \cdot \underline{\nu} + n$. After this final transformation, which makes the Hamiltonian autonomous, the Hamiltonian itself becomes the Nth dynamical invariant. The system is thus integrable, and all compact orbits lie on invariant tori.

It is worth noting that the only essential dependence of this Hamiltonian on the "bare" tunes, $\underline{\nu}$, is contained in the factor $\Delta \equiv \underline{m} \cdot \underline{\nu} + n$, which is, in a directed Euclidean sense, the distance to the resonance line in tune space. This is a remarkable feature of the isolated resonance model: the topological character of the flow depends on the bare tunes only through Δ; different settings of $\underline{\nu}$ lying on the same tune hyperplane, $\Delta =$ constant, will give rise to equivalent dynamics. *Such behavior can both validate and limit the applicability of an isolated resonance model within a particular region of phase space.*

In a perturbation theoretic setting, it is expected that $K_s(\underline{J})$ and $K_r(\underline{J})$ will be "small." For the isolated resonance Hamiltonian to be at all applicable, Δ also must be small enough to be comparable in magnitude to $\underline{\omega}_s(\underline{J})$ and $\underline{\omega}_r(\underline{J})$ within the phase space region of interest. This gives rise to one concept of "resonance width."

Large Amplitude Behavior We examine now the behavior of the flow "at infinity," that is, in the limit $J_1 \longrightarrow \infty$. We have observed already that sum resonances allow the *possibility* of unbounded orbits. To see whether that possibility becomes actual we rearrange Eq. (4.11) as follows:

$$(K - K_l - K_s)\frac{1}{K_r} = \frac{K_r^{(c)}}{K_r} \cos \xi_1 + \frac{K_r^{(s)}}{K_r} \sin \xi_1 .$$

Here we have introduced $K_r^2 \equiv K_r^{(c)\,2} + K_r^{(s)\,2}$. The right-hand side necessarily lies within the interval $(-1, 1)$. This obvious constraint is even more apparent if we introduce a (admittedly ambiguous) phase function, $\phi(\underline{J})$, according to

$$\tan \phi(\underline{J}) = -K_r^{(s)}(\underline{J})/K_r^{(c)}(\underline{J}), \quad \text{with} \quad \operatorname{sgn} \cos \phi = \operatorname{sgn} K_r^{(c)},$$

and write

$$(K - K_l(\underline{J}) - K_s(\underline{J}))\frac{1}{K_r(\underline{J})} = \cos(\xi_1 + \phi(\underline{J})).$$

If $K_r \longrightarrow \infty$ as $J_1 \longrightarrow \infty$, then along an orbit, $K/K_r \longrightarrow 0$. Except for linear resonances, an instance of which we shall examine later, it is also generally true that $K_l/K_r \longrightarrow 0$. This leaves the shear terms to control large amplitude behavior: unbounded motion is only possible for finite, invariant K only along the asymptotic phases, ξ_1^∞, which satisfy

$$\cos(\xi_1^\infty + \phi^\infty) = -\lim_{J_1\to\infty}\left[K_s(\underline{J})/K_r(\underline{J})\right], \tag{4.12}$$

with $\phi^\infty = \lim_{J_1\to\infty}\phi(\underline{J})$. Whether the right-hand side (**rhs**) of this expression lies within $[-1, 1]$ is determined by examining the asymptotic behavior of K_s and K_r. If K_s grows faster than K_r, then **rhs** $\longrightarrow \infty$, and there is no (real) solution for ξ^∞: all orbits are bounded. Physically, this happens because the flow is dominated at large amplitudes by shear terms that produce enough "detuning" to suppress the resonance. If there is not sufficient detuning, so that K_r grows faster than K_s, then **rhs** $\longrightarrow 0$, and unbounded orbits can exist with asymptotic phases,

$$\xi_1^\infty + \phi^\infty \simeq \pm\pi/2.$$

Finally, if K_s and K_r grow at the same rate, so that their ratio approaches some nonzero constant, existence or nonexistence of unbounded orbits and their asymptotic phases are more subtly determined by the parameters of the model, including possibly the values of $J_2, \ldots, J_n$, according to Eq. (4.12).

The term "dynamic aperture" is frequently used. Although ambiguous, it can be interpreted loosely as some measure of the region in phase space occupied by bounded orbits.

Resonant Orbits and Separatrices Resonant, or "critical," orbits of the original Hamiltonian, Eq. (4.7), correspond to fixed points of the projected Hamiltonian, Eq. (4.11). (See Figure 4.3.) They are obtained by setting $\dot{J}_1$ and $\dot{\xi}_1$ to zero.

$$\dot{\xi}_1 = \Delta + \partial_1 K_s + \partial_1 K_r^{(c)}\cos\xi_1 + \partial_1 K_r^{(s)}\sin\xi_1, \tag{4.13}$$

$$\dot{J}_1 = K_r^{(c)}\sin\xi_1 - K_r^{(s)}\cos\xi_1. \tag{4.14}$$

Setting $\dot{J}_1 \equiv 0$ gives us the possibility of two types of critical points: (a) the regular are those for which

$$\sin\xi_1 = K_r^{(s)}/K_r \quad \text{and} \quad \cos\xi_1 = K_r^{(c)}/K_r, \tag{4.15}$$

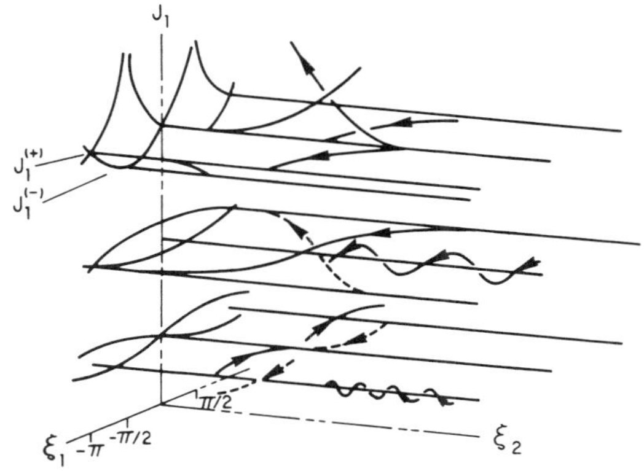

FIGURE 4.3 Conceptual sketch of a constant J_2 slice of the phase space of a four-dimensional isolated resonance model.

and (b) the irregular are those for which $K_r = 0$.[3] Let the coordinates of a fixed point be symbolized as $(\xi_1^{(0)}, J_1^{(0)})$. In the case of a regular fixed point, we substitute Eq. (4.15) back into Eq. (4.13) and set $\dot{\xi}_1 \equiv 0$ to obtain the equation whose solution provides the possible values of $J_1^{(0)}$.

$$\Delta + \frac{\partial}{\partial J_1}(K_s(\underline{J}) + K_r(\underline{J})) \equiv \Delta + \omega_s(\underline{J}) + \omega_r(\underline{J}) = 0. \tag{4.16}$$

This has a wonderful physical interpretation: resonant orbits appear at those amplitudes for which the tune shift represented by $\omega_s + \omega_r$ exactly cancels the resonance offset, Δ, thereby putting the orbit back "on" resonance. In the absence of detailed knowledge of ω_r but reasonable approximations for ω_s, the locations of resonances can sometimes be estimated by setting $\omega_r \approx 0$.

Once the $J_1^{(0)}$ are obtained, they are reinjected into Eq. (4.15) to determine the corresponding $\xi_1^{(0)}$ and finish the task of finding the resonant orbits.

In the case of irregular critical points, we must first solve the equation $K_r = 0$ for allowed values of $J_1^{(0)}$ and then use Eq. (4.13) to get the corresponding values of $\xi_1^{(0)}$. Unlike the regular, the phase of an irregular critical point is not necessarily pinned to a particular value; this must be treated on a case-by-case basis.

Which resonant orbits are elliptic (stable), and which hyperbolic (unstable)? The linearized equations of motion in the tangent space of a regular fixed point are written by expanding Eqs. (4.13) and (4.14) to first order in $d\xi_1$ and dJ_1, using the fixed point solutions to simplify.

$$\begin{pmatrix} \dot{\xi}_1 \\ \dot{J}_1 \end{pmatrix} = \underline{\underline{A}} \cdot \begin{pmatrix} d\xi_1 \\ dJ_1 \end{pmatrix},$$

$$\underline{\underline{A}} = \begin{pmatrix} -K_r^{(s)}K_r^{(c)}\partial_1(K_r^{(s)}/K_r^{(c)}) & \partial_1^2 K_s + (K_r^{(c)}\partial_1^2 K_r^{(c)} + K_r^{(s)}\partial_1^2 K_r^{(s)})/K_r \\ (K_r^{(c)2} - K_r^{(s)2})/K_r & K_r^{(s)}K_r^{(c)}\partial_1(K_r^{(s)}/K_r^{(c)}) \end{pmatrix}_0.$$

[3] I am not aware of any standard nomenclature for this distinction. "Irregular" critical points help to partition phase space but are generally not true resonant orbits.

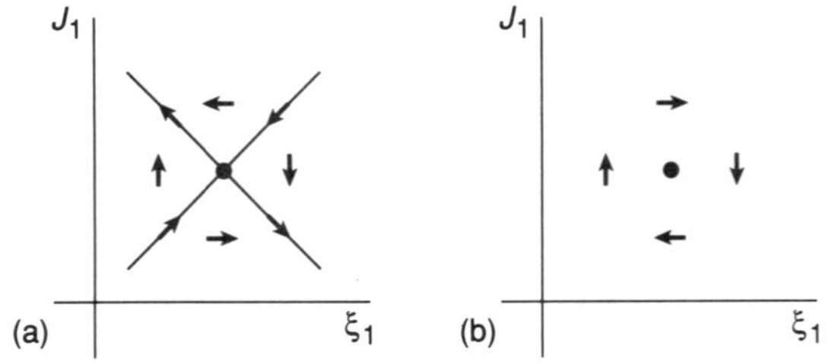

FIGURE 4.4 The nature of a regular fixed point depends on whether the coefficients of $d\xi_1$ and dJ_1 have (a) the same sign or (b) opposite signs.

Of course, symplecticity requires the matrix to be traceless. Evaluating its eigenvalues determines stability of the orbit. Notice that the diagonal terms vanish when $K_r^{(s)} \propto K_r^{(c)}$, $K_r^{(s)} = 0$, or $K_r^{(c)} = 0$. In such cases, the hyperbolicity of a resonant orbit is determined merely by the relative sign of the off-diagonal elements: if they have the same sign then the fixed point is unstable; if they have opposite signs, the fixed point is stable. (See Figure 4.4.)

In analogy with fixed points and periodic orbits, the set of all orbits that approach a hyperbolic resonant orbit as $\theta \longrightarrow +\infty$ is called its "stable" manifold; the set that approach it as $\theta \longrightarrow -\infty$ is its "unstable" manifold. The union of all regular hyperbolic resonant orbits, the irregular critical orbits, and their stable and unstable manifolds is the separatrix of the resonance. *Its dimensionality is sufficient to partition phase space.* To see that this is so, consider that a branch of the unstable manifold existing within a single $J_2, \ldots, J_N$ slice of a $2N$-dimensional phase space is an N-dimensional object; for example, one local chart could comprise the angle coordinates, $\underline{\xi}$ (see Figure 4.3). In addition to joining these branches with the resonant orbits, the union involves letting $J_2, \ldots, J_N$ range over all its possible values, which increases the dimensionality by $N - 1$ to $2N - 1$. A separatrix is therefore obtained by pasting together codimension 1 objects, as required by the boundary of a partition. The union of the (regualar?) resonant orbits alone is of dimension N. Its connected pieces have been given the name "whiskered tori" by Arnold, the "whiskers" referring to the stable and unstable manifolds.

Now this is important: *It is impossible to observe any of this structure by following the orbits continuously* unless one explicitly uses the special resonance coordinates, which are fairly complicated and *depend on the particular resonance of interest.* To visually "reveal" resonance structure in a coordinate-free way, one should sample orbits using a $\phi_{\theta+2\pi\,\theta}$ map[4] in complete analogy with the discussion in Section 2.1.3. Because angle coordinates only appear in the Hamiltonian in linear combination as the arguments of trigonometric functions, the connection between the resonance coordinates and the original, "physical," transverse coordinates is periodic and alters the fixed points of isolated resonance models into true resonant orbits. For example, combine Eq. (4.9) with Eq. (4.10) and apply the map, $\phi_{\theta+2\pi\,\theta}$. The changes in these

[4]This is actually a Poincaré map disguised as a period-advance map: θ marks an azimuthal location in a circular machine.

We now apply the averaging procedure of Eq. (4.3) to the Hamiltonian in Eq. (4.20), noting that $\langle \exp(i\underline{m} \cdot \underline{\varphi})\rangle = \delta_{\underline{m}\,\underline{0}}$.

$$H_{\rm avg}^{(hot)}(\underline{I}) = \frac{B_o R}{|B\rho|} \sum_{\underline{m}} \left(\oint \frac{d\theta}{2\pi} H_{\underline{m}}(\underline{I};\theta) \right) \delta_{\underline{m},\underline{0}}$$
$$= \frac{B_o R}{|B\rho|} \oint \frac{d\theta}{2\pi} H_{\underline{0}}(\underline{I};\theta).$$

This is developed further by employing the expansion given in Eq. (4.21) and the definition in Eq. (4.19).

$$\begin{aligned} H_{\rm avg}(\underline{I}) &= \underline{\nu} \cdot \underline{I} \\ &\quad + \sum_{k,q\geqslant 0} \left[\binom{2k}{k} \binom{2q}{q} \oint \frac{B_o R}{|B\rho|} \frac{d\theta}{2\pi} \beta_1^k(\theta) \beta_2^q(\theta) c_{2k\,2q}(\theta) \right] (I_1/2)^k (I_2/2)^q \\ &\equiv \underline{\nu} \cdot \underline{I} + \sum_{k,q\geqslant 0} h_{kq} I_1^k I_2^q. \end{aligned} \tag{4.22}$$

Among accelerator physicists these nonlinear terms are called "shear terms." The nomenclature is undoubtedly descriptive of nested tori shearing past each other, reflecting the amplitude dependence of the tune. The coefficients h_{kq} are linear functionals of the distributions $c_{2k\,2q}(\theta)$ from the original expansion, Eq. (4.18), in keeping with Eq. (4.22) being a first order approximation.

We shall now specialize by specifically using the $c_{2k\,2q}$ from the multipolar model written in Eq. (3.109) of Section 3.4.3, which produces the following expression for the coefficients h_{kq}.

$$\begin{aligned} h_{kq} &= (-1)^q \frac{1}{k+q} \left(\frac{1}{2}\right)^{k+q+1} \times \binom{2k}{k} \binom{2q}{q} \binom{2k+2q}{2q} \\ &\quad \times \oint \frac{B_o R}{|B\rho|} \frac{d\theta}{2\pi} \beta_1^k(\theta) \beta_2^q(\theta)\, b_{2k+2q-1}(\theta). \end{aligned} \tag{4.23}$$

Three observations:

(a) Only the normal coefficients, b_n, contribute to $H_{\rm avg}$, and then only those for which n is odd: the normal quadrupole, octupole, dodecapole, and so forth. We shall see in the next chapter, however, that the other multipoles will contribute at higher order in the multipole strength.

(b) For a given odd integer j, the normal coefficient b_j is associated with a homogeneous polynomial in I_1 and I_2 of degree $(j + 1)/2$. Put another way, a given normal $4n$-pole ($n = (j + 1)/2$, odd j) will give rise to polynomial terms of degree n.

(c) The combinatorial factors could be written more succinctly: $(2k + 2q)!/(k!\, q!)^2$.

Although written to average the nonlinear terms, Eq. (4.23) can be applied to evaluate the lowest order tune shift arising from normal quadrupole field error. The relevant coefficients are $h_{1\,0}$ and $h_{0\,1}$. The former is evaluated,

$$
\begin{aligned}
h_{1\,0} &= \frac{1}{2} \oint \frac{B_o R}{|B\rho|} \frac{d\theta}{2\pi} \beta_1(\theta) b_1(\theta) \\
&= \frac{1}{4\pi} \oint ds \frac{B'}{|B\rho|} \beta_1,
\end{aligned}
\tag{4.24}
$$

and a similar expression applies to the latter: $h_{0\,1}$ is obtained from this with the replacement $\beta_1 \longrightarrow -\beta_2$. The second line shows the way that this usually appears in most of the accelerator literature, with B' symbolizing $(\partial B_2/\partial x_1)_{\underline{x}=\underline{0}}$. These contributions from the normal quadrupole can be absorbed into the linear part of H_{avg} as a renormalized tune, or a linear tune shift — a tune shift independent of amplitude.

$$
H_{\text{avg}}(\underline{I}) = (\nu_1 + h_{10}) I_1 + (\nu_2 + h_{01}) I_2 + \sum_{k,q} h_{kq} I_1^k I_2^q .
$$

Centered Beam-Beam Tuneshifts Equation (4.22) implies a tune that grows indefinitely large with amplitude, reflecting the unphysical, unbounded nature of the multipole model. An important example in which this does not happen is the "beam-beam" interaction. In our case, this would more appropriately be called the "particle-beam" interaction, as we use it to model the force exerted by a Gaussian bunch on a single test particle. One kick map associated with this process can be obtained from the transverse Hamiltonian,

$$
H(\underline{\varphi}, \underline{I}; \theta) = \underline{\nu} \cdot \underline{I} + \sum_{\theta_{bb}} U(\underline{x}) \delta_{2\pi}(\theta - \theta_{bb}),
$$

where each θ_{bb} marks the location of a beam-beam encounter, treated as impulses, with $\delta_{2\pi}$ being a 2π-periodic delta function. For a Gaussian bunch, the potential U is modeled as follows [72]:

$$
U(\underline{x}) = \lambda \int_0^\infty dt \prod_{k=1}^{2} \frac{1}{\sqrt{t + \sigma_k^2}} \exp\left[-\frac{1}{2} \frac{x_k^2}{t + \sigma_k^2} \right],
\tag{4.25}
$$

$$
\lambda \equiv -\frac{1}{2} \frac{N r_p}{\gamma} (1 + 1/\beta^2) \approx -1.5 \times 10^{-11}\,\text{m} \times \frac{N[10^{10}]}{E[\text{TeV}]} .
$$

Components of the doublet, $\underline{\sigma}$, are the horizontal and vertical standard deviations of the bunch density. These can, *and do*, depend on location, θ_{bb}. N is the number of particles in the bunch, and E is the (lab-frame) energy of a particle in it. The negative sign is chosen to correspond to collisions of a test particle with a bunch of unlike sign, such as p-$\bar{p}$; for p-p it would be positive.[5]

The first order expressions for the tunes are obtained by averaging $\partial H/\partial \underline{I}$ over all angles. The tune *shifts* are given by the following:

$$\Delta\underline{\nu} = \sum_{\theta_{bb}} \left\langle \frac{\partial U}{\partial \underline{I}} \delta_{2\pi}(\theta - \theta_{bb}) \right\rangle_{\theta,\underline{\varphi}} = \sum_{\theta_{bb}} \frac{1}{2\pi} \left\langle \left. \frac{\partial U}{\partial \underline{I}} \right|_{\theta_{bb}} \right\rangle_{\underline{\varphi}} .$$

This is a bit formal, of course: not only does the Hamiltonian contain a delta function, but the integral expression is logarithmically divergent. (The potential energy of a line charge is infinite.) However, this does not present a problem, as only its derivatives are physically significant, and they are finite.

Tune shifts are obtained by differentiating with respect to the "action" coordinates, which means we must substitute for transverse coordinates from Eq. (3.100) into Eq. (4.25). When this is done, the exponential factors take the form

$$\begin{aligned} \exp\left[-\frac{1}{2}\frac{x_k^2}{t + \sigma_k^2} \right] &= \exp\left[-\frac{2\beta_k I_k \sin^2(\psi_k - \nu_k\theta)}{2(t + \sigma_k^2)} \right] \\ &= \exp[\zeta_k(\cos 2\phi_k - 1)], \end{aligned}$$

where $\zeta_k = \beta_k I_k / 2(t + \sigma_k^2)$ and $\phi_k = \psi_k - \nu_k\theta$.

The exponent is now expanded in a Fourier-Bessel series [9]:

$$e^{\zeta(\cos 2\phi - 1)} = Z_0(\zeta) + 2\sum_{n=1}^{\infty} Z_n(\zeta)\cos(n \cdot 2\phi).$$

Here, $Z_n(\zeta) = e^{-\zeta} I_n(\zeta)$, where I_n is the modified Bessel function, $I_n(\zeta) = (-i)^n J_n(i\zeta)$. The integrand contains a product of two such series with differing arguments. Obviously, only the $n = 0$ term will survive angle-averaging, which gives us the result:

$$\Delta\underline{\nu} = \frac{\lambda}{2\pi} \sum_{\theta_{bb}} \int_0^\infty dt \frac{\partial}{\partial \underline{I}} \prod_{k=1}^{2} \frac{1}{\sqrt{t + \sigma_k^2}} Z_0(\zeta_k). \tag{4.26}$$

For purposes of numerical evaluation, it is best to change the integration variable to make the limits of integration finite. A choice that (a) takes the asymptotic behavior of the integrand into account and (b) is symmetric under interchange of

[5]My apologies for not discussing the physics of this expression further. If there is a second edition I will certainly do so. A good derivation of Eq. (4.25) was written by Miguel Furman in "Practical Approximations for the Electric Field of Elliptical Beams," SSC-N312 (April, 1987) in an appendix on Green's functions.

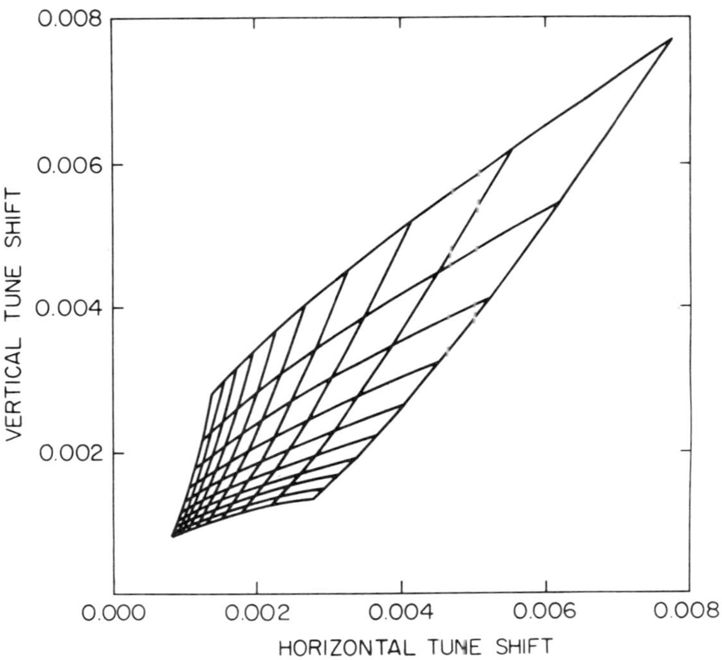

FIGURE 4.5 Tune spread induced by the beam-beam interaction.

indices is $w \equiv 1/(1 + t/\sigma_1\sigma_2)$. With this it is convenient to introduce auxiliary functions, $\xi_1 \equiv 1 - w(1 - \sigma_1/\sigma_2)$, and similarly for ξ_2. It is easy to confirm that $w(t + \sigma_k^2) = \sigma_1\sigma_2\xi_k$ and $\zeta_k = (\beta_k I_k w/2\sigma_1\sigma_2\xi_k)$. The upshot is that Eq. (4.26) is rewritten as follows:

$$\Delta\underline{\nu} = \frac{\lambda}{2\pi}\sum_{\theta_{bb}}\int_0^1 \frac{dw}{w\sqrt{\xi_1\xi_2}}\frac{\partial}{\partial\underline{I}}\left(Z_0(\zeta_1)Z_0(\zeta_2)\right).$$

This equation provides a mapping from the physical $\underline{I}$ quadrant into $\Delta\underline{\nu}$ space, an example of which is shown in Figure 4.5. The point at the upper right is the image of the origin, $\underline{I} = \underline{0}$, and $\Delta\nu \to \underline{0}$ in the limit $\underline{I} \to \infty$.

In particular, evaluating the horizontal tune shift from a single beam-beam interaction gives us

$$\begin{aligned}
\Delta\nu_1 &= \frac{\lambda}{2\pi}\int_0^1 \frac{dw}{w\sqrt{\xi_1\xi_2}}\frac{\partial Z_0(\zeta_1)}{\partial I_1}Z_0(\zeta_2) \\
&= \frac{\lambda}{2\pi}\int_0^1 \frac{dw}{w\sqrt{\xi_1\xi_2}}\frac{\zeta_1}{I_1}[Z_1(\zeta_1) - Z_0(\zeta_1)]Z_0(\zeta_2) \\
&= \frac{\lambda}{4\pi}\frac{\beta_1}{\sigma_1\sigma_2}\int_0^1 \frac{dw}{\sqrt{\xi_1^3\xi_2}}Z_0(\zeta_2)[Z_1(\zeta_1) - Z_0(\zeta_1)].
\end{aligned}$$

The small amplitude limit is obtained by setting $\underline{I} = \underline{0}$.

$$\Delta\nu_1 = -\frac{\lambda}{4\pi}\frac{\beta_1}{\sigma_1\sigma_2}\int_0^1 dw/\sqrt{\xi_1^3\xi_2}.$$

Interestingly enough, this integral can be evaluated in closed form, which results in the small-amplitude beam-beam tune shift.

$$\begin{aligned}\Delta\nu_1 &= -\frac{\lambda}{2\pi}\frac{\beta_1}{\sigma_1(\sigma_1+\sigma_2)}\\ &\approx \frac{1}{2\pi}\left(\frac{Nr_p}{\gamma}\right)\frac{\beta_1}{\sigma_1(\sigma_1+\sigma_2)}.\end{aligned}$$

4.3.2 Two-dimensional Resonances

To first order, the normal form of an isolated resonance model, like that of Eq. (4.6), is obtained by filtering the appropriate harmonic from Eq. (4.20). We shall find the separatrices of the first order, "principal resonances" associated with magnetic multipoles: these are the resonances of highest degree that can be excited to first order by the multipole. Alternatively, a resonance is principal for a multipole if that multipole is the smallest that will excite the resonance, to first order in multipole strength. In this section we shall confine our attention to the two-dimensional, "horizontal" phase space, in which resonant orbits are necessarily periodic. These will be horizontal resonances excited by normal multipoles, and it easy to show that under such conditions $I_2 = 0$ implies $\dot{I}_2 = 0$. In the next section we shall extend the examples to four-dimensional transverse phase space.

Third-Integer Resonance The archetypal resonance model of accelerator physics is the first order exitation of a third-integer — that is, $3\nu_1 + n = 0$ — resonance by sextupoles. The harmonic polynomial representing a sextupolar distribution was written in Eqs. (3.109) and (3.110).

$$A_3/(-B_o)\big|_{\text{normal sextupole}} = \tfrac{1}{3}b_2(\theta)(x_1^3 - 3x_1x_2^2). \tag{4.27}$$

We have simply substituted $\underline{x}$ for $\underline{\zeta}$. From the transformation Eqs. (3.100) we see that $x_k \simeq e^{\pm i\varphi_k}$, so that the first term, x_1^3, will contain harmonic components $e^{\pm i3\varphi_1}$ and $e^{\pm i\varphi_1}$, while the second, $x_1x_2^2$, will contain $e^{\pm i\varphi_1}$ and $e^{\pm i(\varphi_1 \pm 2\varphi_2)}$. The first order resonances excited by normal sextupoles are therefore ν_1, $3\nu_1$, and $\nu_1 \pm 2\nu_2$. We shall ignore the first, study the second here, and save the third for Section 4.3.3. This discussion may be easier to follow if one glances occasionally at Figure 4.6.

Applying Eqs. (4.20) and (4.6), the first order, isolated resonance model for this excitation is

$$
\begin{aligned}
H &= \underline{\nu} \cdot \underline{I} + \left[\left(\oint \frac{d\theta}{2\pi} e^{-in\theta} H_{30}(\underline{I};\theta) \right) e^{i(3\varphi_1 + n\theta)} + \text{c.c.} \right] \\
&\equiv \underline{\nu} \cdot \underline{I} + I_1^{3/2} \left(G e^{i(3\varphi_1 + n\theta)} + G^* e^{-i(3\varphi_1 + n\theta)} \right) \\
&= \underline{\nu} \cdot \underline{I} + I_1^{3/2} \left(2|G| \cos(3\varphi_1 + n\theta + \phi_G) \right), \quad \text{where} \\
G = |G| e^{i\phi_G} &= -\frac{i}{6\sqrt{2}} \oint \frac{d\theta}{2\pi} e^{-in\theta} \left(b_2(\theta) \beta_1^{3/2}(\theta) e^{i3(\psi_1(\theta) - \nu_1 \theta)} \right).
\end{aligned}
$$

As we are assuming a static ring, the phase, ϕ_G, is not essential and can be absorbed into the definition of θ: we merely define $\phi_G \equiv n\theta_0$ and shift the zero of the θ by θ_0. Letting $g \equiv 2|G|$, we thereby obtain the isolated resonance in the form

$$
H = \underline{\nu} \cdot \underline{I} + g I_1^{3/2} \cos(3\varphi_1 + n\theta).
$$

There are no shear terms at this order so, by the arguments of the preceding section, we expect unbounded orbits and a finite "dynamic aperture." Applying the transformations outlined there, we change this into the (projected) resonance normal form Hamiltonian,

$$
K = J_1 \Delta + g J_1^{3/2} \cos \xi_1, \tag{4.28}
$$

where $\Delta = 3\nu + n$ is the tune defect from resonance, and we have dropped the J_2 term, because $J_2 = I_2 = 0$ is a dynamical invariant inessential for our purposes. To ease the notation a little, we shall drop the subscript "1" in what follows. It should be kept in mind that the range of ξ is 6π, not 2π, in accordance with the transformation given in Eq. (4.9), and that the physical phase space comprises three identical copies of the "first Brillouin zone."

Superficially, this looks like a problem with two control parameters, but only the ratio Δ/g is important for studying the orbit space: for example, one or the other can be absorbed in setting a scale for K. Hamilton's equations of motion are written

$$
\begin{aligned}
\frac{dJ}{d\theta} &= g J^{3/2} \sin \xi, \\
\frac{d\xi}{d\theta} &= \Delta + \frac{3}{2} g J^{1/2} \cos \xi,
\end{aligned} \tag{4.29}
$$

from which we obtain the fixed points (regular resonant orbits),

$$
\begin{aligned}
\xi_o &= k\pi, \\
J_o^{1/2} &= (-1)^{k+1} \frac{2}{3} \frac{\Delta}{g}.
\end{aligned} \tag{4.30}
$$

It is easy to check, using the prescription below Fig. 4.4, that these resonant orbits are hyperbolic. Physical solutions require that $J_o > 0$, and we implicitly take $J_o^{1/2} > 0$ as well; accordingly, k must be odd when $\Delta/g > 0$ and even when $\Delta/g < 0$. In either case, there are three inequivalent solutions within the 6π domain of ξ. Physically, this business about signs means that the orientation of the separatrix undergoes a mirror inversion as one passes through resonance, that is, through $\Delta = 0$.

The value of the Hamiltonian at these points is obtained immediately upon multiplying Eq. (4.29) by J and using Eq. (4.28).

$$\Delta J_o + \tfrac{3}{2} g J_o^{3/2} \cos \xi_o = 0 = \tfrac{3}{2} K_o - \tfrac{1}{2} \Delta J_o$$
$$K_o = \tfrac{1}{3} \Delta J_o.$$

The separatrix comprises the level set obtained by setting $K = K_o$.

$$\Delta J + g J^{3/2} \cos \xi = \tfrac{1}{3} \Delta J_o. \tag{4.31}$$

It turns out for this particular problem that this set is most easily visualized in Cartesian coordinates. We define a canonical transformation to coordinates (X, P), exactly as in Eq. (3.52) or Eq. (2.9).

$$\begin{aligned} P &\equiv \sqrt{2J} \cos(\xi/3), \\ X &\equiv \sqrt{2J} \sin(\xi/3). \end{aligned} \tag{4.32}$$

We then can write, with the help of trigonometric identities,

$$\begin{aligned} J &= \tfrac{1}{2}(X^2 + P^2) \\ J^{3/2} \cos \xi &= J^{3/2}(\cos^3(\xi/3) - 3 \cos(\xi/3) \sin^2(\xi/3)) \\ &= \left(\tfrac{1}{2}\right)^{3/2} \left(P^3 - 3PX^2\right). \end{aligned}$$

Equation (4.31), which determines the separatrix, is rewritten as follows:

$$\frac{3}{2} J_o^{1/2}(P^2 + X^2) + (-1)^{k+1} \frac{1}{\sqrt{2}}(P^3 - 3PX^2) - J_o^{3/2} = 0. \tag{4.33}$$

We have used Eq. (4.30) to eliminate Δ/g in favor of J_o. Remarkably enough, this can be factored:

$$\begin{aligned} 0 = &\left(P - (-1)^{k+1} \tfrac{1}{2} \sqrt{2J_o}\right) \\ &\times \left(X + \sqrt{1/3}P + (-1)^{k+1} \sqrt{2J_o/3}\right) \\ &\times \left(X - \sqrt{1/3}P - (-1)^{k+1} \sqrt{2J_o/3}\right). \end{aligned}$$

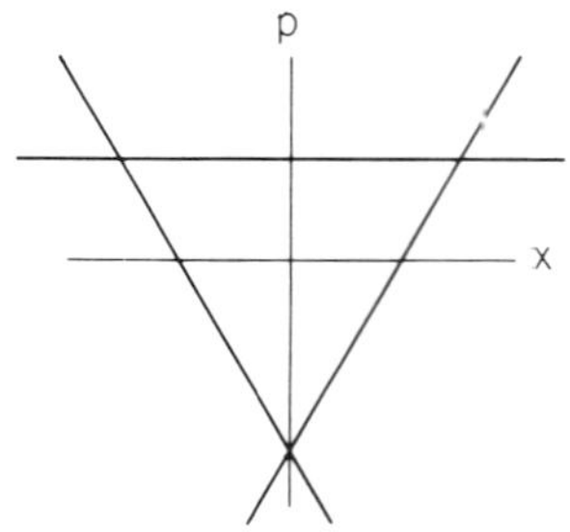

FIGURE 4.6 Sextupole induced third-integer resonance model.

This algebraically expresses the union of three straight (in these coordinates) lines; Figure 4.6 illustrates the situation for odd k. Its partner, Figure 4.7, shows the "trajectories" of the fixed points themselves as a function of the control parameter Δ/g. These also are straight lines.

Remember in interpreting Figure 4.6 physically that the three "fixed points" are actually a single period-three orbit: three traversals of the ring are necessary to return to the same point, and one jumps from one "fixed point" to another under the action of a single Poincaré map, as was discussed in connection with Eq. (4.17). The influence of the third-integer resonance is evident in the Hénon orbit of Figure 5.3(d), which indeed corresponds to a single sextupole kick map. There is, nonetheless, a great apparent difference between this plot and Figure 4.6, which is the isolated resonance model. Getting around that difference is the subject of the next chapter.

Fourth-Integer Resonance Just as the principal resonance excited by sextupoles is the third-integer resonance, octupoles principally excite the fourth-integer resonance. However, unlike the sextupole excitation considered above, octupoles do possess first order shear terms that can detune the resonance sufficiently to close its separatrix and bound orbits. The reader is invited to expend an occasional glance on Figures 4.8 and 4.9 during the following discussion.

Appealing once more to Eq. (3.110) we write the octupolar polynomial,

$$A_3/(-B_o)\big|_{\text{octupole}} = \tfrac{1}{4}b_3(\theta)(x_1^4 - 6x_1^2x_2^2 + x_2^4).$$

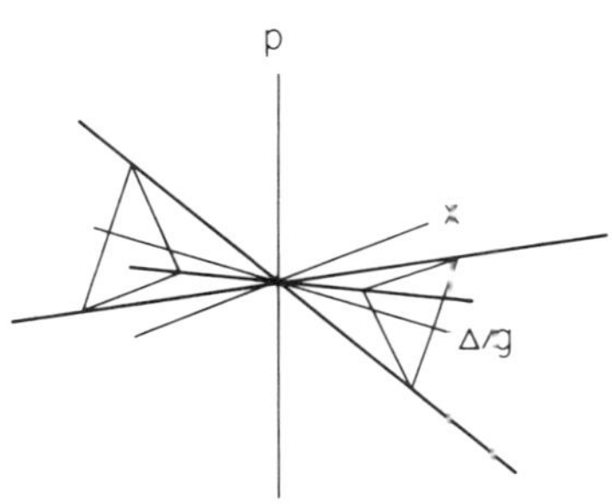

FIGURE 4.7 Trajectories of the resonant orbits as a function of the control parameter, Δ/g.

FIGURE 4.8 Octupole induced fourth-integer resonance model; other topologies are possible.

Using the same argument as before, the first order resonances excited by this polynomial are: $4\nu_1$, by the term x_1^4; $4\nu_2$, by the term x_2^4; $2\nu_1$, with contributions from x_1^4 and $x_1^2x_2^2$; $2\nu_2$, with contributions from x_2^4 and $x_1^2x_2^2$; and $2\nu_1 \pm 2\nu_2$, from the term $x_1^2x_2^2$. Averaging also produces first order shear terms, as in Eq. (4.22). With appropriate choice of the point $\theta = 0$, the isolated resonance model associated with the fourth-integer resonance, $4\nu_1 + n = 0$, then assumes a normal form,

$$K = J_1\Delta + J_2\delta + \kappa J_1^2 + \lambda J_1 J_2 + \mu J_2^2 + gJ_1^2 \cos\xi_1, \tag{4.34}$$

where $\Delta = 4\nu + n$ is the resonance defect. It is left to the reader to write the remaining parameters as functionals of the octupole distribution. The angle ξ_1 is related to original φ_1 according to $\xi_1 = 4\varphi_1 + n\theta$, and so the dynamic range of ξ_1 is 8π.

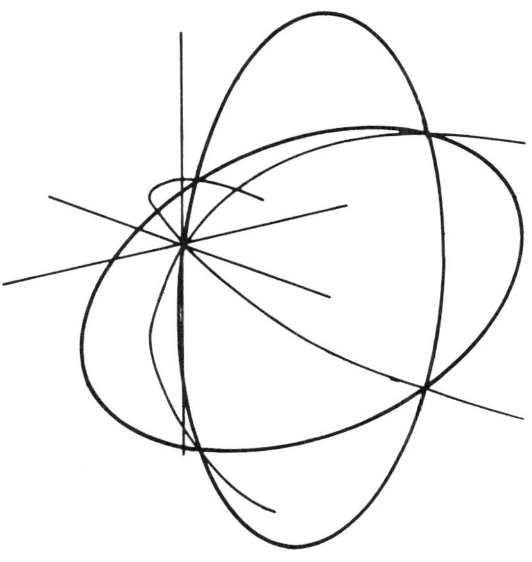

FIGURE 4.9 Scaled trajectories of the resonant orbits as a function of the control parameter, κ/g.

We again concentrate on the horizontal degrees of freedom by setting the invariant $J_2 = I_2 = 0$ and hereafter drop the subscript "1."

$$K = J\Delta + \kappa J^2 + gJ^2 \cos \xi.$$

Again we solve for the regular resonant orbits:

$$\sin \xi = 0 \quad \text{and}$$
$$\Delta + 2J(\kappa + g \cos \xi) = 0$$

As before, we have $\xi_o = n\pi, n \simeq 0, \ldots, 7$, which makes $\cos \xi_o = (-1)^n$, and

$$J_o = -\frac{\Delta}{2(\kappa + (-1)^n g)}.$$

This time, half the resonant orbits are elliptic, the other half, hyperbolic. Only positive solutions for J_o are physically realizable. There are two cases to consider:

Case I: If $|\kappa| > |g|$, then we must have $\text{sgn}\,\Delta = -\text{sgn}\,\kappa$. That is, the resonant orbit exists for ν either above or below resonance, *but not both.* As ν passes through the resonant value, the origin "swallows" the separatrix. Notice the similarity with the pitchfork bifurcation of Figure 3.8; except for the fact that the central resonant orbit is not hyperbolic, the bifurcation upon passing through the fourth-integer resonance is much like a multipronged pitchfork. With very good reasons, this is also like the behavior of the Hénon map in passing through the fourth-integer, as exhibited in Figures 5.3(e) and (g).

Case II: On the other hand, if $|\kappa| < |g|$, then we can pull the same trick as in the third-integer case and compensate the sign change in Δ by a parity change in n. Note, however, that if this is the case, only half the resonant orbits exist at a time and the flow is no longer bounded: the separatrix goes to infinity at the angles satisfying

$$\cos \xi^\infty = -\kappa/g.$$

Thus the origin can "swallow" the resonance without regurgitating it only if the shear term has magnitude large enough to close the separatrix — which is intuitively very reasonable: you cannot swallow something that has strings extending to infinity. Figures 4.8 and 4.9 are the analogs of Figures 4.6 and 4.7.

Half-Integer Resonance One way of extracting particles slowly from an accelerator is to excite a half-integer resonance with quadrupole and octupole magnets. A separatrix is created, which, although open at large amplitudes, nonetheless contains a central region of stable orbits. By manipulating the quadrupoles, the tune slowly approaches the resonant value, and this region is squeezed to a point while particles escape along the separatrix.

The relevant quadrupole and octupole polynomials are $(b_1(\theta)/2)(x_1^2 - x_2^2)$ and $(b_3(\theta)/4)(x_1^4 - 6x_1^2x_2^2 + x_2^4)$. At the outset we shall set $I_2 = 0$ to limit our treatment to the horizontal plane. We then need only consider the polynomial,

$$A_3/(-B_o)\big|_{\text{quadrupole+octupole}} = \tfrac{1}{2}b_1(\theta)x_1^2 + \tfrac{1}{4}b_3(\theta)x_1^4.$$

Their contributions to the shear terms have been written already in Eq. (4.22). Both multipoles excite the half-integer resonance: the quadrupole with a term proportional to I_1, the octupole with a term proportional to I_1^2. As we saw already, the octupole additionally excites a fourth-integer resonance that cannot *a priori* be ignored: if $2\nu_1 + n \approx 0$, there is some possibility that $4\nu_1 + 2n = 2(2\nu_1 + n) \approx 0$. The half-integer resonance terms are

$$\begin{aligned}
&(G_2^{(4)}e^{i(2\varphi_1+n\theta)} + \text{c.c.})(I_1/2) + (G_2^{(8)}e^{i(2\varphi_1+n\theta)} + \text{c.c.})(I_1/2)^2 \\
&\quad = g_2^{(4)}\cos(2\varphi_1 + n\theta + \phi_4)(I_1/2) + g_2^{(8)}\cos(2\varphi_1 + n\theta + \phi_8)(I_1/2)^2, \quad \text{where} \\
&G_2^{(4)} \equiv \tfrac{1}{2}g_2^{(4)}e^{i\phi_4} = -\frac{B_oR}{|B\rho|}\oint\frac{d\theta}{2\pi}e^{-in\theta}\left(\tfrac{1}{2}b_1(\theta)\beta_1(\theta)e^{i2(\psi_1(\theta)-\nu_1\theta)}\right), \quad \text{and} \\
&G_2^{(8)} \equiv \tfrac{1}{2}g_2^{(8)}e^{i\phi_8} = -\frac{B_oR}{|B\rho|}\oint\frac{d\theta}{2\pi}e^{-in\theta}\left(b_3(\theta)\beta_1^2(\theta)e^{i2(\psi_1(\theta)-\nu_1\theta)}\right).
\end{aligned}$$

The corresponding fourth-integer resonance excited by the octupoles is

$$\begin{aligned}
&(G_4^{(8)}e^{i(4\varphi_1+2n\theta)} + \text{c.c.})(I_1/2)^2, \quad \text{where} \\
&G_4^{(8)} = -\frac{B_oR}{|B\rho|}\oint\frac{d\theta}{2\pi}e^{-i2n\theta}\left(\frac{1}{4}b_3(\theta)\beta_1^2(\theta)e^{i4(\psi_1(\theta)-\nu_1\theta)}\right).
\end{aligned} \tag{4.35}$$

In fact, this term is undesirable: (a) practically, in its interference with controlling extraction, and (b) from our immediate point of view, in that it increases the number of control parameters that must be considered. We can expect that a design for half-integer extraction will power octupoles so as to zero $G_4^{(8)}$ while keeping $G_2^{(8)}$ nonzero. At any rate, we shall assume this to be the case.

Under this condition, our isolated resonance Hamiltonian, in one degree of freedom, assumes the normal form,

$$K_o = [\Delta + g_2^{(4)}\cos(\xi_1 + \phi)]J_1 + [g_0^{(8)} + g_2^{(8)}\cos\xi_1]J_1^2,$$

where $I_1 = 2J_1$, $\xi_1 = 2\varphi_1 + n\theta + \phi_8$, and $\phi = \phi_4 - \phi_8$. We also have absorbed the quadrupole-induced tune shift into the definition of Δ. Two of the remaining five control parameters can be eliminated by two rescalings. First, we scale the action coordinate and define $J \equiv (g_2^{(8)}/g_2^{(4)})J_1$. This change is absorbed by multiplying the Hamiltonian by $g_2^{(8)}/g_2^{(4)}$, in accordance with the discussion preceding Eq. (3.37). Second, we rescale the orbit parameter according to $d\theta' = (g_2^{(4)2}/g_2^{(8)})\,d\theta$, which is absorbed by multiplying the Hamiltonian by an additional factor of $g_2^{(8)}/g_2^{(4)2}$. The combined result is a Hamiltonian,

$$K = \frac{g_2^{(8)\,2}}{g_2^{(4)\,3}} K_o = [\delta + \cos(\xi + \phi)]J - [\kappa + \cos\xi]J^2, \tag{4.36}$$

where $\delta = \Delta/g_2^{(4)}$ and $\kappa = g_0^{(8)}/g_2^{(8)}$. In words, δ is the difference between the tune and the half-integer $n/2$ scaled by the nth harmonic quadrupole driving term, κ is the ratio of 0th to n^{th} harmonic octupole driving terms, and ϕ is the relative phase between the octupole and quadrupole harmonics.

Unlike the trivial examples considered so far, this dynamic possesses a variety of inequivalent flows, which we will classify by mapping its transition boundary, the surface of points in the three-dimensional (δ, κ, ϕ) control space on which the system undergoes bifurcations. These are of two kinds:

(a) Local bifurcations, or catastrophes, are characterized by the coalescing of hyperbolic and elliptic resonant orbits. When the control parameters assume critical values, a cusp is formed in phase space, reflecting the fact that the local, linearized dynamic vanishes. The bifurcation is "local" in that the existence of a cusp can be noted using local observations. The set of control parameters at which catastrophes occur is called the "catastrophe surface," following Thom [89].

(b) Global bifurcations, whose existence cannot be discerned by any local observation. Rather than resonant orbits coalescing, entire branches of the separatrix merge and emerge in different ways, as was illustrated in Fig. (3.2): the separatrix as a whole undergoes a transition; locally, nothing interesting happens. Following Gilmore [40], we shall call the set of control parameters at which these global bifurcations occur the "Maxwell surface."

The transition boundary partitions control space, and we shall sketch a separatrix associated with each of its regions. (The point being that within each region all the flows are topologically equivalent.) This will be done using the $(x, p) = (\sqrt{2J}\sin(\xi/2), \sqrt{2J}\cos(\xi/2))$ coordinates, exactly as in Eq. (4.32). We are aided by two conspicuous symmetries.

(a) $\phi \to -\phi,\ \xi \to -\xi$: The flow at $-\phi$ is obtained from that at ϕ through a reflection across the p axis.

(b) $\xi \to \xi + \pi,\ \kappa \to -\kappa,\ \delta \to -\delta,\ H \to -H$: The flow at $(-\delta, -\kappa)$ is obtained from that at (δ, κ) by a 90° phase space rotation and "time"-reversal. Because of the last two we need explore only half of the control space explicitly.

As in our previous examples, the first step in sketching the separatrices is to find hyperbolic resonant orbits, which are critical points of the projected Hamiltonian, Eq. (4.36). The fixed point equations, obtained by applying Hamilton's equations and setting derivatives to zero, can be written as follows:

$$\begin{aligned} J^{(0)} &= -\sin(\xi^{(0)} + \phi)/\sin\xi^{(0)}, \\ \sin\xi^{(0)}[\delta + \cos(\xi^{(0)} + \phi)] &= 2\sin(\xi^{(0)} + \phi)[\kappa + \cos\xi^{(0)}]. \end{aligned} \tag{4.37}$$

We are especially interested in finding those solutions that are degenerate, or multiple, because they are associated with catastrophes, that is, resonant orbits that fall on top of each other. The procedure for finding these solutions is somewhat involved and yields no physical insights. For these reasons, its explanation is relegated to Appendix A. We shall satisfy ourselves here with displaying its results.

The topography of the transition boundary can be envisaged most easily by sectioning it at constant values of ϕ. Eight such sections are shown in Figure 4.10; the solid lines indicate the catastrophe surface, and the dashed lines, the Maxwell surface. The latter lies in the elliptically cross-sectioned tube

$$\delta^2 - 2\kappa\delta\cos\phi + \kappa^2 = \sin^2\phi. \tag{4.38}$$

The former is a more complicated surface consisting of two major pieces. The first piece is independent of ϕ and contains the planes $\delta = \pm 1$ and $\kappa = \pm 1$. These determine the flow characteristics at the origin and at infinity, as we shall see shortly. The second grows out of the corners $\delta = \kappa = \pm 1$ as ϕ increases. It remains confined to the unit square when $\phi \in [0, \pi/2]$ but leaks into the regions $|\kappa| > 1$, forming a cusp at its extremes. It begins to intersect itself at $\phi = 5\pi/6$ and eventually attaches itself to the two straight lines $2\kappa + \delta = \pm 1$ when $\phi = \pi$.

Figures 4.11 and 4.12 show how the dynamical possibilities are organized on the sections at $\phi = 0$ and $\phi = \pi$. The former encloses nine regions possessing four different classes of behavior; the latter, eighteen regions, with seven classes. The origin is stable or unstable accordingly as $|\delta| > 1$ or $|\delta| < 1$ (this expresses the half-integer stop band in our units); separatrices are closed or open in phase space accordingly as $|\kappa| > 1$ or $|\kappa| < 1$. These transitions occur for all values of ϕ, but at $\phi = 0$, they are the only ones possible. On the Maxwell boundary at $\phi = \pi$ the separatrix consists of two straight lines intersecting at and a circle centered at the origin. No separatrix is associated with the regions $\delta, \kappa > 1$ and $\delta, \kappa < -1$ because there are no unstable fixed points there.

In the intervening region, for $0 < \phi < \pi$, the dynamical effects of crossing transition boundaries are appreciated best by following paths in the control space. Figure 4.13, for example, illustrates what happens along part of the path $(\delta, \kappa, \phi) = (-0.6, 0.4, 0 \ldots \pi)$.

In going from $\phi = 0$ to $\phi = 150°$, the system does not cross a transition boundary, so the separatrix in Figure 4.13(a) is still equivalent to two intersecting lines. Passage through the catastrophe surface in Figures 4.13(b) and 4.13(d) is signaled in phase space by the creation of a symmetric pair[6] of degenerate, critical points, each

[6]Please recognize that this "pair" of "fixed points" correspond to *one* orbit, as discussed with Eq. (4.17).

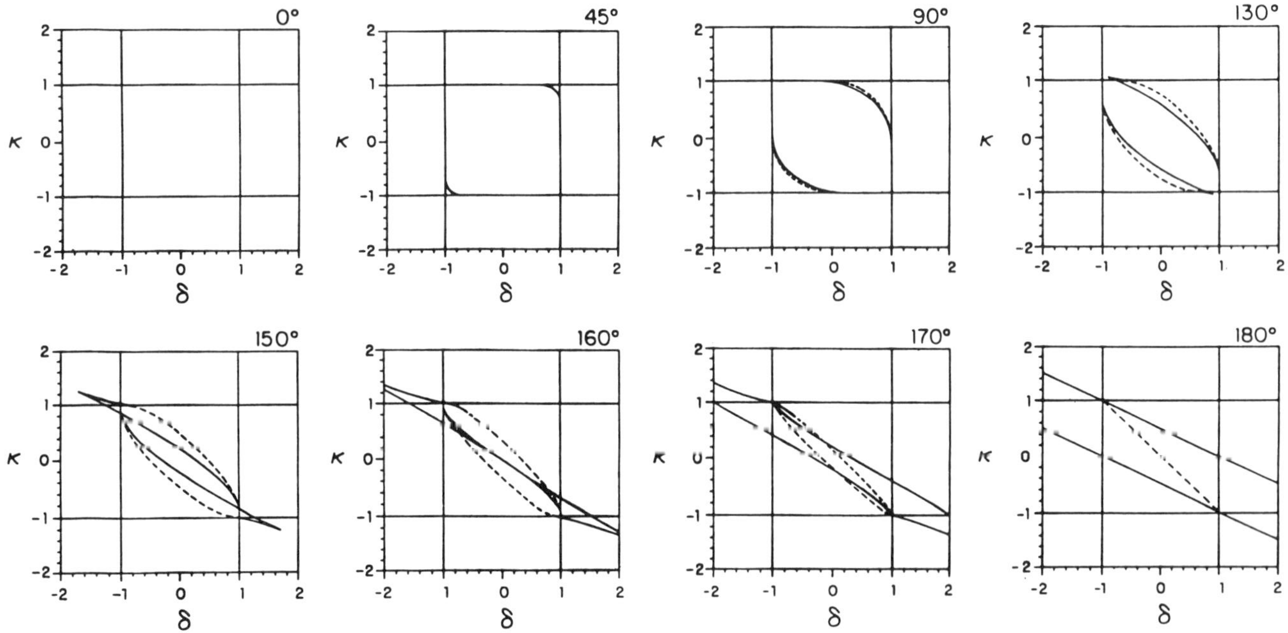

FIGURE 4.10 Sections of the catastrophe and Maxwell surfaces at constant values of ϕ.

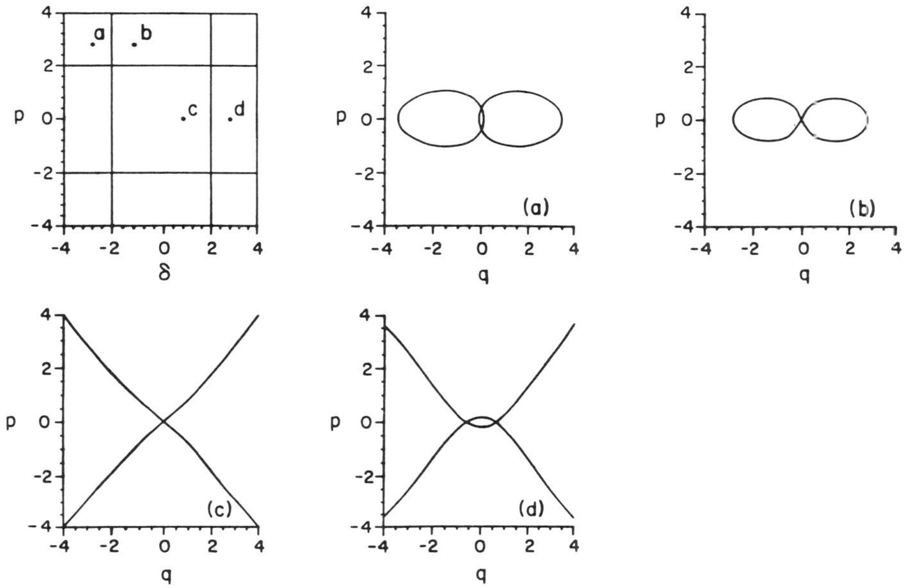

FIGURE 4.11 Generic separatrices at the control plane $\phi = 0$.

appearing as a cusp in some trajectory. Those trajectories become new branches of the separatrix as each degeneracy splits into a stable-unstable pair of normal fixed points.

A global saddle-switch bifurcation, like that of Figure 3.2, occurs when the system crosses the Maxwell surface in Figure 4.13(f). The Hamiltonian at the unstable critical orbits created in Figure 4.13(b) takes on the value zero, and those branches merge with the one passing through the origin. They fall apart again in Figure 4.13(g), but a transition has occurred: the flows in Figures 4.13(e) and 4.13(g) are not equivalent. In particular, the latter contains orbits that wind around the origin — even though the origin itself is unstable — while the former does not. However, this transition has taken place without passing through a catastrophe. The critical points are always nonsingular: they are always the same in number and of the same character. This kind of transition has a nonlocal character. That is, it cannot be characterized by what happens at one point of phase space. A minimal characterization requires equating the value of the Hamiltonian at two separated, unstable critical points, which leads to Eq. (4.38).

As the system control point approaches $\phi = \pi$, the stable areas created in Figure 4.13(d) continue to expand until another merger takes place and we end up with the separatrix of Figure 4.12(h).

Another example, this time along a path at constant ϕ, is presented in Figure 4.14. At the endpoints the flows are equivalent to that of Figure 4.11(a), because all three of these points can be connected without crossing the transition surface. Crossing the boundary in Figure 4.14(b) creates degenerate fixed points and new separatrix branches that encircle stable critical orbits, as shown. As the system moves from Figure 4.14(b) to 4.14(f) each unstable critical orbit separates itself from its stable

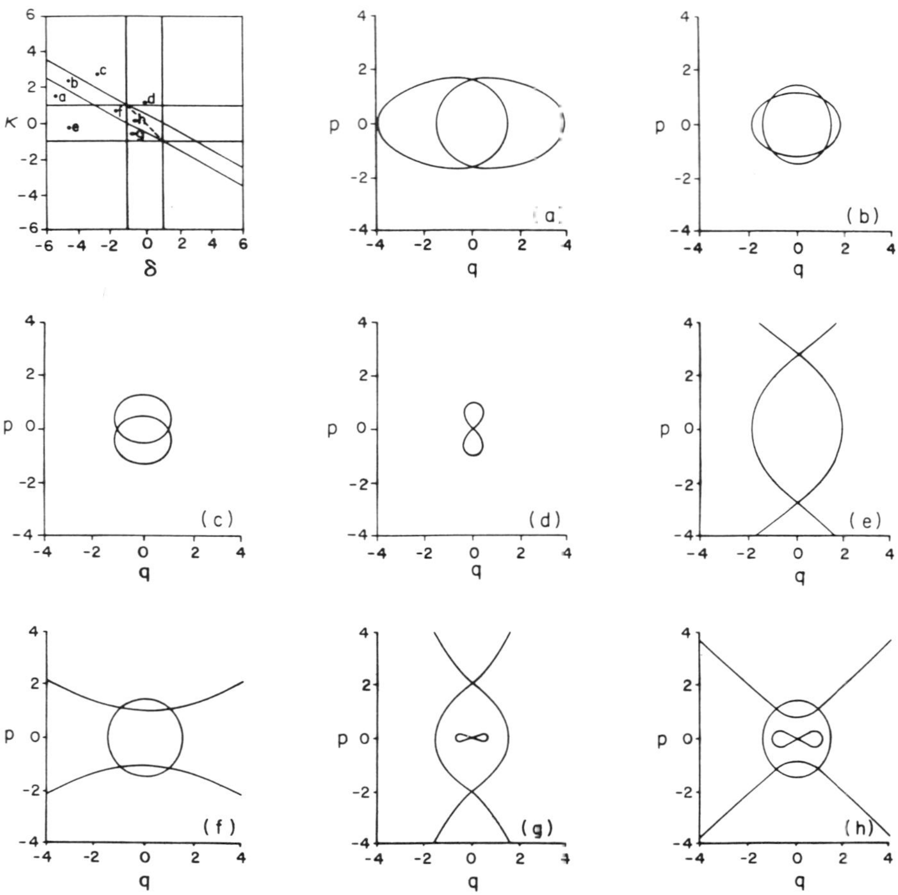

FIGURE 4.12 Generic separatrices at the control plane $\phi = \pi$.

partner and travels across to attach itself to the other one, which of course annihilates it, the cusp being formed upon crossing the Maxwell surface.

4.3.3 Four-dimensional Resonances

Although useful, the intuition developed from resonance models in two-dimensional phase spaces can be extremely misleading: resonant orbits in higher dimensions are *not* necessarily periodic and their separatrices possess much more complicated structure. To study something of this complexity, we shall sketch the separatrices of a $\nu_1 + \nu_2$ resonance, as principally excited by skew quadrupoles, and a $\nu_1 + 2\nu_2$ resonance, as principally excited by normal sextupoles.

Visualizing objects in four dimensions is a little involved, but not impossible. One method is to take a sequence of three-dimensional slices, much as one might present a cube to a two-dimensional creature by slicing it from bottom to top. Of

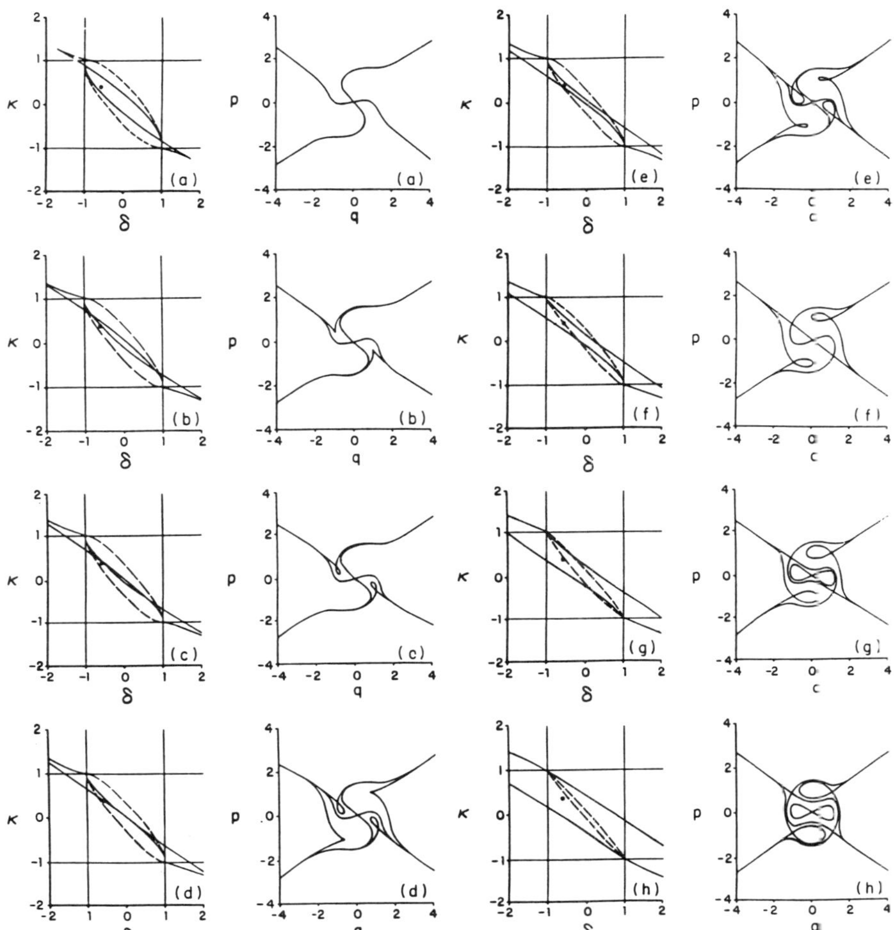

FIGURE 4.13 Following the dynamics along a control path at constant $(\delta, \kappa) = (-0.6, 0.4)$.

course, we must take some care in arranging the slices; our two-dimensional friend may form a distorted concept of a cube were it presented sliced along a diagonal. We shall obtain a good representation of the four-dimensional flow by drawing the separatrix within the three-dimensional dynamically invariant surfaces specified by the condition J_2 = constant. By patching these slices together we obtain the full separatrix, which is a three-dimensional object that legitimately partitions the four-dimensional phase space.

$\nu_1 + \nu_2$ *Resonance* From the monomial

$$A_3/(-B_o)\big|_{\text{skew quadrupole}} = -a_1(\theta)x_1x_2,$$

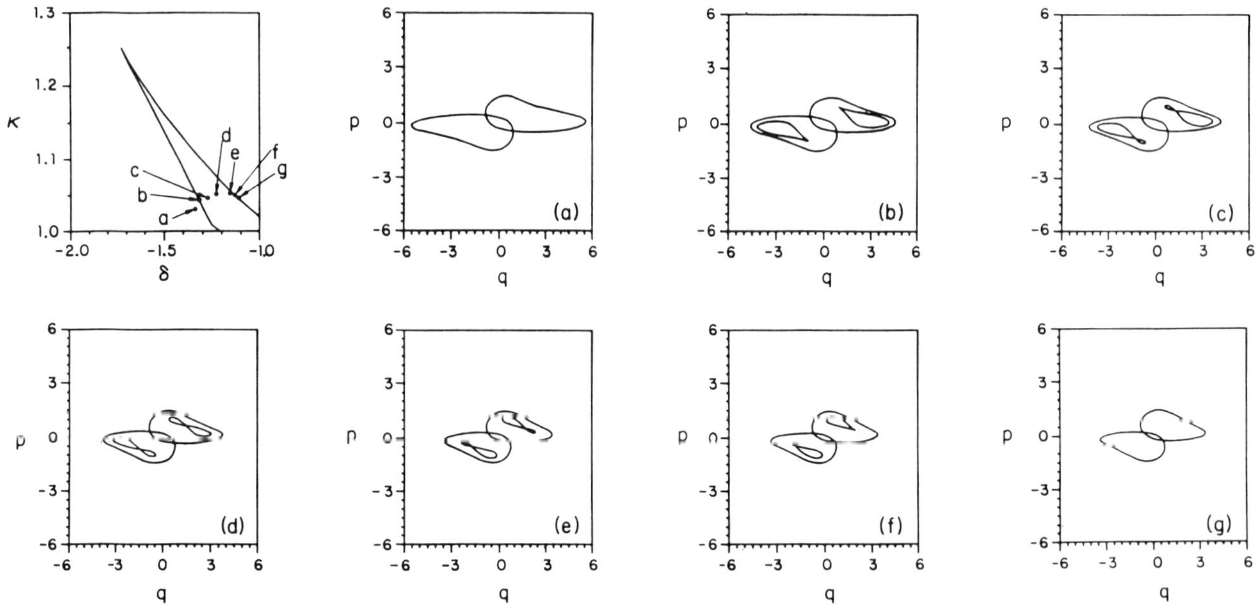

FIGURE 4.14 Following the dynamics along a control path at constant $\phi = 150°$. (Note the enlarged scale of the control space diagram.)

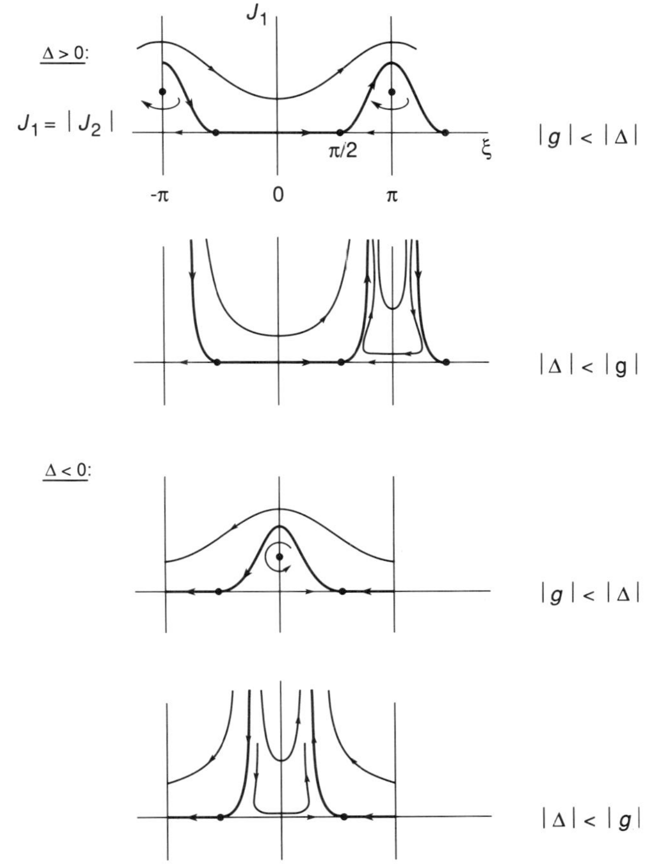

FIGURE 4.15 Flow diagrams for the projected Hamiltonian of the $\nu_1 + \nu_2$ resonance assuming $0 < g$. For $g < 0$, reverse the direction of the flow and interchange $0 < \Delta \leftrightarrow \Delta < 0$.

we see that skew quadrupoles principally excite the $\nu_1 \pm \nu_2$ resonances. We shall study the sum resonance here. Bypassing a discussion that should be familiar by now, the model associated with this excitation is

$$H = \underline{\nu} \cdot \underline{I} + g(I_1 I_2)^{1/2} \cos(\varphi_1 + \varphi_2 + n\theta + \phi), \quad \text{where}$$

$$g e^{i\phi} \equiv -\frac{1}{4\pi} \frac{B_o R}{|B\rho|} \oint d\theta e^{-in\theta} \left(a_1(\theta) \sqrt{\beta_1 \beta_2} e^{i(\tilde{\psi}_1 + \tilde{\psi}_2)} \right).$$

We can assume $g > 0$ without losing generality because of the symmetry: $g \to -g, \phi \to \phi + \pi$. The linearity of this dynamic is expressed in these coordinates by the scaling symmetry: $\underline{I} \to \lambda \underline{I}, H \to \lambda H$.

Referring to Figure 4.15 as needed will make it easier to follow the analysis. First we apply Eq. (4.9) and Eq. (4.10), with $\underline{m} = (1, 1)$, to rewrite this as an autonomous, projected Hamiltonian. One possible transformation is

$$\begin{pmatrix} I_1 \\ I_2 \end{pmatrix} = \begin{pmatrix} 1 & 1 \\ 1 & -1 \end{pmatrix} \cdot \begin{pmatrix} J_1 \\ J_2 \end{pmatrix},$$

$$\xi_1 = \varphi_1 + \varphi_2 + n\theta + \phi,$$

$$\xi_2 = \varphi_1 - \varphi_2.$$

The coordinate J_2 is free to assume any value, but J_1 is constrained by the requirement that both I_1 and I_2 be non-negative, from which we infer that the condition $J_1 \geqslant |J_2|$ describes the physical region. The new representation of Hamiltonian is $K = H + nJ_1$.

$$K = (\nu_1 - \nu_2)J_2 + (\nu_1 + \nu_2 + n)J_1 + g(J_1^2 - J_2^2)^{1/2} \cos \xi_1.$$

It is expected that $\Delta = \nu_1 + \nu_2 + n$ is a small quantity. Indeed, for this Hamiltonian to be at all interesting Δ must be small enough so that Eq. (4.16) is almost satisfied. Invariant manifolds must run parallel to ξ_2, since ξ_2 does not appear in K. (See Fig. 4.3.) As expected, the observable J_2 is dynamically invariant and

$$d\xi_2/d\theta = \partial K/\partial J_2 = \nu_1 - \nu_2 - g\frac{J_2}{(J_1^2 - J_2^2)^{1/2}} \cos \xi_1.$$

The more relevant vector field components are given by projection onto the "1" coordinates:

$$\frac{dJ_1}{d\theta} = -\frac{\partial K}{\partial \xi_1} = g(J_1^2 - J_2^2)^{1/2} \sin \xi_1,$$

$$\frac{d\xi_1}{d\theta} = \frac{\partial K}{\partial J_1} = \Delta + g\frac{J_1}{(J_1^2 - J_2^2)^{1/2}} \cos \xi_1,$$

where we have identified the resonance defect, $\Delta = \nu_1 + \nu_2 + n$.

We proceed to find the resonant orbits, corresponding to the critical points of this projected flow. Setting dJ_1 to zero provides two possibilities:

$$dJ_1 = 0 \Rightarrow J_1 = |J_2|, \quad \text{or}$$

$$\xi_1 \simeq 0 \text{ or } \pi.$$

The former condition identifies the "irregular" critical points; the latter, the "regular." Notice that $J_1 = |J_2|$ means that either I_1 or I_2 must vanish, so that the transverse motion is instantaneously planar. Now setting $d\xi_1$ to zero produces the final conditions.

$$d\xi_1 = 0 \Rightarrow (J_1^2 - J_2^2)^{1/2} = -\frac{g}{\Delta} J_1 \cos \xi_1.$$

Irregular critical points must therefore be pinned at angles, $\xi_1 \simeq \pm\pi/2$, while the regular resonant orbits occur at

$$J_1 = |J_2|/\sqrt{1-(g/\Delta)^2},$$

$$\xi_1 \simeq \begin{cases} 0, & -1 < g/\Delta < 0 \\ \pi, & 0 < g/\Delta < 1 \end{cases}.$$

Behavior of the orbits at large amplitudes is studied by employing Eq. (4.12):

$$K/J_1 = (\nu_1 - \nu_2)J_2/J_1 + \Delta + g(1 - J_2^2/J_1^2)^{1/2}\cos\xi_1.$$

As K is a constant of the motion, J_1 can become arbitrarily large along an orbit only by approaching the asymptotic angle, ξ_1^∞, where

$$\cos\xi_1^\infty = -\Delta/g.$$

As a consequence, all orbits are bounded when $1 < |\Delta/g|$. One speaks of the tune as being within the "stop band" of the resonance if this condition is violated. That the entire phase space is either stable or unstable — rather than having a central region of bounded orbits, as in the examples of the previous section — is merely a consequence of the underlying linearity: a central region of stable orbits would define a scale for phase space that linear systems cannot possess. Put another way, the existence of unbounded orbits in a linear system implies that any neighborhood of the origin must intersect unbounded orbits.

Testing the ellipticity of the "regular" resonant orbits is left as an exercise.

$\nu_1 + 2\nu_2$ Resonance Slicing phase space along J_2 isosurfaces was not useful in the previous example because of its linearity. The situation is very different for a $\nu_1 + 2\nu_2$ resonance, which is principally excited by normal sextupoles via the $x_1x_2^2$ term appearing in the polynomial written in Eq. (4.27). The isolated resonance model for this excitation is

$$H = \underline{\nu}\cdot\underline{I} + gI_1^{1/2}I_2\cos(\varphi_1 + 2\varphi_2 + n\theta + \phi),$$

where we leave it to the reader to write $ge^{i\phi}$ as a functional of the sextupole distribution. The transformation to the resonance projected coordinates, using Eqs. (4.9) and (4.10), yields

$$\begin{aligned} J_1 &= (I_1 + 2I_2)/5, \\ J_2 &= (2I_1 - I_2)/5, \\ \xi_1 &= \varphi_1 + 2\varphi_2 + n\theta + \phi, \\ \xi_2 &= 2\varphi_1 - \varphi_2. \end{aligned} \quad (4\ 39)$$

On this chart the Hamiltonian is represented as

$$K = J_1\Delta + J_2\Gamma + gI_1^{1/2}I_2\cos\xi_1, \tag{4.40}$$

where $\Delta \equiv \nu_1 + 2\nu_2 + n$ and $\Gamma \equiv 2\nu_1 - \nu_2$. The Hamiltonian flow, projected along ξ_2, is given by the vector field

$$\begin{aligned} \dot{J}_1 &= gI_1^{1/2}I_2\sin\xi_1, \\ \dot{\xi}_1 &= \Delta + gI_1^{-1/2}(\tfrac{1}{2}I_2 + 2I_1)\cos\xi_1. \end{aligned} \tag{4.41}$$

Resonant, or critical, orbits of K are projected into fixed points of Eqs.(4.41). The regular ones are those for which $\sin\xi_1 = 0$; the irregular ones are those for which either $I_1 = 0$ or $I_2 = 0$.

Symmetries of the projected flow will allow us to confine our attention to the parameter quadrant: $\Delta > 0$, $g > 0$. Clearly, if we simultaneously change the signs of both these quantities, the flow simply changes direction. Changing the sign of g alone can be compensated for by the transformation $\xi_1 \to \xi_1 + \pi$. Finally, changing the sign of Δ alone amounts to performing both previous transformations in succession. In fact there are no essential parameters in this problem: both Δ and g can be made to vanish by a simple scaling transformation. Let us define $\kappa \equiv \Delta/g$, and scale the amplitude variables by κ^2.

$$j_{1,2} \equiv J_{1,2}/\kappa^2, \quad i_{1,2} \equiv I_{1,2}/\kappa^2. \tag{4.42}$$

Then the level sets — which determine the topology of the flow — of the function

$$\mathcal{K} \equiv g^2(K - J_2\Gamma)/\Delta^3 = j_1 + i_1^{1/2}i_2\cos\xi_1$$

are identical to those of K. Furthermore, $\mathcal{K}$ can act as a true Hamiltonian for the scaled variables provided we simultaneously rescale $\theta \to \theta\Delta^3/g^2$.

For reference purposes, we present the answer before its development. The separatrix is sketched in Figure 4.16. Each frame shows its intersection with a single three-dimensional J_2 leaf, projected along the ξ_2 direction onto the (ξ_1, J_1) plane. A few points should be kept in mind while scanning these pictures. First, the ξ_1 axis corresponds *not* to $J_1 = 0$ but to $J_1 = -2J_2$ $(I_1 = 0)$, when $J_2 < 0$, and to $J_1 = \frac{1}{2}J_2$ $(I_2 = 0)$, when $J_2 > 0$. Second, the dynamical range of ξ_1 is 6π: we are viewing only one-third of the full projection; each picture is repeated twice. Third, remember that a "fixed point" in the diagram is the projection of a resonant orbit, which is a 1-torus, a closed curve corresponds to a 2-torus, and an open (unbounded) curve corresponds to a two-dimensional surface.

The fixed point equation, Eq. (4.16), which we use to find the resonant orbits, is written

$$\Delta/g \pm I_1^{-1/2}(\tfrac{1}{2}I_2 + 2I_1) = 0. \tag{4.43}$$

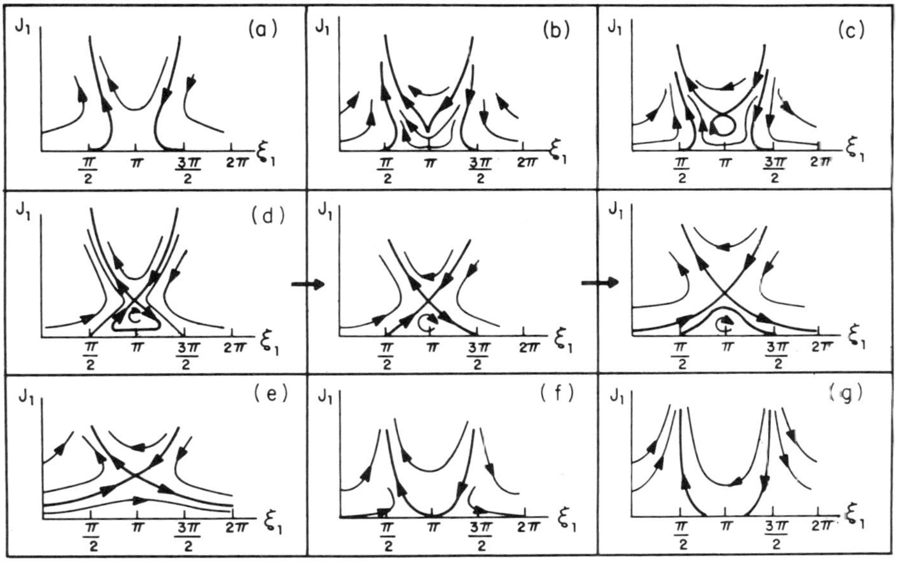

FIGURE 4.16 Flow diagrams for the projected Hamiltonian of the (1,2) resonance.

Clearly if $\Delta/g > 0$ ($\Delta/g < 0$) then the $-$ $(+)$ sign is indicated, which means that $\xi_1^{(0)} \simeq \pi$ ($\xi_1^{(0)} \simeq 0$). To be definite, let us assume in everything that follows that $\Delta > 0$ and $g > 0$, and rely on the previously mentioned symmetries to extrapolate results to other regions. In any case we shall set $\kappa \equiv |\Delta/g|$. The fixed point equation (4.43) can be written

$$\tfrac{1}{2}I_2 + 2\left(I_1^{1/2} - \frac{\kappa}{4}\right)^2 = \frac{\kappa^2}{8},$$

which describes an ellipse in $(\sqrt{I_1}, \sqrt{I_2})$ space. We want to express this in terms of the constant of motion J_2. Substituting from Eqs. (4.39) we get the following:

$$3I_1 - \kappa I_1^{1/2} - \tfrac{5}{2}J_2 = 0, \tag{4.44}$$

with solution

$$\sqrt{I_{1\pm}^{(0)}} = \tfrac{1}{6}(\kappa \pm \sqrt{\kappa^2 + 30J_2}). \tag{4.45}$$

This is sketched in Fig. (4.17), with + and − branches identified.

Applying the tangent space Eq. (4.17) to Eq. (4.41) provides us with the localized flow.

$$\dot{J}_1 = gI_1^{(0)\,1/2}I_2^{(0)} \cos \xi_1^{(0)} d\xi_1, \tag{4.46}$$

$$\dot{\xi}_1 = g/4\,I_1^{(0)\,-3/2}(6J_1^{(0)} + 17J_2) \cos \xi_1^{(0)} dJ_1. \tag{4.47}$$

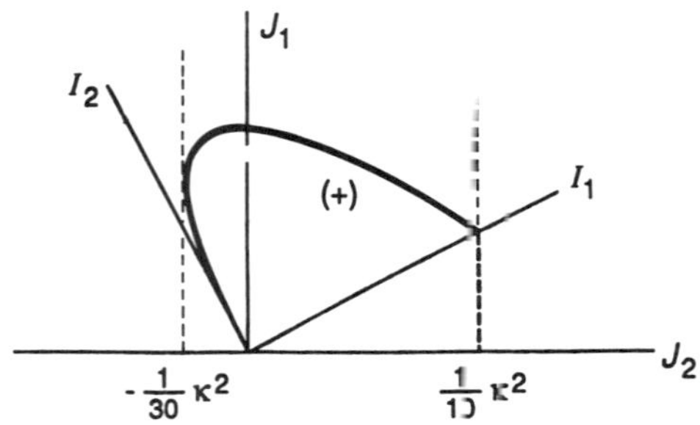

FIGURE 4.17 Track of the projected resonant orbits (fixed points).

Therefore unstable resonant orbits are those for which $6J_1^{(0)} + 17J_2 > 0$. (See the discussion of Fig. 4.4.) This inequality must be satisfied within the region $J_2 > 0$, and since the + branch is the only one to exist in this region, this is enough to establish that the + branch is the unstable one and the − branch the stable one. As a double-check, we examine the region $J_2 < 0$. First note that

$$6J_{1\pm}^{(0)} + 17J_2 = \tfrac{1}{3}\left[\kappa^2 + 30J_2 \pm \kappa\sqrt{\kappa^2 + 30J_2}\right],$$

so, as expected, since $\kappa^2 + 30J_2 > 0$, it automatically follows that the + branch is unstable. Furthermore, since $\sqrt{\kappa^2 + 30J_2} < \kappa$, we have immediately that

$$\kappa^2 + 30J_2 < \kappa\sqrt{\kappa^2 + 30J_2},$$

which means that the − branch traces the path of a stable fixed point.

Recall that the irregular critical points are found by setting K_r to zero in Eq.(4.14). This provides the value of the amplitude variable at the resonant orbit, after which the phase value is obtained from Eq.(4.13) upon setting $\dot{\xi}_1$ to zero. In our model, $K_r = 0$ means that either I_1 or I_2 must vanish. Consider first approaching the surface $I_1 = 0$. Equation (4.41) is then dominated by the $I_1^{-1/2}$ term:

$$I_1^{1/2}\dot{\xi}_1 \approx \tfrac{1}{2}gI_2\cos\xi_1.$$

$\dot{\xi}_1$ is indefinite at $\xi_1 \simeq \pm\pi/2$. This is the location of the irregular critical point, whose phase, in this case, happens to be pinned. (See Figures 4.16(a)–(d).) At any other value of ξ_1,

$$\dot{\xi}_1 = \mathrm{sgn}[g\cos\xi_1]\cdot\infty.$$

This type of behavior will obviously occur whenever K_r contains a term whose highest power of I_1 is $\frac{1}{2}$, and similarly for I_2.

Now consider the surface $I_2 = 0$. In this case Eq.(4.41) simplifies to the following:

$$\dot{\xi}_1 = \Delta + 2gI_1^{1/2}\cos\xi_1.$$

The resonant orbit thus sits at

$$\cos \xi_1^* = -\tfrac{1}{2}\kappa I_1^{1/2}$$

and is not pinned, but varies with I_1. Real solutions are possible only for $I_1 > \kappa^2/4$. Notice also that $\xi_1^* \to \pm\pi/2$ as $I_1 \to \infty$. In Figure 4.17, the point $(I_1 = \kappa^2/4, I_2 = 0)$ corresponds to the intersection of the track of the "regular fixed point" (resonant orbit, actually) with the I_2 axis. What happens is that the unstable regular fixed point is pushed down into the surface $I_2 = 0$ where it splits into two irregular critical points. (See Figure 4.16.)

Note that $\dot{\xi}_1$ remains finite for $I_2 = 0$, whereas $\dot{\xi}_1$ was infinite on $I_1 = 0$. Nearby orbits approach the instability in more leisurely fashion. This is intuitively appealing: one would expect a normal sextupole resonance to destabilize an orbit with zero vertical emittance more slowly than one having zero horizontal emittance.

The J_1 amplitudes at which the regular resonant orbits occur on a particular J_2 surface are found by intersecting the fixed point track of Figure 4.17 with the corresponding line, $J_2 =$ constant. The merger that occurs at $J_2 = -\frac{1}{30}\kappa^2$, $J_1 = \frac{17}{180}\kappa^2$ represents a local bifurcation.[7] As exhibited in Figures 4.16(a)–(c), the topological character of the flow changes when κ^2 crosses this threshold: below it, all orbits diverge; above it, there is an "island" of stable orbits. Precisely at catastrophe, stable and unstable resonant orbits merge to form a cusp, which annihilates both of them. If one thinks of J_2 as a control variable for the projected Hamiltonian — so that the control space is the set of doublets (J_2, κ) — then the subset $\{ (J_2, \kappa) | J_2 = -\kappa^2/30 \}$, which marks control points at which bifurcation occurs, is the catastrophe surface.

A global, saddle-switch bifurcation can occur in our example when the Hamiltonian takes on the same value at the unstable regular and irregular resonant orbits, so that a branch of the separatrix can connect them. Assigning $\xi_1 = \pi/2$ and $I_1 = 0$ in Eq. (4.40) gives us the value of the reduced Hamiltonian at the irregular critical point.

$$K_{\text{irr f p}} = -2J_2\Delta + J_2 \underline{m} \times \underline{\nu}.$$

Here, $\underline{m} \times \underline{\nu} \equiv m_1\nu_2 - m_2\nu_1$.Its value at a regular resonant orbit is

$$K_{\text{reg f p}} = J_1^{(0)}\Delta + J_2\underline{m} \times \underline{\nu} - g\sqrt{I_1^{(0)} I_2^{(0)}}.$$

Now the algebra is made simpler by expressing everything in terms of I_1 and J_2.

$$I_2^{(0)} = 2I_1^{(0)} - 5J_2,$$
$$J_1^{(0)} = I_1^{(0)} - 2J_2. \qquad (4.43)$$

[7] It is a bifurcation of the projected Hamiltonian; it is a fold in the separatrix of the original, four-dimensional system.

Making these substitutions gives us the following:

$$K_{\text{reg fp}} = (I_1^{(0)}\Delta - 2J_2) + J_2 \underline{m} \times \underline{\nu} - 2gI_1^{(0)\,3/2} + 5gI_1^{(0)\,1/2}J_2.$$

Now use the fixed point equation, Eq. (4.44), to write

$$5gI_1^{(0)\,1/2}J_2 = 6gI_1^{(0)\,3/2} - 2I_1^{(0)}\Delta.$$

Making this substitution and simplifying a little, we obtain

$$K_{\text{reg fp}} = K_{\text{irr fp}} + 4gI_1^{(0)\,3/2} - I_1^{(0)}\Delta.$$

The necessary condition for a global bifurcation, $K_{\text{reg fp}} = K_{\text{irr fp}}$, can thus be written as follows:

$$\sqrt{I_1^{(0)}} = \tfrac{1}{4}\kappa.$$

Substituting for $I_1^{(0)}$ from Eq. (4.45) gives us the value of J_2 where the bifurcation occurs.

$$J_2 = -\tfrac{1}{40}\kappa^2$$

The "control space" surface $\{ (J_2, \kappa) | J_2 + \frac{1}{40}\kappa^2 = 0 \}$ is the "Maxwell surface" of this example.

The actual separatrix of the $(1, 2)$ resonance model is obtained by pasting together all the iso-J_2 phase space slices that we have sketched. In doing so we obtain an object of the correct dimensionality; the separatrix is indeed a three-dimensional object (remember Fig. 4.3), and it does partition the four-dimensional phase space. To recapitulate (refer to Figure 4.16): (a) For J_2 large and negative all orbits are unbounded except the irregular resonant orbits, which are pinned to the surface $I_1 = 0$ at phases $\xi_1 \simeq \pm\pi/2$. (b) As J_2 increases, a local bifurcation, or catastrophe, occurs on the leaf $J_2 = -\frac{1}{30}\kappa^2$. It is heralded by the appearance of a new branch of the separatrix connected nontransversally (forming a cusp) to a new resonant orbit, a 1-torus. (c) That orbit splits, and for $-\frac{1}{30}\kappa^2 < J_2 < -\frac{1}{40}\kappa^2$ there is a single class of bounded orbits. (d) A global bifurcation, a saddle-switch, occurs on the surface $J_2 = -\frac{1}{40}\kappa^2$. At this precise value, the surface $I_1 = 0$ is stable for phases that are 2π-equivalent to the range $\pi/2 < \xi_1 < 3\pi/2$. On the surfaces $-\frac{1}{40}\kappa^2 < J_2 < 0$ there are two classes of bounded orbits. The first, say, Class A, is as before and is characterized by a bounded phase, $\pi/2 < \xi_1 < 3\pi/2$. The second, Class B, has an unboundedly increasing phase ξ_1. (Another way of saying this: Class A orbits exist in "islands.") The entire surface $I_1 = 0$ is now locally stable. (e) For $0 < J_2 < \frac{1}{10}\kappa^2$, the Class A orbits have disappeared; Class B orbits are still bounded. (f) When $\frac{1}{10}\kappa^2 < J_2$, Class B has disappeared as well. All orbits are once more unbounded, except the two unpinned irregular resonant orbits in the plane $I_2 = 0$, which begin at $\xi_1 \simeq \pi$ at $J_2 = \frac{1}{10}\kappa^2$ and (g) wander to $\xi_1 \simeq \pm\pi/2$ as $J_2 \to \infty$.

PROBLEMS

4.1 Why does averaging the Van der Pol equation produce (reasonably) correct results in the limit of small γ but not large γ?

4.2 Show that $I_2 = 0$ implies $\dot{I}_2 = 0$ when the nonlinearities are excited by normal multipoles.

4.3 Express all the control parameters (g, κ, and so forth) which appear in the isolated resonance models of this chapter as functionals of multipole distributions.

4.4 Confirm that the expression for $H^{(hot)}$ given in Eq. (4.21) is real.

4.5 Why do skew multipoles not contribute to the first order tune shifts?

4.6 Explicitly write the first order shear terms arising from normal octupoles, decapoles, and duodecapoles.

4.7 Give a simple physical argument why the sign of the tune shift arising from a normal quadrupole field error must be as written in Eq. (4.24).

4.8 Sketch how a weak, octupole-induced shear term, of the form κJ^2, will close the separatrix of the isolated third-integer resonance. Consider both cases: $\kappa > 0$ and $\kappa < 0$.

4.9 The closed spearatrix sketched in Figure 4.8 is not the only one possible. Find the other. How does the topology depend on the sign of κ?

4.10 Suppose we do *not* set $J_2 = 0$ in Eq. (4.34). How would this affect the fourth-integer resonance?

4.11 Show that the first order projected form of the $\nu_1 = \nu_2$ resonance Hamiltonian can be written as follows:

$$H = \nu_1 I_1 + \nu_2 I_2 - \sqrt{I_1 I_2}\left(g e^{i(\varphi_1 - \varphi_2)} + \text{c.c.}\right),$$

$$g \equiv \frac{1}{4\pi}\int ds \frac{B'}{|B\rho|}\sqrt{\beta_1\beta_2}\exp i[\tilde{\psi}_1 - \tilde{\psi}_2].$$

Here, of course, B' refers to $\partial B_1/\partial x_1$, that is, the skew component, and $\tilde{\psi}_k = \psi_k - \nu_k\theta$. Write the components of the linear vector field in terms of the normal form coordinates, a and a^*, of Eq. (2.9), and show that the eigentunes are

$$\nu_\pm = (\nu_1 + \nu_2)/2 \pm \sqrt{[(\nu_1 - \nu_2)/2]^2 + |g|^2}.$$

4.12 Write the isolated resonance models for all resonances excited by an octupole distribution, to first order in octupole strength. Evalute the models' parameters as functionals of the octupole distribution.

4.13 Using Eq. (4.17), show that the regular resonant orbits of the $\nu_1 + \nu_2$ principal resonance, excited by skew quadrupoles, are elliptic.

4.14 What are the physical meanings of an infinite phase rate, $\xi_1 = \infty$, which occurs in several examples in this chapter? (As a heuristic guide, consider a point in phase space passing through "the origin" at constant velocity.)

CHAPTER 5

Perturbation Theories

PEANUTS reprinted by permission of UFS, Inc.

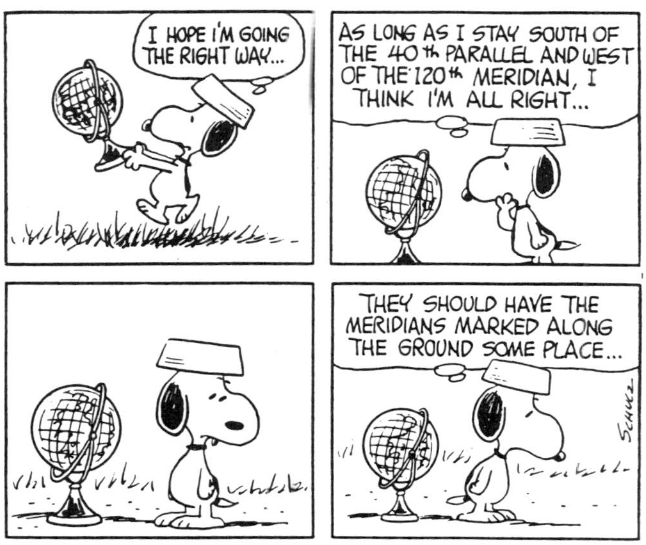

— Charles Schultz
Peanuts

A perturbation theory is a quest for useful coordinates.[1] In quantum mechanics, it leads to the construction of bases that diagonalize operators; in classical mechanics the objective is to find coordinate charts on which a vector field is represented "as simply as possible." When a simple representation has been found, at least on some

[1] And like all quests, it leads to failure if pushed to the extreme, but is a worthwhile endeavor if not.

local chart, we call it the vector field's "normal form," a term that unfortunately, and incorrectly, suggests uniqueness. The twist (or shear) and resonance models of the previous chapter are examples of such normal forms. They were motivated heuristically by averaging accelerator Hamiltonians, but a more correct derivation will be produced in this chapter. We shall find that these normal forms legitimately apply in coordinates that are *distorted* relative to the original. That distortion is the major piece missing from naive averaging, which otherwise would be equivalent to first order perturbation theory.

We can view the role of perturbation theory in another manner: it produces analogies. Very few problems can be solved exactly; the overwhelming majority, only approximately. Analogy asserts that, if one adopts an appropriate viewpoint, something that is unknown can be made to look like something else, about which we perhaps know a great deal. A perturbation theory asserts that two systems really belong to a class of systems that behave similarly, *at least locally and macroscopically,* and that it is possible to find a "path" through this equivalence class that takes us from one to the other. It gives us a way to construct that path parametrically using an "order parameter" natural to the problem at hand.

The subject is vast, and we cannot begin to do it justice. We shall concentrate on three particular perturbation theories that are eminently suited for automation. The first is a straightforward, "follow-your-nose" method, which, although having a long history, is not as well known as it should be. It will be introduced in Section 5.1 with an easy one-dimensional example that motivates the concept of normal forms. It is quickly generalized to many dimensions, and a systematic procedure is specified for performing the perturbation theoretic transformations. The Lie bracket will appear *naturally* as the fundamental operation driving the procedure. In Section 5.2 we use the Lie bracket to construct a vector field whose orbits are the "paths" through the equivalence class of dynamics mentioned above. We shall also detail its association with the resonant normal forms of the last chapter. Section 5.3 is devoted to a more recent method, which proceeds assuming that the transformation is built up from a composition of exponential maps. All three procedures are driven by Lie algebraic operations, and we shall point out their similarities as we go.

Do not be discouraged by the *apparently* tedious nature of the examples we shall encounter: perturbation theoretic calculations should not be carried out by hand beyond the first or second order anyway. We have chosen to stress Lie algebraic methods over others because *they translate immediately into finite, recursive, or iterative algorithms* that can be implemented transparently on computers. All three procedures that we shall study can be and have been incorporated into a number of programs. However, we are not going to discuss those here: our goal is to understand the physics and mathematics behind the programs.

5.1 BASIC PROCEDURE

Whether classical or quantum, perturbation theory has a tendency to become computationally intense. To help mark the way and not lose sight of physics peering through the symbols, we shall work a simple problem in a rather naive way before

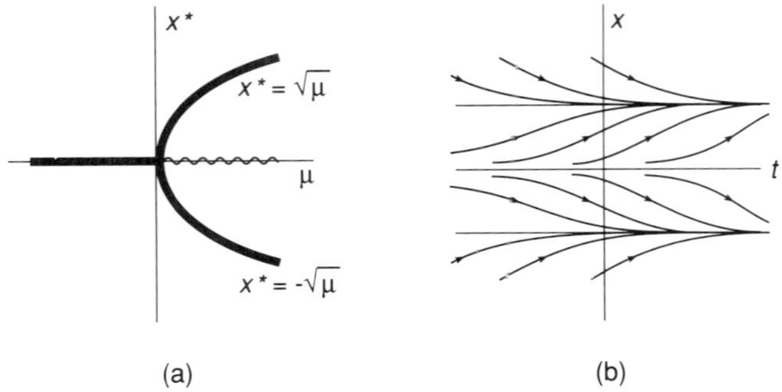

FIGURE 5.1 (a) Bifurcation diagram illustrating a "pitchfork" bifurcation at $\mu = 0$, (b) Extended flow field in the case $\mu > 0$.

plunging into the more formal aspects of the theory. Although this problem is trivial it possesses essential features and properties that carry over to and can be identified in the more complicated formalisms that follow.

5.1.1 A One-dimensional Model

Averaged Van der Pol Dynamic Let us return to the averaged Van der Pol dynamic we encountered in Eq. (4.2). With a suitable redefinition of parameters and rescaling it can be cast into the form

$$\dot{x} = x(\mu - x^2). \tag{5.1}$$

The flow on extended phase space is sketched in Figure 5.1 in the case $\mu > 0$; the fixed points at $x_c = \pm\sqrt{\mu}$ correspond to the limit cycle of the original Van der Pol oscillator.Our objective is to develop a change of coordinates, $x \longrightarrow y$, that simplifies the dynamics. One might argue that this vector field is already pretty simple: only a cubic term keeps it from complete triviality. We shall try to choose a new chart so as to make that term vanish.

The first step in this process is crucial. Naively, one might expect to write something like $y = F(x)$ and search for F. In fact, it turns out to be more fruitful if we do this in the other direction: search for F where $x = F(y)$. We can narrow this further by specifying that (a) we want the origin to remain invariant under the transformation: $0 \longmapsto 0$, and (b) we require that the *linearized* dynamics near $x = y = 0$ be expressed identically. These sensible, though not necessary, conditions yield a form that is commonly called a "near-identity" transformation,[2]

$$x = y + F(y), \tag{5.2}$$

where F must be $o(y)$ to satisfy condition (b): that is, $F(y)/y \longrightarrow 0$ as $y \longrightarrow 0$.

[2]Usually with a perhaps misplaced sense that F is "small"

Combining Eq. (5.2) with Eq. (5.1) provides us with the new representation.

$$\begin{aligned}(1 + F'(y))\dot{y} &= \mu(y + F(y)) - (y + F(y))^3 \\ \dot{y} &= (1 + F'(y))^{-1} \cdot [\mu(y + F(y)) - (y + F(y))^3]. \end{aligned} \tag{5.3}$$

At this point it is easy to formulate strategies that do not work well, the most obvious being to substitute an arbitrary polynomial form for F and attempt to find its coefficients. What *does* work is to perform the transformation in stages. Concentrate first on eliminating the cubic term. Let us assume that F is not an *arbitrary* polynomial but simply a single term: $F(y) = \kappa y^3$, with κ to be determined. Equation (5.3) becomes

$$\begin{aligned}\dot{y} &= (1 + 3\kappa y^2)^{-1} \cdot [\mu(y + \kappa y^3) - (y + \kappa y^3)^3] \\ &= (1 - 3\kappa y^2 + 9\kappa^2 y^4 + O(y^6)) \cdot [\mu y + (\mu\kappa - 1)y^3 - 3\kappa y^5 + O(y^7)] \\ &= \mu y - (1 + 2\mu\kappa)y^3 + 6\kappa^2\mu y^5 + O(y^7).\end{aligned}$$

We have identified the leading order behavior of $\dot{y}$ and buried everything else in a rather complicated (algebraic) function, which we do not bother to write explicitly. Now set $\kappa \equiv -1/2\mu$ so as to eliminate the coefficient of y^3.

$$\dot{y} = \mu y + \frac{3}{2\mu} y^5 + O(y^7).$$

On the surface, it does not appear as though we have made much progress: we have replaced a rather simple system containing only one cubic term with one containing a much more complicated function that is $O(y^5)$. Let us assume, appearances notwithstanding, that there is some advantage to doing this. Having done it once, we can repeat the process in order to eliminate the y^5 term. Consider another transformation, $y \longrightarrow w$, of the form

$$y = w + \gamma w^5,$$

where γ will be chosen so as to kill fifth order terms. Manipulations proceed as before.

$$\begin{aligned}\dot{y} = (1 + 5\gamma w^4)\dot{w} &= \mu(w + \gamma w^5) + \frac{3}{2\mu}(w + \gamma w^5)^5 + O(w^7) \\ &= \mu w + (\mu\gamma + \frac{3}{2\mu})w^5 + O(w^7) \\ \dot{w} &= \mu w + (4\mu\gamma - \frac{3}{2\mu})w^5 + O(w^7).\end{aligned}$$

We set $\gamma \equiv 3/8\mu^2$ to zero the coefficient of w^5, which leaves

$$\dot{w} = \mu w + O(w^7). \tag{5.4}$$

For completeness, we list the sequence of transformations found so far and their composition.

$$x = y - \frac{1}{2\mu} y^3,$$

$$y = w + \frac{3}{8\mu^2} w^5,$$

$$x = w - \frac{1}{2\mu} w^3 + \frac{3}{8\mu^2} w^5 + O(w^7). \tag{5.5}$$

Scale and Convergence What has been accomplished? We started with a simple cubic nonlinearity and, by means of two successive transformations, Eq. (5.5), produced an alternate representation with a much more complicated nonlinearity that is $O(w^7)$. Presumably, we could continue this process and keep pushing the nonlinear terms to arbitrarily high order. The payoff is buried in the *assumption* that accumulated higher order terms are smaller than retained lower order terms. At some point we neglect them and substitute the normal form, $\dot{w} = \mu w$, for the original system. This is then easy to solve, $w(t) = w(0)e^{\mu t}$ after which we transform back to the original coordinates. However, note well that the dynamical system of Eq. (5.1) is definitely *not* topologically equivalent to a linear one when $\mu > 0$: a nontrivial one-dimensional linear system has only one fixed point, while the dynamical system that we started with has three, at $x = 0$ and $\pm\sqrt{\mu}$. Therefore no transformation can be expected to work globally; it must fail when x approaches one of the "extra" fixed points.

This topological fact is reflected in the analytical form of the expansion. We have proceeded under the assumption that terms of higher "order" are smaller than those of smaller "order" without touching upon what the order parameter actually is. Clearly, in this problem it is determined by the length scale. What is important is the dimensionless quantity, $w/\sqrt{\mu}$: rather than write Eq. (5.5) and Eq. (5.4) as we have, it might have been more informative to express them as follows:

$$x = w \cdot [1 - \tfrac{1}{2}(w/\sqrt{\mu})^2 + \tfrac{3}{8}(w/\sqrt{\mu})^4 + O((w/\sqrt{\mu})^6)], \tag{5.6}$$
$$\dot{w} = \mu w \cdot [1 + O((w/\sqrt{\mu})^6)].$$

We cannot depend on the coefficients to make the series converge globally. If it does — and there is no reason to assume this[3] — convergence will come from requiring that

[3] In a sense, convergence may be a red herring. A series can sometimes be useful for computational purposes if it is asymptotic rather than convergent.

$x/\sqrt{|\mu|} \approx w/\sqrt{|\mu|} \ll 1$.[4] This holds true for negative as well as positive μ even though only the latter is associated with a triplet of fixed points. (This is reminiscent of the observation that the Taylor series for $1/(1+x)$ diverges near $x \approx 1$ even though — and *because* — the pole is located at $x \approx -1$.)

Near the bifurcation point, $\mu \approx 0$, the entire programme breaks down, as the reference point for expansion and the positions of the "extra" fixed points approach coincidence. This too is reflected analytically in the coefficients, which contain factors of $1/\sqrt{\mu}$. As $\mu \to 0$ these become "small denominators," an expression we shall encounter repeatedly. Let me repeat for emphasis, however, that this analytic phenomenon is really a reflection of what is happening geometrically. Fundamentally, to expect the expansion to work beyond $x \approx \sqrt{|\mu|}$ is unreasonable on topological grounds. The analytic phenomena — divergence, small denominators, and the like — merely attest to this impossibility. They are the *effects*, not the *cause*.

Renormalization procedures in quantum field theory are sometimes characterized as a very effective method for disposing of serious difficulties by "sweeping them under a rug," this being either criticism or high praise, depending on the speaker. In a similar vein, we may say of this perturbative, normal form strategy that we are eliminating undesirable features of a system by "pushing them over a cliff," the abyss of "higher order terms." Whether this be a worthy objective under all circumstances may be open to debate. It is, nonetheless, the fundamental strategy behind the manipulations in this chapter.

Important Features As mentioned before, the mathematics of perturbation procedures can, at times, obscure their underlying structure. It would be well to keep in mind that the more sophisticated manipulations we will encounter in the following sections are introduced as nothing more than efficient ways to achieve in higher dimensional problems the same objectives as those of this simple example. I wrote already that the example possessed *essential* features which carry over into more complicated problems. It is worthwhile to list the more important so that we can recognize them later. Taken together, they really outline a strategy for doing perturbation theory.

(a) In studying either periodic or autonomous dynamics, we begin by representing the vector field—or the map, as the case may be—on a chart whose origin is a critical (or recurring) orbit: most typically, a fixed point; less frequently, a periodic orbit; possibly, a point on a resonant, nonperiodic orbit.[5]

(b) Require that the perturbation procedure not affect the representation of the linear vector field, obtained by linearizing the problem about the origin. This leads to a "near-identity" transformation. In multi-dimensional problems it is best to begin,

[4] The situation may even be worse, as each step in the perturbation procedure may reduce the interval of convergence.

[5] In this edition we will only study the first case explicitly. The other two are useful when studying such things as the detailed behavior of a separatrix or the "renormalization group." The maps of nonperiodic dynamics are important for problems like examining the "optical" aberrations of a spectrometer. An atlas of two or more charts is used: the origin of the first being the initial state and that of the last, the final state of some fiducial orbit.

if possible, with a chart that diagonalizes the linear field; these are then the linear "normal form" coordinates, the coordinates that make the representation of the linear dynamic as simple as possible. (We will get to all that shortly.)

(c) Tentatively accepting the linearized dynamic as a target normal form for the vector field (or map), we next seek to eliminate only the lowest order nonlinear terms from the representation via a transformation whose reverse specification contains only terms of that order. The simplicity of this key idea[6] masks its brilliance. The new chart provides another set of (normal form) coordinates and, as required, the representation of the linearized dynamics has not been affected.

(d) *Represent the vector field on the new chart.* We then have the identical problem as in (c) but at higher order; that is, the lowest order nonlinear terms that appear are now of higher order than before. Other than that, we are confronted with the same problem and solve it in the same way. Step (c) is repeated as many times as desired. At the nth stage, the representations of the lower order vector fields (or maps) remain the same, terms of order n are (if possible) eliminated, [7] and those at even higher order get rewritten in preparation for the next step.

(e) The complete transformation from the first chart to the last is obtained by composing the individual, much simpler ones together.

There is, in fact, another important feature which has been obscured somewhat by doing a problem in only one dimension. Before writing it, we will generalize this process of finding normal form coordinates to higher dimensional phase spaces. For the most part, however, this describes the strategy behind the perturbation theories we will study.

5.1.2 Multidimensional Dynamics

Although we did not demonstrate this, the procedure employed to reduce the problem of the last section could have been continued indefinitely. Within some neighborhood of the origin the representation of the dynamics could be made to look arbitrarily close to that of a linear system. The two issues we want to address here are: (a) whether this is always the case, and (b) how to generalize the procedure to arbitrary dimensions. We shall find that the Lie bracket plays the central role in studying both of these issues. Although we did not call attention to it, the bracket was, in fact, a key player in the manipulations of the preceding section.

Reduction Procedure We repeat the strategy of the previous calculation in more dimensions. To lay out the problem quite generally, suppose that we have a dynamic that is expressed locally by an nth degree polynomial plus an arbitrary function of degree $n + 1$.

$$\dot{\underline{z}} = \underline{v}(\underline{z}) = \sum_{k=1}^{n-1} \tilde{\underline{v}}^{(k)}(\underline{z}) + \underline{v}^{(n)}(\underline{z}) + O(\underline{z}^{n+1}). \tag{5.7}$$

[6]Which may be due to Poincaré, like so many others.

[7]If both possible *and desired.* In our example, all the terms were made to vanish except the linear, but we shall see in the next section that this will *not generally be true.*

The polynomial has been split into homogeneous pieces; any $\underline{v}^m$ or $\underline{\tilde{v}}^m$ stands for a homogeneous polynomial of degree m. In particular, $\underline{\tilde{v}}^{(1)}$ is the linear term. We assume that all polynomials of degree less than n either vanish or have been reduced as much as possible. That is, they are in "normal form," which is emphasized by using the symbol $\underline{\tilde{v}}$, rather than $\underline{v}$, to represent these terms. Our goal is to reduce the nth degree polynomial, thus bringing the problem to the next stage.

To accomplish this, we introduce a transformation in the same form as before:

$$\underline{z} = \underline{y} + \underline{F}^{(n)}(\underline{y}). \tag{5.8}$$

The superscript on $\underline{F}$ signals that it is a homogeneous polynomial of degree n. The dynamical equation for y is written, recursively,

$$\underline{\dot{y}} = -\underline{\dot{y}} \cdot \underline{\underline{\partial F}}^{(n)}(\underline{y}) + \underline{v}(\underline{z}), \tag{5.9}$$

and now we go about identifying terms of various degrees. First consider the term $\underline{\tilde{v}}^{(1)}(\underline{z})$, which appears in $\underline{v}(\underline{z})$.

$$\underline{\tilde{v}}^{(1)}(\underline{z}) = \underline{\tilde{v}}^{(1)}(\underline{y} + \underline{F}^{(n)}(\underline{y})) = \underline{\tilde{v}}^{(1)}(\underline{y}) + \underline{F}^{(n)}(\underline{y}) \cdot \underline{\underline{\partial v}}^{(1)}(\underline{y}) + O(y^{n+1}).$$

The other terms, being of higher degree, are handled more easily.

$$\underline{\tilde{v}}^{(k)}(\underline{z}) = \underline{\tilde{v}}^{(k)}(\underline{y} + \underline{F}^{(n)}(\underline{y})) = \underline{\tilde{v}}^{(k)}(\underline{y}) + O(y^{n+k-1}).$$

This leaves the recursive term, $\underline{\dot{y}} \cdot \underline{\underline{\partial F}}^{(n)}(\underline{y})$. Since differentiation lowers the degree of a polynomial by one, and since $\underline{\tilde{v}}^{(1)}(\underline{y})$ is the only term of degree one on the right-hand side of Eq. (5.9), we write

$$\underline{\dot{y}} \cdot \underline{\underline{\partial F}}^{(n)}(\underline{y}) = (\underline{\tilde{v}}^{(1)}(\underline{y}) + O(y^2)) \cdot \underline{\underline{\partial F}}^{(n)}(\underline{y}) = \underline{\tilde{v}}^{(1)}(\underline{y}) \cdot \underline{\underline{\partial F}}^{(n)}(\underline{y}) + O(y^{n+1}).$$

This is the "special role" played by $\underline{\tilde{v}}^{(1)}$: it effectively replaces $\underline{\dot{y}}$ on the right-hand side of Eq. (5.9).

Putting these pieces back into Eq. (5.9), and rearranging terms a little, we reap the result,

$$\begin{aligned} \underline{\dot{y}} &= \sum_{k=1}^{n-1} \underline{\tilde{v}}^{(k)}(\underline{y}) \\ &\quad + \underline{F}^{(n)}(\underline{y}) \cdot \underline{\underline{\partial \tilde{v}}}^{(1)}(\underline{y}) - \underline{\tilde{v}}^{(1)}(\underline{y}) \cdot \underline{\underline{\partial F}}^{(n)}(\underline{y}) + \underline{v}^{(n)}(\underline{y}) \\ &\quad + O(y^{n+1}) \\ &= \sum_{k=1}^{n-1} \underline{\tilde{v}}^{(k)}(\underline{y}) + \left([\underline{F}^{(n)}, \underline{\tilde{v}}^{(1)}](\underline{y}) + \underline{v}^{(n)}(\underline{y})\right) + O(y^{n+1}). \end{aligned} \tag{5.10}$$

Coefficients of polynomials with degrees smaller than n remain unchanged. The nth degree polynomials, which we are attempting to eliminate, include a Lie bracket between $\underline{\tilde{v}}^{(1)}$ and $\underline{F}^{(n)}$. In order to make these terms vanish, we would like to choose $\underline{F}^{(n)}$, so that

$$[\underline{F}^{(n)}, \underline{\tilde{v}}^{(1)}] + \underline{v}^{(n)} \equiv \underline{\tilde{v}}^{(n)} \overset{?}{\equiv} 0. \tag{5.11}$$

We shall see this equation and its variations appearing again and again in all kinds of perturbation theories. They express the key order-by-order relationships between the new representation of the vector field and the transformation which connects the two charts; they are a systematic, explicit working out of Eq. (2.54). Following Arnold [5], we will call them the theory's "homological equations."

Regrettably, it is not always possible to solve Eq. (5.11) as it is written, with a zero on the right. To get a feeling for the type of problem that occurs, we write $\underline{\tilde{v}}^{(1)}(\underline{y}) = \underline{\underline{\Lambda}} \cdot \underline{y}$ and, noting that $\underline{\underline{\partial \tilde{v}}}^{(1)} = \Lambda^T$, write Eq. (5.11) as follows:

$$\underline{F}^{(n)} \cdot \Lambda^T - \underline{y} \cdot \Lambda^T \cdot \underline{\underline{\partial F}}^{(n)} = \underline{\tilde{v}}^{(n)} - \underline{v}^{(n)}.$$

To simplify further, suppose that the linear dynamic has been diagonalized,[8] so that $\underline{\underline{\Lambda}}_{ij} = \lambda_i \delta_{ij}$, and the component form of Eq. (5.11) simplifies to

$$\left(\lambda_k - \sum_j \lambda_j y_j \frac{\partial}{\partial y_j}\right) F_k^{(n)}(\underline{y}) = \tilde{v}_k^{(n)}(\underline{y}) - v_k^{(n)}(\underline{y}). \tag{5.12}$$

Now, each $F_k^{(n)}$ is a homogeneous polynomial of degree n. Let $\psi_{\underline{m}}^{(n)}$ represent the homogeneous monomials of degree n:

$$\psi_{\underline{m}}^{(n)}(\underline{y}) \equiv \prod_j y_j^{m_j}, \quad \text{where each} \quad m_j \geqslant 0 \quad \text{and} \quad \sum m_j = n.$$

We expand the homogeneous polynomials in this basis,

$$F_k^{(n)}(\underline{y}) = \sum_{\{\underline{m}\}} \mathrm{f}_{\underline{m}} \psi_{\underline{m}}^{(n)}(\underline{y}),$$

$$v_k^{(n)}(\underline{y}) = \sum_{\{\underline{m}\}} \mathrm{v}_{\underline{m}} \psi_{\underline{m}}^{(n)}(\underline{y}), \quad \text{and}$$

$$\tilde{v}_k^{(n)}(\underline{y}) = \sum_{\{\underline{m}\}} \tilde{\mathrm{v}}_{\underline{m}} \psi_{\underline{m}}^{(n)}(\underline{y}).$$

[8] With eigenvectors and eigenvalues that may be complex.

The importance of this representation is that the $\psi_{\underline{m}}^{(n)}$ are eigenstates of the operator $y_j \partial/\partial y_j$ appearing in Eq. (5.12).

$$\left(y_j \frac{\partial}{\partial y_j} \right) \psi_{\underline{m}}^{(n)}(\underline{y}) = m_j \psi_{\underline{m}}^{(n)}(\underline{y})$$

Combining this with the observation that Eq. (5.12) can be satisfied identically only if it is satisfied *by each monomial term* leads to the family of equations,

$$\left(\lambda_k - \sum_j \lambda_j m_j \right) \mathrm{f}_{\underline{m}} = \tilde{\mathrm{v}}_{\underline{m}} - \mathrm{v}_{\underline{m}}. \tag{5.13}$$

Our task is to choose the $\mathrm{f}_{\underline{m}}$ and $\tilde{\mathrm{v}}_{\underline{m}}$ such that this is satisfied. Ideally we would like $\tilde{\mathrm{v}}_{\underline{m}} = 0$, so we set

$$\mathrm{f}_{\underline{m}} \equiv -\mathrm{v}_{\underline{m}}/(\lambda_k - \sum_j \lambda_j m_j). \tag{5.14}$$

This will work *unless* there is an array of integers, say, $\underline{l}$, such that

$$\sum_j l_j \lambda_j = 0. \tag{5.15}$$

In that case, the denominator vanishes for monomials $\psi_{\underline{m}}^{(n)}$ for which $m_j = l_j - \delta_{kj}$. Those terms must be absorbed into the $\tilde{\mathrm{v}}_{\underline{m}}$ of Eq. (5.13) in order to make the right-hand side vanish. This provides the prescription for the "normal form," the "simplest possible" transformed vector field. It will contain only the residual terms that arise because of the resonance condition written in Eq. (5.15).

Of course, in practice Eq. (5.15) is seldom satisfied exactly. If one is close to resonance, in the sense that the denominator in Eq. (5.14) is small, then it may be wise to absorb the offending term into $\tilde{\mathrm{v}}_{\underline{m}}$ nonetheless.

Example: Quadratic Nonlinearity To illustrate the process we consider adding a quadratic force term to the harmonic oscillator: a continuous version of the Hénon system. We take as our local dynamical equation, on a two-dimensional phase space,

$$\dot{\underline{z}} = \begin{pmatrix} 0 & \omega \\ -\omega & 0 \end{pmatrix} \cdot \underline{z} + \begin{pmatrix} 0 \\ g_s z_1^2 \end{pmatrix}.$$

The symbols g_s is used for the nonlinear coupling to reflect the fact that, in the context of accelerator theory, this would be a sextupole-like term.

The goal of our first transformation is to nullify the quadratic, sextupole term. This is a matter of personal choice, but I find it convenient to carry out the Lie bracket calculations in terms of commutators of vector fields, relying on the morphism of

Eq. (2.67) and Eq. (2.68). Accordingly, let us install the transformation functions, $\underline{F}^{(2)}(\underline{y})$, as the components of a vector field,

$$\mathbf{F}^{(2)} = (f_1 y_1^2 + f_2 y_1 y_2 + f_3 y_2^2) \frac{\partial}{\partial y_1} + (f_{3'} y_1^2 + f_{2'} y_1 y_2 + f_{1'} y_2^2) \frac{\partial}{\partial y_2}, \tag{5.16}$$

where the coefficients are to be determined. The basis for this representation are the six vector fields whose coefficients are the three homogeneous polynomials of degree two. In this same notation, the linear part of the vector field is written

$$\mathbf{V}^{(1)} = \underline{v}^{(1)}(\underline{y}) \cdot \frac{\partial}{\partial \underline{y}} = \omega \left(y_2 \frac{\partial}{\partial y_1} - y_1 \frac{\partial}{\partial y_2} \right).$$

We seek to choose coefficients such that Eq. (5.11) is satisfied.

$$\left[\frac{1}{\omega} \mathbf{V}^{(1)}, \mathbf{F}^{(2)} \right] = \frac{g_s}{\omega} y_1^2 \frac{\partial}{\partial y_2}. \tag{5.17}$$

To accomplish this, we first study the action of $\mathbf{V}^{(1)}$ on the six basis fields in Eq. (5.16). Consider first its action on $y_1^2 \partial / \partial y_1$.

$$\begin{aligned} \left[y_2 \frac{\partial}{\partial y_1} - y_1 \frac{\partial}{\partial y_2}, y_1^2 \frac{\partial}{\partial y_1} \right] &= y_2(2y_1) \frac{\partial}{\partial y_1} - \left(-y_1^2 \frac{\partial}{\partial y_2} \right) \\ &= 2y_1 y_2 \frac{\partial}{\partial y_1} + y_1^2 \frac{\partial}{\partial y_2}. \end{aligned}$$

The corresponding result on $y_2^2 \partial / \partial y_2$ is obtained from this by changing the signs and interchanging indices.

$$\left[y_2 \frac{\partial}{\partial y_1} - y_1 \frac{\partial}{\partial y_2}, y_2^2 \frac{\partial}{\partial y_2} \right] = -2y_2 y_1 \frac{\partial}{\partial y_2} - y_2^2 \frac{\partial}{\partial y_1}.$$

The others are done in like fashion.

$$\begin{aligned} \left[y_2 \frac{\partial}{\partial y_1} - y_1 \frac{\partial}{\partial y_2}, y_1 y_2 \frac{\partial}{\partial y_1} \right] &= y_2^2 \frac{\partial}{\partial y_1} - y_1^2 \frac{\partial}{\partial y_1} + y_1 y_2 \frac{\partial}{\partial y_2}, \\ \left[y_2 \frac{\partial}{\partial y_1} - y_1 \frac{\partial}{\partial y_2}, y_1 y_2 \frac{\partial}{\partial y_2} \right] &= -y_1^2 \frac{\partial}{\partial y_2} + y_2^2 \frac{\partial}{\partial y_2} - y_1 y_2 \frac{\partial}{\partial y_1}, \\ \left[y_2 \frac{\partial}{\partial y_1} - y_1 \frac{\partial}{\partial y_2}, y_2^2 \frac{\partial}{\partial y_1} \right] &= -2y_1 y_2 \frac{\partial}{\partial y_1} + y_2^2 \frac{\partial}{\partial y_2}, \\ \left[y_2 \frac{\partial}{\partial y_1} - y_1 \frac{\partial}{\partial y_2}, y_1^2 \frac{\partial}{\partial y_2} \right] &= 2y_1 y_2 \frac{\partial}{\partial y_2} - y_1^2 \frac{\partial}{\partial y_1}. \end{aligned}$$

Applying these to Eq. (5.17), we get the following result:

$$\begin{aligned}\left[\frac{1}{\omega}\mathbf{V}^{(1)}, \mathbf{F}^{(2)}\right] &= (-f_2 - f_{3'})y_1^2 \frac{\partial}{\partial y_1} \\ &\quad +(2f_1 - f_{2'} - 2f_3)y_1y_2 \frac{\partial}{\partial y_1} \\ &\quad +(-f_{1'} + f_2)y_2^2 \frac{\partial}{\partial y_1} \\ &\quad +(f_1 - f_{2'})y_1^2 \frac{\partial}{\partial y_2} \\ &\quad +(-2f_{1'} + f_2 + 2f_{3'})y_1y_2 \frac{\partial}{\partial y_2} \\ &\quad +(f_{2'} + f_3)y_2^2 \frac{\partial}{\partial y_2} \\ &\equiv \frac{g_s}{\omega} y_1^2 \frac{\partial}{\partial y_2}.\end{aligned}$$

This yields six easily solved equations for the six unknown coefficients, whose solutions are

$$\begin{aligned}f_1 &= g_s/3\omega, \\ f_3 &= -f_{2'} = 2g_s/3\omega,\end{aligned}$$

all others being zero. The function of transformation is thus written

$$\mathbf{F}^{(2)} = \frac{g_s}{3\omega}\left(y_1^2 \frac{\partial}{\partial y_1} + 2y_2^2 \frac{\partial}{\partial y_1} - 2y_1y_2 \frac{\partial}{\partial y_2}\right),$$

which, upon returning to array notation, is equivalent to

$$\underline{z} = \underline{y} + \frac{g_s}{3\omega}\begin{pmatrix} y_1^2 + 2y_2^2 \\ -2y_1y_2 \end{pmatrix}.$$

By construction, the vector field representation on the $\underline{y}$ chart is linear up to terms of $O(\underline{y}^3)$.

Now, in fact, we have done this the hard way, if only to illustrate that, regardless of chart, solving homological equations requires nothing more complicated than solving a set of linear equations. This will be true at any order, because, as seen in Eq. (5.11), the homological equations always involve a bracket *with the linear part* of the vector field (or map). *The degree of a polynomial is invariant under such an operation, and this is what makes it possible to solve the homological equations in such a simple fashion, one degree at a time.* However, there is yet an easier way, which is: to use a (possibly complex) chart that diagonalizes the linear part of the vector field.[9] If this is done, then the homological equations decouple not only in degree but term by term. It is left as an exercise to the reader to repeat the calculations of this section but using the normal form coordinates described in Equations (2.8), (2.9), and (2.11).

Maps An equation analogous to Eq. (5.10) can be derived for writing a perturbation series for a time-evolution map rather than a vector field. We rewrite Eq. (5.7) to represent a polynomial map,

$$\underline{z}' = \underline{P}(\underline{z}) = \sum_{k=1}^{n-1} \underline{\tilde{P}}^{(k)}(\underline{z}) + \underline{P}^{(n)}(\underline{z}) + O(\underline{z}^{n+1}),$$

where each $\underline{P}^m$ or $\underline{\tilde{P}}^m$ is a homogeneous polynomial of degree m. As before, we seek to put the nth degree term into normal form via the substitution in Eq. (5.8). Since these are polynomials, we can isolate this term readily.

$$\begin{aligned}\underline{y}' = \underline{z}' - \underline{F}^{(n)}(\underline{y}') &= -\underline{F}^{(n)}(\underline{y}') + \sum_{k=1}^{n-1} \underline{\tilde{P}}^{(k)}(\underline{y} + \underline{F}^{(n)}(\underline{y})) + \underline{P}^{(n)}(\underline{y} + \underline{F}^{(n)}(\underline{y})) \\ &\quad + O(\underline{y}^{n+1}) \\ &= \sum_{k=1}^{n-1} \underline{\tilde{P}}^{(k)}(\underline{y}) \\ &\quad + \underline{\tilde{P}}^{(1)}(\underline{F}^{(n)}(\underline{y})) - \underline{F}^{(n)}(\underline{\tilde{P}}^{(1)}(\underline{y})) + \underline{P}^{(n)}(\underline{y}) \\ &\quad + O(y^{n+1}).\end{aligned}$$

The center line becomes the nth degree normal form, which is written functionally using the composition operator:

$$\underline{\tilde{P}}^{(n)} \equiv \underline{P}^{(n)} + \underline{\tilde{P}}^{(1)} \circ \underline{F}^{(n)} - \underline{F}^{(n)} \circ \underline{\tilde{P}}^{(1)}. \tag{5.18}$$

This formally looks very similar to Eq. (5.11), with a composition commutator replacing the Lie bracket. However, composition *is not a Lie algebraic operation* (it is not even bilinear), so the analogy ends there. (However, in the final section of this chapter we will consider an algorithm for exponential maps based on Lie brackets.)

As a minor aside, we note that Eqs.(5.11) and (5.18) can be made to look even more similar by mixing the operations. Because $\underline{\tilde{v}}^{(1)}$ is linear,

$$\underline{F}^{(n)}(\underline{y}) \cdot \underline{\underline{\partial \tilde{v}}}^{(1)}(\underline{y}) = \underline{\tilde{v}}^{(1)}(\underline{F}^{(n)}(\underline{y})) = (\underline{\tilde{v}}^{(1)} \circ \underline{F}^{(n)})_{(\underline{y})}.$$

[9]Or, if this is impossible (consider a particle floating freely in space), at least puts it into Jordan normal form.

Therefore Eq. (5.11) can be rewritten as follows:

$$\underline{\tilde{v}}^{(n)} \equiv \underline{v}^{(n)} + \underline{\tilde{v}}^{(1)} \circ \underline{F}^{(n)} - \underline{\tilde{v}}^{(1)} \cdot \underline{\underline{\partial F}}^{(n)}.$$

Similarly, by reversing this argument, the homological equation for maps, Eq. (5.18), is written

$$\underline{\tilde{P}}^{(n)} \equiv \underline{P}^{(n)} + \underline{F}^{(n)} \cdot \underline{\underline{\partial \tilde{P}}}^{(1)} - \underline{F}^{(n)} \circ \underline{\tilde{P}}^{(1)}.$$

Important Features (continued) We are now ready to write the final important feature of the perturbation theories under study, the one postponed from our earlier discussion on pages 245 and 246.

(f) At each stage of the iterative (recursive?) procedure we solve a linear system of equations, the homological equations, for both the next transformation and the normal form vector field at that order. The homological equations *have the same structure at all orders* — and thus are solved in the same manner — a structure that results from applying a Lie derivative (i.e., a bracket) with respect to the linear part of the vector field (or map). One generally chooses to make the normal form as simple as possible; typically, one tries to zero it entirely, but topologicial barriers (i.e., recurrent sets: periodic orbits, resonances, separatrices) — which manifest themselves analytically by making the linear system singular or near-singular (the regrettably named "small denominator" problem) — may prevent this or make it inconvenient. In such cases, the troublesome terms — which really are those belonging to the kernel of the Lie operator; you cannot invert a nontrivial projection, — are absorbed by the normal form at that order and the procedure is continued.

Because the homological equations do indeed have the same structure at all orders they are *ideally* suited to programming on a computer using iterative algorithms.

5.2 VECTOR FIELDS OF TRANSFORMATIONS

The basic procedure of the previous section was predicated on an order parameter that essentially expressed a gauge for phase space coordinates. This in itself is interesting, as phase space does not possess a natural scale: in general, the coordinates do not even have the same dimensions. Indeed, most analysts will automatically rescale in the vicinity of elliptic fixed points by working with the eigencoordinates of the linearized dynamic; in two dimensions, this makes the orbits look circular as one approaches the fixed point. Apart from that, by choosing scale as the order parameter we lose a handle on studying bifurcations of the system, as was done for the half-integer resonance in the last chapter, for example.

We consider in this section a perturbation procedure in which the order parameter can be associated either with phase space scale, as before, or with any control parameter of the system. This will be done by the remarkable trick of writing two dynamics simultaneously: the first being the one we are interested in studying, the second, a formal vector field that identifies the *order* parameter with its *orbit* parameter. Like that of the previous section, it is suitable for automation and has been (repeatedly) automated. Our special interest in it here is in its ability to generate rigorously, and naturally, the resonance normal forms motivated via averaging in the last chapter. The original form of this procedure was developed for autonomous Hamiltonian dynamics by Deprit [28] and is called "Deprit's algorithm." It should come as no surprise by this time that this perturbation series also is based on Lie brackets.

5.2.1 General Dynamics

Preliminaries We begin by yet again setting up the equations that express our problem. Suppose that $\dot{\underline{z}} = \underline{v}(\underline{z}, \epsilon)$ and $\dot{\underline{y}} = \underline{u}(\underline{y}, \epsilon)$ represent the same dynamic on two charts, that is, using two sets of coordinates *now related by a continuous family of transformations indexed by the order parameter.*

$$\underline{z} = \underline{F}(\underline{y}, \epsilon), \qquad \underline{y} = \underline{G}(\underline{z}, \epsilon), \qquad G = F^{-1}. \tag{5.19}$$

We require that $\lim_{\epsilon\to 0} \underline{F} = \lim_{\epsilon\to 0} \underline{G} = \underline{I}$ — where, recall, $\underline{I} : \underline{z} \longmapsto \underline{z}$ is the identity function, with components, $I_k(\underline{z}) = z_k$ — so that these are near-identity transformations. If it is necessary to work with nonautonomous vector fields (and it is), then we accept that everything is to be referred to an extended phase space, so that we can formally proceed as though the fields were autonomous. The differential forms of Eqs. (5.19) are written

$$\begin{aligned} d\underline{z} &= d\epsilon\; \partial_\epsilon \underline{F}(\underline{y}, \epsilon) + d\underline{y} \cdot \underline{\underline{\partial F}}(\underline{y}, \epsilon), \\ d\underline{y} &= d\epsilon\; \partial_\epsilon \underline{G}(\underline{z}, \epsilon) + d\underline{z} \cdot \underline{\underline{\partial G}}(\underline{z}, \epsilon). \end{aligned} \tag{5.20}$$

These differential forms should be taken literally as *defining* $\partial_\epsilon \underline{G}$, $\underline{\underline{\partial G}}$, and so forth. Setting $d\underline{y} = \underline{0}$ in the first of these provides us with the relation

$$\left(\frac{d\underline{z}}{d\epsilon}\right)_{d\underline{y}=\underline{0}} = \partial_\epsilon \underline{F}(\underline{y}, \epsilon). \tag{5.21}$$

This is the vector field that will be used to generate the transformation. The transformed coordinate, $\underline{z}$, is viewed as propagating along an "orbit," with ϵ as the orbit parameter, with "initial condition" given by $\underline{y}$, as confirmed by the condition $d\underline{y} = \underline{0}$. The difficulty is that the right-hand side is represented as a function of $\underline{y}$, not $\underline{z}$. We correct that formally by defining the $\underline{z}$ representation,

$$\underline{w}(\underline{z}; \epsilon) \equiv \left(\frac{d\underline{z}}{d\epsilon}\right)_{d\underline{y}=\underline{0}} = (\partial_\epsilon \underline{F})_{(\underline{y}, \epsilon)} = (\partial_\epsilon \underline{F})_{(\underline{G}(\underline{z}, \epsilon), \epsilon)}, \tag{5.22}$$

or written functionally as a composition of functions,

$$\underline{w} = (\partial_\epsilon \underline{F}) \circ \underline{G}. \tag{5.23}$$

We can write this in one more way, using the facts that (a) $\underline{F}$ and $\underline{G}$ are inverse functions and (b) composition is associative.

$$\underline{w} \circ \underline{F} = \partial_\epsilon \underline{F} \circ \underline{G} \circ \underline{F} = \partial_\epsilon \underline{F}. \tag{5.24}$$

This will turn out to be useful later. Applying Eq. (5.22) to the second of Eqs. (5.20) and setting $d\underline{y} = \underline{0}$ produces

$$\partial_\epsilon \underline{G}(\underline{z}, \epsilon) = -\underline{w}(\underline{z}, \epsilon) \cdot \underline{\underline{\partial G}}(\underline{z}, \epsilon) = -(\underline{w} \cdot \underline{\underline{\partial G}})_{(\underline{z}, \epsilon)}.$$

Since the arguments are the same on both sides we can write this functionally in terms of a dot product:

$$\partial_\epsilon \underline{G} = -\underline{w} \cdot \underline{\underline{\partial G}}. \tag{5.25}$$

We might have anticipated this result. Equation (5.25) is the inversion of Eqs.(5.21) and (5.23): if $\underline{y} \longmapsto \underline{z}$ along the "transformation orbit," whose evolution is governed by $\underline{w}$, then we would expect the inverse transformation $\underline{z} \longmapsto \underline{y}$ to evolve along the ϵ-reversed "transformation orbit," as governed by $-\underline{w}$. The extra factor of $\underline{\underline{\partial G}}$ serves to convert the representation of $-\underline{w}$ into the appropriate chart, according to the rule written in Eqs.(2.54) and (2.56).

Equation (2.54) provides us with the required connection between $\underline{u}$ and $\underline{v}$, but now we put in the arguments, according to the respective charts.

$$\begin{aligned}
\dot{\underline{y}} &= \dot{\underline{z}} \cdot \underline{\underline{\partial G}} \\
&= \underline{v}(\underline{z}) \cdot \underline{\underline{\partial G}}(\underline{z}) \\
&= \underline{v}(\underline{F}(\underline{y})) \cdot \underline{\underline{\partial G}}(\underline{F}(\underline{y})) \\
&= (\underline{v} \cdot \underline{\underline{\partial G}})_{(\underline{F}(\underline{y}))} \\
&\equiv \underline{u}(\underline{y}).
\end{aligned}$$

This can be written in functional form by eliminating the argument from the fourth line and using the composition operator:

$$\underline{u} = (\underline{v} \cdot \underline{\underline{\partial G}}) \circ \underline{F}. \tag{5.26}$$

Of course, a symmetric relation holds for $\underline{v}$ written in terms of $\underline{u}$.

Perturbation Series The program now is to find a vector field, $\underline{w}$, such that $\underline{u}$ is in normal form, that is, "as simple as possible." We do this perturbatively by deriving a homological equation to bootstrap the perturbation series. Functional notation will be used from the outset to avoid clutter; if this becomes too confusing, simply replace the formal arguments.

We assume that $\underline{v}$ is given to us as a polynomial or a power series in ϵ. (This does not necessarily mean it is a polynomial in the phase space coordinates.) Indeed, $\underline{v}$ is frequently a first degree polynomial in ϵ, so we are taking a very general case. Our objective is to construct the corresponding series for u; that is, if we write a formal series,

$$\underline{v}(\underline{z}, \epsilon) = \sum_{n=0}^{\infty} \frac{\epsilon^n}{n!} \underline{v}_n(\underline{z}) \quad \text{and} \quad \underline{u}(\underline{y}, \epsilon) = \sum_{n=0}^{\infty} \frac{\epsilon^n}{n!} \underline{u}_n(\underline{y}), \tag{5.27}$$

we seek to construct all the $\underline{u}_n$'s from the $\underline{v}_n$'s. This will be done through the identification

$$\underline{u}_n(\underline{y}) = (\partial_\epsilon^n \underline{u}(\underline{y}, \epsilon))_{d\underline{y}=\underline{0}, \epsilon=0}. \tag{5.28}$$

As the derivation is rather tedious, the impatient may skip to Eqs.(5.39) and (5.42), which contain the desired result. (But do not miss Eq. (5.30), Eq. (5.31), and Eq. (5.29), which are beautiful results on their own.) At the other extreme, a reader with extra time at his disposal may want to examine the elegant integral formulation of perturbation theory contained in Section 32 of Abraham and Robbin's *Transversal Mappings and Flows* (W.A. Benjamin, 1967). See especially the integro-differential equation which expresses their "32.1 Perturbation Theorem." I have strong suspicion that the algorithm to be developed in this section is nothing more or less than the detailed working out of this equation for its practical application.

We could short-circuit calculating the right-hand side of Eq. (5.28) by a simple argument. If we (a) accept that $\underline{u}$ and $\underline{v}$ are the same geometric vector field simply represented on two different charts and (b) recognize $(\partial_\epsilon)_{d\underline{y}=\underline{0}}$ as the local representation of the extended Lie derivative operator (see Eq. (2.70)), $D_\epsilon = \partial_\epsilon + L_{\underline{w}}$, for vector fields using the $\underline{y}$ coordinate chart, then we have immediately that

$$\underline{u}_n = \left(D_\epsilon^n \underline{v}\right)_{\epsilon=0}. \tag{5.29}$$

Coefficients of the new representation are obtained by applying D_ϵ repeatedly to the original one. This is now a relationship between functions whose domain is a single phase space chart; that is, the arguments are the same on both sides of this equation.

If this argument seems a little too pat and you are uncomfortably suspicious of having been swindled, take heart: here is a longer, more detailed proof, which we write in the form of a lemma.

Lemma 5.1 Eq. (5.29) is correct: $\underline{u}_n = \left(D_\epsilon^n \underline{v}\right)_{\epsilon=0}$, where D_ϵ is the extended Lie derivative operator, $D_\epsilon = \partial_\epsilon + L_{\underline{w}}$.

Proof: As a first step we calculate $\partial_\epsilon \underline{u}$, which is done by differentiating the right-hand side of Eq. (5.26).

$$\begin{aligned}\partial_\epsilon \underline{u} &= \partial_\epsilon((\underline{v} \cdot \underline{\underline{\partial G}}) \circ \underline{F}) \\ &= (\partial_\epsilon \underline{v} \cdot \underline{\underline{\partial G}}) \circ \underline{F} + (\underline{v} \cdot \partial_\epsilon \underline{\underline{\partial G}}) \circ \underline{F} \\ &\quad + \partial_\epsilon \underline{F} \cdot (\underline{\partial}(\underline{v} \cdot \underline{\underline{\partial G}}) \circ \underline{F}).\end{aligned}$$

In the last term we substitute from Eq. (5.24) for $\partial_\epsilon \underline{F}$, while in the second term we use Eq. (5.25) to express $\partial_\epsilon \underline{\underline{\partial G}} = \underline{\partial} \partial_\epsilon \underline{G} = -\underline{\partial}(\underline{w} \cdot \underline{\underline{\partial G}})$.

$$\begin{aligned}\partial_\epsilon((\underline{v} \cdot \underline{\underline{\partial G}}) \circ \underline{F}) &= (\partial_\epsilon \underline{v} \cdot \underline{\underline{\partial G}}) \circ \underline{F} - (\underline{v} \cdot \underline{\partial}(\underline{w} \cdot \underline{\underline{\partial G}})) \circ \underline{F} + (\underline{w} \cdot \underline{\partial}(\underline{v} \cdot \underline{\underline{\partial G}})) \circ \underline{F} \\ &= (\partial_\epsilon \underline{v} \cdot \underline{\underline{\partial G}} - \underline{v} \cdot \underline{\partial}(\underline{w} \cdot \underline{\underline{\partial G}}) + \underline{w} \cdot \underline{\partial}(\underline{v} \cdot \underline{\underline{\partial G}})) \circ \underline{F} \\ &= (\partial_\epsilon \underline{v} \cdot \underline{\underline{\partial G}} - \underline{v} \cdot (\underline{\underline{\partial w}} \cdot \underline{\underline{\partial G}}) - \underline{v} \cdot (\underline{w} \cdot \underline{\underline{\underline{\partial\partial G}}}) \\ &\quad + \underline{w} \cdot (\underline{\underline{\partial v}} \cdot \underline{\underline{\partial G}}) + \underline{w} \cdot (\underline{v} \cdot \underline{\underline{\underline{\partial\partial G}}})) \circ \underline{F}.\end{aligned}$$

The third and fifth terms cancel, since $\underline{\underline{\partial\partial}}$ is symmetric. We factorize what remains and write it as follows:

$$\begin{aligned}\partial_\epsilon((\underline{v} \cdot \underline{\underline{\partial G}}) \circ \underline{F}) &= ((\partial_\epsilon \underline{v} + \underline{w} \cdot \underline{\underline{\partial v}} - \underline{v} \cdot \underline{\underline{\partial w}}) \cdot \underline{\underline{\partial G}}) \circ \underline{F} \\ &= ((\partial_\epsilon \underline{v} + [\,\underline{w}, \underline{v}\,]) \cdot \underline{\underline{\partial G}}) \circ \underline{F}.\end{aligned}$$

We recognize our old friend, the extended Lie derivative, seen here with ϵ as the orbit parameter,

$$D_\epsilon \underline{v} \equiv \partial_\epsilon \underline{v} + [\,\underline{w}, \underline{v}\,]. \tag{5.30}$$

It is worth repeating the final result in terms of D_ϵ.

$$\partial_\epsilon((\underline{v} \cdot \underline{\underline{\partial G}}) \circ \underline{F}) = (D_\epsilon \underline{v} \cdot \underline{\underline{\partial G}}) \circ \underline{F}. \tag{5.31}$$

$D_\epsilon \underline{v}$ essentially looks like the total derivative of $\underline{v}$ under a flow generated by the vector field $\underline{w}$ and using ϵ as the orbit parameter ("time"). This interpretation is also consistent with Eq. (5.21) and Eq. (5.23). The operators ∂_ϵ and D_ϵ are seen to be equivalent under the linear map, $\underline{v} \longmapsto (\underline{v} \cdot \underline{\underline{\partial G}}) \circ \underline{F}$. (See Figure 5.2.) In particular, this means that we can iterate the relation, so that in general,

$$\partial_\epsilon^n((\underline{v} \cdot \underline{\underline{\partial G}}) \circ \underline{F}) = (D_\epsilon^n \underline{v} \cdot \underline{\underline{\partial G}}) \circ \underline{F}. \tag{5.32}$$

Matters now simplify considerably. According to Eq. (5.28) and Eq. (5.26) we only need evaluate Eq. (5.32) at $\epsilon = 0$. This reduces both $\underline{F}$ and $\underline{G}$ to the identity

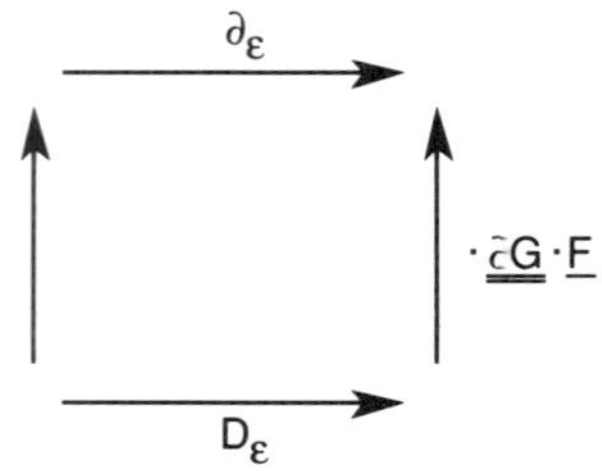

FIGURE 5.2 Commutative diagram of the equivalence between ∂_ϵ and D_ϵ.

map, and $(\underline{\underline{\partial G}})_{\epsilon=0} = \underline{\underline{1}}$, which provides us with an extraordinarily compact result:

$$\underline{u}_n = \left(D_\epsilon^n \underline{v}\right)_{\epsilon=0}. \qquad \square$$

To write that this is a beautiful result is to understate the matter seriously.

Systematization Given a generator, $\underline{w}$, of the transformation, Eq. (5.29) provides a rule for writing components of the vector field on the new chart. To make this procedure systematic, and indeed automatic, and to help us find the "most useful" generator, we shall write w itself as a formal power series.

$$\underline{w}(\underline{z}, \epsilon) = \sum_{n=0}^{\infty} \frac{\epsilon^n}{n!} \underline{w}_{n+1}(\underline{z}). \tag{5.33}$$

The shifted subscript is for later convenience. Do not be confused: $\underline{z}$ is simply a dummy argument now, indicating a point in $\mathcal{R}^n$. We could replace $\underline{z}$ with $\underline{y}$ in the above equation, and in fact shall do so below, without changing its meaning.

We now proceed to evaluate $D_\epsilon \underline{v}$ using this expansion, Eq. (5.27) and Eq. (5.30).

$$D_\epsilon \underline{v} = \sum_{n=1}^{\infty} \frac{\epsilon^{n-1}}{(n-1)!} \underline{v}_n + \sum_{n=0}^{\infty} \sum_{m=0}^{\infty} \frac{\epsilon^{m+n}}{m!n!} [\underline{w}_{m+1}, \underline{v}_n]. \tag{5.34}$$

Shift the index on the first sum, rewrite the second using an index $k \equiv m + n$, and collect terms.

$$\begin{aligned} D_\epsilon \underline{v} &= D_\epsilon \left(\sum_{k=0}^{\infty} \frac{\epsilon^k}{k!} \underline{v}_k \right) \\ &= \sum_{k=0}^{\infty} \frac{\epsilon^k}{k!} \left(\underline{v}_{k+1} + \sum_{m=0}^{k} \binom{k}{m} [\underline{w}_{m+1}, \underline{v}_{k-m}] \right) \\ &\equiv \sum_{k=0}^{\infty} \frac{\epsilon^k}{k!} \underline{v}_k^{(1)}. \end{aligned}$$

Effectively, D_ϵ replaces a series whose coefficients are $\underline{v}_k$ with one whose coefficients are $\underline{v}_k^{(1)}$, the relationship between the two being given by the second line above. Since D_ϵ is a linear operator we can iterate this to obtain coefficients $\underline{v}_k^{(2)}$, $\underline{v}_k^{(3)}$, and so forth.

$$D_\epsilon^n \underline{v} = \sum_{k=0}^{\infty} \frac{\epsilon^k}{k!} \underline{v}_k^{(n)},$$

$$\underline{v}_k^{(n)} \equiv \underline{v}_{k+1}^{(n-1)} + \sum_{m=0}^{k} \binom{k}{m} [\underline{w}_{m+1}, \underline{v}_{k-m}^{(n-1)}]. \tag{5.35}$$

This will hold for all n if we make the natural identification, $\underline{v}_k^{(0)} \equiv \underline{v}_k$.

According to Eq. (5.29), we only need the coefficients, $\underline{v}_0^{(n)}$; as we shall see, the rest serve as "auxiliary functions" to aid in their computation. Putting the pieces together, we rewrite Eq. (5.27) as follows:

$$\underline{u}(\underline{y}, \epsilon) = \sum_{n=0}^{\infty} \frac{\epsilon^n}{n!} \underline{v}_0^{(n)}(\underline{y}). \tag{5.36}$$

We will perform one final set of manipulations in order to remove the auxiliary functions, $\underline{v}_k^{(n)}, k \neq 0$, from consideration. Begin by bringing the sum to the other side of Eq. (5.35), and replacing n with $n + 1$ and k with $k - 1$.

$$\underline{v}_k^{(n)} = \underline{v}_{k-1}^{(n+1)} - \sum_{m=0}^{k-1} \binom{k-1}{m} [\underline{w}_{m+1}, \underline{v}_{k-1-m}^{(n)}]. \tag{5.37}$$

This expresses the auxiliary function, $\underline{v}_k^{(n)}$, as a superposition of linear functionals of auxiliary functions with smaller subscripts. By repeating this on the right-hand side, we eventually would be able to write $\underline{v}_k^{(n)}$ as a superposition of linear functionals of $\underline{v}_0^{(m)}, m = n, \ldots, n + k$. Formalize this by writing

$$\underline{v}_k^{(n)} = \sum_{j=0}^{k} \binom{k}{j} \underline{\mathcal{K}}_j(\underline{v}_0^{(n+k-j)}), \tag{5.38}$$

where the $\underline{\mathcal{K}}_j$ are linear operators to be determined. Clearly, by setting $k = 0$, we must have $\underline{\mathcal{K}}_0(\underline{v}) = \underline{v}$, which will serve as the seed in the recursion about to be written. That such $\underline{\mathcal{K}}_j$ exist is not completely trivial, as *they do not depend on n*. We establish their existence by construction, and write the result as a lemma.

Lemma 5.2 The linear operators sought by Eq.(5.38) are constructed recursively according to

$$\underline{\mathcal{K}}_0 = \underline{I},$$

$$\underline{\mathcal{K}}_j() = - \sum_{m=0}^{j-1} \binom{j-1}{m} [\underline{w}_{j-m}, \underline{\mathcal{K}}_m()]. \tag{5.39}$$

Proof: We work somewhat backwards. Consider the difference, from Eq. (5.38),

$$\underline{v}_k^{(n)} - \underline{v}_{k-1}^{(n+1)} = \underline{\mathcal{K}}_k(\underline{v}_0^{(n)}) + \sum_{j=0}^{k-1}\left[\binom{k}{j} - \binom{k-1}{j}\right]\underline{\mathcal{K}}_j(\underline{v}_0^{(n+k-j)})$$

$$= \sum_{j=1}^{k}\binom{k-1}{j-1}\underline{\mathcal{K}}_j(\underline{v}_0^{(n+k-j)}), \tag{5.40}$$

the last line following from the basic, Pascal triangle recursion among binomial coefficients. With an eye to inserting this into Eq. (5.37), we write

$$\underline{v}_{k-1-m}^{(n)} = \sum_{j=0}^{k-1-m}\binom{k-1-m}{j}\underline{\mathcal{K}}_j(\underline{v}_0^{(n+k-1-m-j)})$$

$$= \sum_{j=m+1}^{k}\binom{k-1-m}{j-1-m}\underline{\mathcal{K}}_{j-1-m}(\underline{v}_0^{(n+k-j)}).$$

Combine this with Eq. (5.40) and Eq. (5.37).

$$\sum_{j=1}^{k}\binom{k-1}{j-1}\underline{\mathcal{K}}_j(\underline{v}_0^{n+k-j})$$

$$= -\sum_{m=0}^{k-1}\sum_{j=m+1}^{k}\binom{k-1}{m}\binom{k-1-m}{j-1-m}[\underline{w}_{m+1}, \underline{\mathcal{K}}_{j-1-m}(\underline{v}_0^{(n+k-j)})].$$

We now interchange the order of summations and use the combinatorial identity,

$$\binom{k-1}{m}\binom{k-1-m}{j-1-m} = \binom{k-1}{j-1}\binom{j-1}{m},$$

to write

$$\sum_{j=1}^{k}\binom{k-1}{j-1}\left(\underline{\mathcal{K}}_j(\underline{v}_0^{(n+k-j)}) + \sum_{m=0}^{j-1}\binom{j-1}{m}[\underline{w}_{m+1}, \underline{\mathcal{K}}_{j-1-m}(\underline{v}_0^{(n+k-j)})]\right) = 0.$$

This will clearly be satisfied provided that we define the $\underline{\mathcal{K}}_j$'s according to the rule

$$\underline{\mathcal{K}}_j(\underline{v}) = -\sum_{m=0}^{j-1}\binom{j-1}{m}[\underline{w}_{m+1}, \underline{\mathcal{K}}_{j-1-m}(\underline{v})]$$

$$= -\sum_{m=0}^{j-1}\binom{j-1}{m}[\underline{w}_{j-m}, \underline{\mathcal{K}}_m(\underline{v})],$$

which is the recursion that we sought.[10] The proof is now formally completed by running through these manipulations in reverse. □

Considered as operators, the $\underline{\mathcal{K}}_m$'s are clearly polynomials in Lie derivatives, as defined in Eq. (2.70). Recast as an operator equation, Eq. (5.39) becomes

$$\mathbf{K}_j = -\sum_{m=0}^{j-1} \binom{j-1}{m} \mathbf{L}_{\underline{w}_{j-m}} \mathbf{K}_m,$$

the first few steps of which are

$$\begin{aligned}
\mathbf{K}_0 &= \mathbf{1}, \\
\mathbf{K}_1 &= -\mathbf{L}_{\underline{w}_1}, \\
\mathbf{K}_2 &= -\mathbf{L}_{\underline{w}_2} + \mathbf{L}^2_{\underline{w}_1}, \\
\mathbf{K}_3 &= -\mathbf{L}_{\underline{w}_3} + 2\mathbf{L}_{\underline{w}_2}\mathbf{L}_{\underline{w}_1} + \mathbf{L}_{\underline{w}_1}\mathbf{L}_{\underline{w}_2} - \mathbf{L}^3_{\underline{w}_1}.
\end{aligned}$$

However, while they appear sophisticated and elegant, there is (probably) no advantage to writing these operator polynomials explicitly.

A direct connection between the $\underline{u}_k$'s and $\underline{v}_k$'s is obtained, at last, by setting $n = 0$ in Eq. (5.38).

$$\underline{v}_k^{(0)} = \sum_{j=0}^{k} \binom{k}{j} \underline{\mathcal{K}}_j(\underline{v}_0^{(k-j)}) = \underline{v}_0^{(k)} + \sum_{j=1}^{k} \binom{k}{j} \underline{\mathcal{K}}_j(\underline{v}_0^{(k-j)}). \tag{5.41}$$

Making the identifications $\underline{v}_k^{(0)} = \underline{v}_k$ and $\underline{v}_0^{(k)} = \underline{u}_k$ (see Eq. (5.36)) and rearranging terms slightly, we obtain our final expression,

$$\underline{u}_k = \underline{v}_k - \sum_{j=1}^{k} \binom{k}{j} \underline{\mathcal{K}}_j(\underline{u}_{k-j}). \tag{5.42}$$

To get some idea of how this all works, we pause to consider a few illustrations, after which we will once again pick up the thread of this development.

Example: General We have, from Eq. (5.42) and Eq. (5.39),

$$\begin{aligned}
\underline{u}_0 &= \underline{v}_0, \\
\underline{u}_1 &= \underline{v}_1 + [\,\underline{w}_1, \underline{v}_0]. \qquad (5.43)
\end{aligned}$$

[10] I am tempted to write the mnemonic, $\mathcal{K}^j = w(w + \mathcal{K})^{j-1}$, in analogy with the one for Bernoulli numbers.

We have substituted $\underline{v}_0$ for $\underline{u}_0$ in the second line. The strategy is to choose $\underline{w}_1$ so that $\underline{u}_1$ is as simple as possible, that is, to bring $\underline{u}_1$ into "normal form." To eliminate this first degree term entirely would require $\underline{w}_1$ to satisfy

$$[\underline{v}_0, \underline{w}_1] = \underline{v}_1.$$

As in the preceding section (see Eq. (5.11)) it is not guaranteed that we can do this, as $\underline{v}_1$ may contain terms not in the range of $\mathbf{L}_{\underline{v}_0}$. Such terms, if they exist, must necessarily be swallowed by $\underline{u}_1$ before the equation admits a solution. The simplest $\underline{u}_1$ will then consist of only those terms. (Of course, we always have the freedom to make $\underline{u}_1$ *more* complicated.)

The second degree term is written similarly.

$$\begin{aligned}
\underline{u}_2 &= \underline{v}_2 - 2\mathcal{K}_1(\underline{u}_1) - \mathcal{K}_2(\underline{u}_0) \\
&= \underline{v}_2 + 2[\underline{w}_1, \underline{u}_1] + [\underline{w}_2, \underline{u}_0] + [\underline{w}_1, \mathcal{K}_1(\underline{u}_0)] \\
&= \underline{v}_2 + 2[\underline{w}_1, \underline{u}_1] + [\underline{w}_2, \underline{v}_0] - [\underline{w}_1, [\underline{w}_1, \underline{v}_0]] \\
&= \underline{v}_2 + 2[\underline{w}_1, \underline{v}_1] - [\underline{w}_1, [\underline{w}_1, \underline{v}_0]] - [\underline{v}_0, \underline{w}_2].
\end{aligned} \tag{5.44}$$

Again our task is to suppress $\underline{u}_2$, which is done by limiting it to the minimum necessary in order that the equation be solvable, that is, to the component of $\underline{v}_2 + 2[\underline{w}_1, \underline{v}_1] - [\underline{w}_1, [\underline{w}_1, \underline{v}_0]]$ not in the range of $\mathbf{L}_{\underline{v}_0}$.

The pattern is obvious. At the nth stage of the sequence we expect, and obtain, a homological equation of the form

$$L_{v_0}\underline{w}_n = [\underline{v}_0, \underline{w}_n] = (\underline{v}_n + \text{ other stuff}) - \underline{u}_n.$$

(See Eq. (5.52) below.) The minimal $\underline{u}_n$ will swallow only those terms on the right-hand side within the null space of $\mathbf{L}_{\underline{v}_0}$, which, put back into the series, becomes the "normal form" vector field, $\underline{u}$. In essence, the mathematical problem is identical to that exhibited in Eq. (5.11). And like Eq. (5.11) or Eq. (5.5), what gives rise to these terms is the existence of critical orbits, producing topological structures that cannot be transformed away, and expressed analytically by the presence of "small denominators." We shall see shortly that, in the contexts of accelerator Hamiltonians and of the present method, these give rise to the classic resonance and shear models presented in the last chapter, *but now understood to be applied to the transformed coordinates.*

Example: Linear Dynamics In applying the perturbation procedure to a linear dynamic, all the functions that appear are themselves linear, as is the final transformation. We associate with each of them a corresponding matrix, as

$$\underline{v}_n(\underline{z}) = \underline{\underline{V_n}} \cdot \underline{z}, \qquad \underline{w}_n(\underline{z}) = \underline{\underline{W_n}} \cdot \underline{z}, \quad \text{and so forth.}$$

In what follows, recall from Problem 2.29 that the matrix commutator algebra of linear dynamics is an antimorphism of the Lie bracket algebra: if $\underline{a}$ and $\underline{b}$ are linear dynamics with matrix representations $\underline{\underline{A}}$ and $\underline{\underline{B}}$, then their Lie bracket, $[\underline{a}, \underline{b}]$, is a linear dynamic with matrix $-[\underline{\underline{A}}, \underline{\underline{B}}]$. From Eq. (5.43) we have immediately

$$\begin{aligned} \underline{\underline{U_0}} &= \underline{\underline{V_0}}, \\ \underline{\underline{U_1}} &= \underline{\underline{V_1}} + [\underline{\underline{V_0}}, \underline{\underline{W_1}}]. \end{aligned} \tag{5.45}$$

For the sake of making a correspondence between what we do here and quantum perturbation theory, we shall assume that the unperturbed vector field is diagonal.

$$(\underline{\underline{V_0}})_{ij} = \lambda_i \delta_{ij}, \quad \text{or} \quad (\underline{v}_0(\underline{z}))_i = \lambda_i z_i.$$

We leave ambiguous the question of whether these eigenvalues are real or complex; those whom complex phase space makes uncomfortable may simply assume that they are real. In light of this, we can evaluate the commutator in Eq. (5.45) and rewrite it in component form as follows:

$$\left(\underline{\underline{U_1}}\right)_{ij} = \left(\underline{\underline{V_1}}\right)_{ij} + (\lambda_i - \lambda_j)\left(\underline{\underline{W_1}}\right)_{ij}. \tag{5.46}$$

Our "normal form" mentality asserts that we choose $\underline{\underline{W_1}}$ so as to zero $\underline{\underline{U_1}}$. A problem arises if $\underline{\underline{V_0}}$ is degenerate, so that $\lambda_i = \lambda_j$ for some $i \neq j$. Perturbation theory alone cannot zero the matrix elements in the degenerate subspaces, and the matrix elements of $\underline{\underline{U}}$ and $\underline{\underline{V}}$ will be identical in those sectors. Otherwise, if there is no degeneracy, or in the nondegenerate blocks, we can choose

$$\left(\underline{\underline{W_1}}\right)_{ij} = -\frac{1}{\lambda_i - \lambda_j}\left(\underline{\underline{V_1}}\right)_{ij}, \quad \text{for } i \neq j. \tag{5.47}$$

This has the effect of zeroing all off-diagonal elements of $\underline{\underline{U_1}}$. The residual terms are the diagonal elements:

$$\left(\underline{\underline{U_1}}\right)_{ij} = \left(\underline{\underline{V_1}}\right)_{ii} \delta_{ij}. \tag{5.48}$$

Diagonal elements, $\underline{\underline{W_1}}_{ii}$, are not at all determined by this procedure. We are free to take $\underline{\underline{W_1}}_{ii} = 0$.

The second degree matrices we obtain from Eq. (5.44).

$$\underline{\underline{U_2}} = \underline{\underline{V_2}} + 2[\underline{\underline{V_1}}, \underline{\underline{W_1}}] + [\underline{\underline{W_1}}, [\underline{\underline{V_0}}, \underline{\underline{W_1}}]] + [\underline{\underline{V_0}}, \underline{\underline{W_2}}].$$

We treat the most difficult term first. Components of the commutator, $[\underline{\underline{V_0}}, \underline{\underline{W_1}}]$, were written in Eq. (5.46), and $[\underline{\underline{V_0}}, \underline{\underline{W_2}}]$ will look similar. Combine this with $\underline{\underline{W_1}}$ from

Eq. (5.47) to obtain

$$\begin{aligned}
([\underline{\underline{W_1}}, [\underline{\underline{V_0}}, \underline{\underline{W_1}}]])_{ij} &= \sum_k \bar{\delta}_{ik} (\underline{\underline{V_1}})_{ik} \bar{\delta}_{kj} \frac{1}{\lambda_k - \lambda_j} (\underline{\underline{V_1}})_{kj} \\
&\quad - \sum_k \bar{\delta}_{ik} \frac{1}{\lambda_i - \lambda_k} (\underline{\underline{V_1}})_{ik} \bar{\delta}_{kj} (\underline{\underline{V_1}})_{kj} \\
&= \bar{\delta}_{ij} \frac{1}{\lambda_i - \lambda_j} ((\underline{\underline{V_1}})_{ij} (\underline{\underline{V_1}})_{jj} - (\underline{\underline{V_1}})_{ii} (\underline{\underline{V_1}})_{ij}) \\
&\quad + \sum_k (\underline{\underline{V_1}})_{ik} \left(\bar{\delta}_{kj} \frac{1}{\lambda_k - \lambda_j} - \bar{\delta}_{ik} \frac{1}{\lambda_i - \lambda_k} \right) (\underline{\underline{V_1}})_{kj},
\end{aligned}$$

where $\bar{\delta}_{ij} = 1 - \delta_{ij}$. We also have

$$([\underline{\underline{V_1}}, \underline{\underline{W_1}}])_{ij} = - \sum_k (\underline{\underline{V_1}})_{ik} \left(\bar{\delta}_{kj} \frac{1}{\lambda_k - \lambda_j} - \bar{\delta}_{ik} \frac{1}{\lambda_i - \lambda_k} \right) (\underline{\underline{V_1}})_{kj},$$

which promises to combine nicely with the previous expression. Putting the pieces together provides us with the final result.

$$\begin{aligned}
(\underline{\underline{U_2}})_{ij} &= (\underline{\underline{V_2}})_{ij} + \sum_k (\underline{\underline{V_1}})_{ik} \left(\bar{\delta}_{ik} \frac{1}{\lambda_i - \lambda_k} - \bar{\delta}_{kj} \frac{1}{\lambda_k - \lambda_j} \right) (\underline{\underline{V_1}})_{kj} \\
&\quad + \bar{\delta}_{ij} \frac{1}{\lambda_i - \lambda_j} ((\underline{\underline{V_1}})_{ij} (\underline{\underline{V_1}})_{jj} - (\underline{\underline{V_1}})_{ii} (\underline{\underline{V_1}})_{ij}) + (\lambda_i - \lambda_j)(\underline{\underline{W_2}})_{ij}.
\end{aligned}$$

Again, for $\lambda_i \neq \lambda_j$ we choose $\underline{\underline{W_2}}$ so that $(\underline{\underline{U_2}})_{ij} = 0$. Residual elements exist in the degenerate subspaces, if there are any, and certainly on the diagonals. To compare with quantum theory, we recall that $\frac{1}{2!}\underline{\underline{U_2}}$ and $\frac{1}{2!}\underline{\underline{V_2}}$ are the coefficients that appear in our perturbation series, and we write

$$\frac{1}{2!} (\underline{\underline{U_2}})_{ii} = \frac{1}{2!} (\underline{\underline{V_2}})_{ii} + \sum_k \bar{\delta}_{ik} \frac{(\underline{\underline{V_1}})_{ik} (\underline{\underline{V_1}})_{ki}}{\lambda_i - \lambda_k}. \tag{5.49}$$

Equations (5.48) and (5.49) are precisely those obtained when studying the perturbed Schrödinger equation in quantum mechanics, where usually $\underline{\underline{V_2}} = \underline{\underline{0}}$. Notice that the goal here was *not explicitly stated* as making the new representation diagonal but bringing it to "normal form," making it "as simple as possible." Sensibly enough, it turns out that the normal form coordinates are indeed the eigencoordinates of the linear problem, the residual vector field providing the perturbed (renormalized?) eigenvalues. The "small denominator" problem is here associated with near-degeneracy: eigenvalues of the unperturbed system that are "almost" equal.

While this formalism is not a reasonable substitute for the more elegant and computationally succinct Lippman-Schwinger equation [60, 75], its utility is not really evident until one considers nonlinear dynamics.

Example: Van der Pol Oscillator We consider Van der Pol's equation one final time and apply the perturbation procedure to it. The vector field is represented in action-angle coordinates by employing Eq. (2.9).

$$\begin{aligned} \dot{a}/a &= \dot{I}/2I - i\dot{\varphi}, \\ \dot{I} &= 2I\,\mathrm{Re}(\dot{a}/a), \\ \dot{\varphi} &= -\mathrm{Im}(\dot{a}/a). \end{aligned}$$

The expression for $\dot{a}$ was previously established in Eq. (4.1), from which we obtain

$$\dot{a}/a = -i + \tfrac{1}{2}\gamma(1 + e^{i2\varphi})(\mu - 3I(1 - \cos 2\varphi)).$$

Plugging this in above and rearranging terms into harmonic order we find

$$\begin{aligned} \dot{\varphi} &= 1 + \gamma[\tfrac{1}{2}(3I - \mu)\sin 2\varphi - \tfrac{3}{4}I\sin 4\varphi] \\ \dot{I} &= \gamma[I(\mu - \tfrac{3}{2}I) + \mu I\cos 2\varphi + \tfrac{3}{2}I^2\cos 4\varphi]. \end{aligned}$$

In applying perturbation theory to this we treat γ as the order parameter, which means we are considering the case of weak nonlinearity. Decompose the original vector field as follows:

$$\begin{aligned} \mathbf{V} &\equiv \mathbf{V}_0 + \gamma\mathbf{V}_1, \\ \mathbf{V}_0 &= \frac{\partial}{\partial\varphi}, \\ \mathbf{V}_1 &= (I(\mu - \tfrac{3}{2}I) + \mu I\cos 2\varphi + \tfrac{3}{2}I^2\cos 4\varphi)\frac{\partial}{\partial I} \\ &\quad + (\tfrac{1}{2}(3I - \mu)\sin 2\varphi - \tfrac{3}{4}I\sin 4\varphi)\frac{\partial}{\partial\varphi}. \end{aligned}$$

The first degree homological equation is very simple.

$$[\mathbf{W}_1, \mathbf{V}_0] = -\frac{\partial W_{1I}}{\partial\varphi}\frac{\partial}{\partial I} - \frac{\partial W_{1\varphi}}{\partial\varphi}\frac{\partial}{\partial\varphi} = \mathbf{U}_1 - \mathbf{V}_1.$$

Ideally we would like to set $\mathbf{U}_1 = \mathbf{0}$. However, $\mathbf{V}_1$ contains the term $I(\mu - \frac{3}{2}I)\partial/\partial I$. If this is retained, then W_{1I} will contain a term that grows linearly in φ. This is not an impossibility: it simply means that our transformation would be multiple-valued, and keeping track of the orbit properly would require an infinite atlas. Most people prefer the option of allowing the field $\mathbf{U}_1$ to swallow this term, absorbing the limit cycle into the dynamics rather than leaving it in the transformation,

$$\mathbf{U}_1 \equiv I(\mu - \tfrac{3}{2}I)\frac{\partial}{\partial I}.$$

The resultant Lie generator field is then periodic in φ.

$$\mathbf{W}_1 = (\tfrac{1}{4}(3I - \mu)\cos 2\varphi - \tfrac{3}{16}I\cos 4\varphi)\frac{\partial}{\partial\varphi} - (\tfrac{1}{2}\mu I\sin 2\varphi + \tfrac{3}{8}I^2\sin 4\varphi)\frac{\partial}{\partial I}. \tag{5.50}$$

We recognize, of course, the emergence of the limit cycle, as in Eqs.(4.2) and (5.1). If it did not appear here, the transformation would have to accomplish it via a multivalued term. That is the trade-off one makes by electing to swallow these terms: it is a choice between burying the limit cycle in the dynamics, in which case the transformation is single-valued, or in the transformation, in which case the dynamics is made more complicated. However, Eq. (5.1) illustrates how a second, independent perturbation series can be developed in tandem to simplify even further.

Homological Equations As we saw in these examples, at each order, k, we must solve a homological equation for $\underline{w}_k$ and $\underline{u}_k$, with the objective of making $\underline{u}_k$ as simple as possible. The general homological equation is buried in Eq. (5.42), rather opaquely because of its brevity. Somewhat less elegant but more useful expressions can be obtained by singling out the term $[\underline{w}_k, \underline{v}_0]$. That term appears in $\underline{\mathcal{K}}(\underline{u}_0)$, which we first isolate.

$$\underline{u}_k = \underline{v}_k - \underline{\mathcal{K}}_k(\underline{u}_0) - \sum_{m=1}^{k-1}\binom{k}{m}\underline{\mathcal{K}}_m(\underline{u}_{k-m}). \tag{5.51}$$

From Eq. (5.39) we have that

$$\begin{aligned}\underline{\mathcal{K}}_k(\underline{u}_0) &= -\sum_{m=0}^{k-1}\binom{k-1}{m}[\underline{w}_{k-m}, \underline{\mathcal{K}}_m(\underline{u}_0)] \\ &= -[\underline{w}_k, \underline{u}_0] - \sum_{m=1}^{k-1}\binom{k-1}{m}[\underline{w}_{k-m}, \underline{\mathcal{K}}_m(\underline{u}_0)],\end{aligned}$$

where we have used $\underline{\mathcal{K}}_0(\underline{u}_0) = \underline{u}_0$. Substituting this back into Eq. (5.51), employing the identity, $\underline{u}_0 = \underline{v}_0$, and rearranging terms provide the homological equation for general degree, k.

$$[\underline{w}_k, \underline{v}_0] = \underline{u}_k - \underline{v}_k + \sum_{m=1}^{k-1}\left(\binom{k}{m}\underline{\mathcal{K}}_m(\underline{u}_{k-m}) - \binom{k-1}{m}[\underline{w}_{k-m}, \underline{\mathcal{K}}_m(\underline{v}_0)]\right). \tag{5.52}$$

The presence of $\underline{\mathcal{K}}_m(\underline{v}_0)$ would prove extremely inconvenient later, when we specialize to nonautonomous Hamiltonian systems. Another expression can be derived using an ingenious little trick, due to Kamel [51, 52]. From Eq. (5.38) we have

$$\underline{v}_{k-1}^{(1)} = \sum_{j=0}^{k-1} \binom{k-1}{j} \underline{\mathcal{K}}_j(\underline{u}_{k-j}) = \underline{u}_k + \sum_{j=1}^{k-1} \binom{k-1}{j} \underline{\mathcal{K}}_j(\underline{u}_{k-j}).$$

On the other hand, using Eq. (5.35) produces the following:

$$\underline{v}_{k-1}^{(1)} = \underline{v}_k + \sum_{m=0}^{k-1} \binom{k-1}{m} [\underline{w}_{m+1}, \underline{v}_{k-1-m}].$$

Equating these two, isolating the term $[\underline{w}_k, \underline{v}_0]$, shifting dummy indices, and rearranging slightly result in the homological equation,

$$\begin{aligned} [\underline{w}_k, \underline{v}_0] &= \underline{u}_k - \underline{v}_k \\ &- \sum_{m=1}^{k-1} \left\{ \binom{k-1}{m-1} [\underline{w}_m, \underline{v}_{k-m}] - \binom{k-1}{m} \underline{\mathcal{K}}_m(\underline{u}_{k-m}) \right\}. \end{aligned} \tag{5.53}$$

In addition to having $\underline{v}_0$ $(= \underline{u}_0)$ appear only in the commutator on the left-hand side, this homological equation has the additional advantage over Eq. (5.52) that it is written explicitly in terms of the sequence of $\underline{v}_k$'s; in many problems, all of them beyond $\underline{v}_1$ vanish.

Coordinate Transformations Having established the homological equations whose solution provides the generator, $\underline{w}(\underline{z}, \epsilon)$, we now turn to the problem of actually writing the transformation functions, $\underline{F}$ and $\underline{G}$, themselves. As with all else, this will be accomplished by expansions in the order parameter.

$$\underline{z} = \underline{F}(\underline{y}, \epsilon) \equiv \sum_{n=0}^{\infty} \frac{\epsilon^n}{n!} \underline{F}_n(\underline{y}), \underline{y} = \underline{G}(\underline{z}, \epsilon) \equiv \sum_{n=0}^{\infty} \frac{\epsilon^n}{n!} \underline{G}_n(\underline{z}), .$$

The coefficients, $\underline{G}_n$, are constructed easily using a recursion relation based on Eq. (5.25). Inserting the expansions, we obtain

$$\begin{aligned} \sum_{k=0}^{\infty} \frac{\epsilon^k}{k!} \underline{G}_{k+1}(\underline{z}) &= -\left(\sum_{n=0}^{\infty} \frac{\epsilon^n}{n!} \underline{w}_{n+1}(\underline{z}) \right) \cdot \left(\sum_{m=0}^{\infty} \frac{\epsilon^m}{m!} \underline{\underline{\partial G}}_m(\underline{z}) \right) \\ &= -\sum_{k=0}^{\infty} \frac{\epsilon^k}{k!} \left(\sum_{m=0}^{k} \binom{k}{m} \underline{w}_{k+1-m}(\underline{z}) \cdot \underline{\underline{\partial G}}_m(\underline{z}) \right). \end{aligned}$$

Since the arguments are the same on both sides, we can write the function recursion, for all $k \geqslant 1$,

$$\underline{G}_k = -\sum_{m=0}^{k-1} \binom{k-1}{m} \underline{w}_{k-m} \cdot \underline{\underline{\partial G}}_m. \tag{5.54}$$

The seed of this recursion is $\underline{G}_0$, as is required by $\underline{G}$ being a near-identity transformation, so that the linearized dynamics are represented identically in both variables.

$$\underline{G}_0(\underline{z}) = \underline{z}.$$

Finding the coefficients, $\underline{F}_n$, is a little more involved. We begin from Eq. (5.22), which defines the vector field, $\underline{w}(\underline{z}, \epsilon)$. Inserting the expansions into this relation produces

$$\sum_{n=0}^{\infty} \frac{\epsilon^n}{n!} \underline{w}_{n+1}(\underline{z}) = \sum_{n=0}^{\infty} \frac{\epsilon^n}{n!} \underline{F}_{n+1}(\underline{y}).$$

Regrettably, we cannot simply deduce that $\underline{w}_{n+1} = \underline{F}_{n+1}$, as their arguments are different. We can, however, express

$$\begin{aligned} \underline{F}_{n+1}(\underline{y}) &= \left(\left(\frac{d}{d\epsilon} \right)^n_{d\underline{y}=\underline{0}} \sum_{k=0}^{\infty} \frac{\epsilon^k}{k!} \underline{w}_{k+1}(\underline{z}) \right)_{\epsilon=0} \\ &\equiv \left(\sum_{k=0}^{\infty} \frac{\epsilon^k}{k!} \underline{w}^{(n)}_{k+1}(\underline{z}) \right)_{\epsilon=0} \\ &= \underline{w}^{(n)}_1(\underline{y}), \end{aligned}$$

where we recall, from Eq's (5.21) and (5.23), that on the $\underline{z}$ chart, $d/d\epsilon = \partial_\epsilon + \underline{w} \cdot \underline{\partial}$. The calculations we need look formally identical to Eq. (5.34), except that the Lie bracket has been replaced by $\underline{w} \cdot \underline{\partial}$. By the same sort of manipulations as before we obtain an equation that is the analog of Eq. (5.35).

$$\underline{w}^{(n)}_k = \underline{w}^{(n-1)}_{k+1} + \sum_{m=0}^{k-1} \binom{k-1}{m} \underline{w}_{m+1} \cdot \underline{\underline{\partial w}}^{(n-1)}_{k-m}.$$

That the indices have slipped a little relative to Eq. (5.35) is an artifact of having defined $\underline{w}$ in terms of a shifted index (see Eq. (5.33)). In fact, if we replace $\underline{\tilde{w}}^{(n)}_k \equiv \underline{w}^{(n)}_{k+1}$ everywhere, these differences would be eliminated. We can then follow the same path that took us from Eq. (5.35) to Eq. (5.41). Accordingly, we define linear operators $\underline{\mathcal{L}}_j$, using the recursive scheme,

$$\begin{aligned} \underline{\mathcal{L}}_0(\underline{v}) &= \underline{v}, \\ \underline{\mathcal{L}}_j(\underline{v}) &= -\sum_{m=0}^{j-1} \binom{j-1}{m} \underline{w}_{j-m} \cdot \underline{\partial}(\underline{\mathcal{L}}_m(\underline{v})), \end{aligned} \tag{5.55}$$

in terms of which we obtain

$$\underline{w}^{(0)}_{k+1} = \underline{w}^{(k)}_1 + \sum_{j=1}^{k} \binom{k}{j} \underline{\mathcal{L}}_j(\underline{w}^{(k-j)}_1).$$

Now identify $\underline{w}_1^{(k)} = \underline{F}_{k+1}$ and $\underline{w}_{k+1}^{(0)} = \underline{w}_{k+1}$ to get the final result,

$$\underline{F}_k = \underline{w}_k - \sum_{j=1}^{k-1} \binom{k-1}{j} \underline{\mathcal{L}}_j(\underline{F}_{k-j}), \tag{5.56}$$

the seed of which is, of course, $\underline{F}_0(\underline{y}) = \underline{y}$. We note in passing that this can be written more symmetrically by using $\underline{\mathcal{L}}_0(\underline{F}_k) = \underline{F}_k$.

$$\underline{w}_k = \sum_{j=0}^{k-1} \binom{k-1}{j} \underline{\mathcal{L}}_j(\underline{F}_{k-j}).$$

To emphasize the algorithmic, procedural nature of these rather obscure looking equations, we explicitly write the first four orders below.

$$\begin{aligned}
\underline{F}_1 &= \underline{w}_1, \\
&\quad \underline{\mathcal{L}}_1(\underline{F}_1) = -\underline{w}_1 \cdot \underline{\partial}(\underline{F}_1); \\
\underline{F}_2 &= \underline{w}_2 - \underline{\mathcal{L}}_1(\underline{F}_1), \\
&\quad \underline{\mathcal{L}}_1(\underline{F}_2) = -\underline{w}_1 \cdot \underline{\partial}(\underline{F}_2), \\
&\quad \underline{\mathcal{L}}_2(\underline{F}_1) = -\underline{w}_2 \cdot \underline{\partial}(\underline{F}_1) - \underline{w}_1 \cdot \underline{\partial}(\underline{\mathcal{L}}_1(\underline{F}_1)); \qquad (5.57) \\
\underline{F}_3 &= \underline{w}_3 - 2\underline{\mathcal{L}}_1(\underline{F}_2) - \underline{\mathcal{L}}_2(\underline{F}_1), \\
&\quad \underline{\mathcal{L}}_1(\underline{F}_3) = -\underline{w}_1 \cdot \underline{\partial}(\underline{F}_3), \\
&\quad \underline{\mathcal{L}}_2(\underline{F}_2) = -\underline{w}_2 \cdot \underline{\partial}(\underline{F}_2) - \underline{w}_1 \cdot \underline{\partial}(\underline{\mathcal{L}}_1(\underline{F}_2)), \\
&\quad \underline{\mathcal{L}}_3(\underline{F}_1) = -\underline{w}_3 \cdot \underline{\partial}(\underline{F}_1) - 2\underline{w}_2 \cdot \underline{\partial}(\underline{\mathcal{L}}_1(\underline{F}_1)) - \underline{w}_1 \cdot \underline{\partial}(\underline{\mathcal{L}}_2(\underline{F}_1)); \\
\underline{F}_4 &= \underline{w}_4 - 3\underline{\mathcal{L}}_1(\underline{F}_3) - 3\underline{\mathcal{L}}_2(\underline{F}_2) - \underline{\mathcal{L}}_3(\underline{F}_1). \qquad (5.58)
\end{aligned}$$

Clearly these are in a form ready for programming.

5.2.2 Hamiltonian Dynamics

Because the iterative steps are written in terms of Lie brackets, and because of the antimorphism between Lie and Poisson brackets expressed in Eqs.(3.7) through (3.9), their specialization to Hamiltonian systems is immediate and easy. Looking ahead to accelerator applications, we perform the reduction in a form suitable for nonautonomous as well as autonomous Hamiltonian systems.

Autonomous We begin, however, with the autonomous case, as it is straightforward, and the other is treated more easily with it in hand. Accordingly, we write the vector field representations, $\underline{u}$ and $\underline{v}$, *and* $\underline{w}$ as well, in Hamiltonian form:

$$\underline{v} = \underline{\underline{J}} \cdot \underline{\partial} H, \quad \underline{u} = \underline{\underline{J}} \cdot \underline{\partial} K, \quad \underline{w} = \underline{\underline{J}} \cdot \underline{\partial} S.$$

These relations are understood to include the ϵ-ordered expansions of these objects, such as $\underline{v}_n = \underline{\underline{J}} \cdot \partial H_n$, and so forth. It now follows that each $\underline{\mathcal{K}_j}(\underline{v})$ also represents a Hamiltonian vector field, which proposition we prove in the form of a lemma.

Lemma 5.3 Let $\underline{\mathcal{K}_j}(\underline{v})$ be defined as in Eq.(5.39), with $\underline{v}$ and $\underline{w}_n, n = 1, 2, \ldots,$ representing arbitrary Hamiltonian vector fields. Then for every j, $\underline{\mathcal{K}_j}(\underline{v})$ also represents a Hamiltonian vector field.

Proof: The lemma is demonstrated inductively on j. First, it is certainly true if $j = 0$, since $\underline{\mathcal{K}_0}(\underline{v}) = \underline{v}$, and the hypothesis gives us that $\underline{v}$ is Hamiltonian. Now, suppose that it is true for $0 \leqslant m < j$. Let $\underline{v} = \underline{\underline{J}} \cdot \partial H$ and $\underline{w}_n = \underline{\underline{J}} \cdot \partial S_n, \forall n$. We shall write $\underline{\mathcal{K}_m}(\underline{v}) = \underline{\underline{J}} \cdot \partial \mathcal{K}(H)$ for $0 \leqslant m < j$. Using the morphism between Lie and Poisson brackets, expressed in Eqs.(3.7) through (3.9), and using Eq. (5.39), we have the following:

$$
\begin{aligned}
\underline{\mathcal{K}_j}(\underline{v}) &= -\sum_{m=0}^{j-1} \binom{j-1}{m} [\underline{w}_{j-m}, \underline{\mathcal{K}_m}(\underline{v})] \\
&= \underline{\underline{J}} \cdot \partial \left(\sum_{m=0}^{j-1} \binom{j-1}{m} [S_{j-m}, \mathcal{K}_m(H)] \right) \\
&\equiv \underline{\underline{J}} \cdot \partial \mathcal{K}_j(H).
\end{aligned}
$$

This is the inductive step, and the lemma is proved. □

In addition to establishing the important property of being Hamiltonian, this lemma also provides the recursion to be used for calculating the functions $\mathcal{K}_m$. With this in hand we proceed to rewrite Eq. (5.42) as follows:

$$
\underline{\underline{J}} \cdot \partial \left(K_k - H_k + \sum_{j=1}^{k} \binom{k}{j} \mathcal{K}_j(K_{k-j}) \right) = 0.
$$

This has the usual ambiguity always associated with Hamiltonian systems. Typically, this is never used; rather we write the scalar equation,

$$
K_k \equiv H_k - \sum_{j=1}^{k} \binom{k}{j} \mathcal{K}_j(K_{k-j}).
$$

Again, while this expression is succinct, it is not the most useful for calculations. As before, it is better to write the homological equation explicitly by isolating the bracket, $[S_k, H_0]$. The result is written by inspection from Eq. (5.53).

$$
[S_k, H_0] = H_k - K_k - \sum_{m=1}^{k-1} \left\{ \binom{k-1}{m} \mathcal{K}_m(K_{k-m}) + \binom{k-1}{m-1} [S_m, H_{k-m}] \right\}. \quad (5.59)
$$

We shall see shortly that an extra term must be added for nonautonomous Hamiltonians.

The coordinate transformations, Eq. (5.54) and Eq. (5.56), are translated into a delightfully simple Hamiltonian form. First, Eq. (5.54) is manipulated by substituting from the definition of S, $\underline{w} = \underline{\underline{J}} \cdot \underline{\partial} S$.

$$\begin{aligned}
\underline{G}_k &= -\sum_{m=0}^{k-1} \binom{k-1}{m} \underline{w}_{k-m} \cdot \underline{\underline{\partial G}}_m \\
&= -\sum_{m=0}^{k-1} \binom{k-1}{m} \left(\underline{\underline{J}} \cdot \underline{\partial} S_{k-m}\right) \cdot \underline{\underline{\partial G}}_m \\
&= \sum_{m=0}^{k-1} \binom{k-1}{m} [S_{k-m}, \underline{G}_m]
\end{aligned}$$

Here, as in Eq. (3.11) we interpret the commutator between a scalar and a vector field representation as a Poisson bracket evaluated component-wise, and we have employed the definition of Poisson brackets given in Eq. (3.8).

The key issue in rewriting Eq. (5.56) is actually the translation of Eq. (5.55), which defines the $\mathcal{L}_j$'s.

$$\underline{\mathcal{L}}_j(\underline{v}) = -\sum_{m=0}^{j-1} \binom{j-1}{m} \left(\underline{\underline{J}} \cdot \underline{\partial} S_{j-m}\right) \cdot \underline{\partial}(\underline{\mathcal{L}}_m(\underline{v})), \tag{5.60}$$

$$= \sum_{m=0}^{j-1} \binom{j-1}{m} [S_{j-m}, \underline{\mathcal{L}}_m(\underline{v})]. \tag{5.61}$$

That the perturbative recursions could be written in terms of Poisson brackets was immediately evident from the morphism between Lie and Poisson brackets; that these transformation equations also can be written in this form comes as something of a bonus.

Nonautonomous As long as we were dealing with arbitrary vector fields we were able to bury the issue of autonomy with the trick of making nonautonomous systems formally autonomous by referring the coordinates $\underline{z}$ and $\underline{y}$ to extended phase space. We now need to make that more explicit, as the phase space of Hamiltonian systems is necessarily even dimensional. To accomplish the translation we mentally adjust all of our expressions from the previous sections by appending a superscript "e," meaning "extended," and proceed to separate the extra coordinate explicitly.

$$\underline{z}^e = \begin{pmatrix} \underline{z} \\ \theta \end{pmatrix}, \underline{v}^e = \begin{pmatrix} \dot{\underline{z}} \\ \dot{\theta} \end{pmatrix} = \begin{pmatrix} \dot{\underline{z}} \\ 1 \end{pmatrix}.$$

The assignment $\dot{\theta} = 1$ is, of course, a choice — the choice of identifying the extra coordinate with the orbit parameter. We could easily have left it completely general.

There is, however, no perceived advantage to doing so, and it would clutter the development. The extended generator, $\underline{w}^e$, is written similarly, but with a major difference.

$$\underline{w}^e = \begin{pmatrix} d\underline{z}/d\epsilon \\ d\theta/d\epsilon \end{pmatrix} = \begin{pmatrix} d\underline{z}/d\epsilon \\ 0 \end{pmatrix}.$$

In fact, only the coefficient $\underline{v}_0^e$ has a nonzero extra component, and so the consequences of extended phase space will only be felt where $\underline{v}_0^e$ appears in the homological equation. From Eq. (5.52), this is in the commutator, $[\underline{w}_k^e, \underline{v}_0^e]$, which we evaluate below.

$$\begin{aligned}
[\underline{w}_k^e, \underline{v}_0^e] &= \begin{pmatrix} \underline{w}_k \cdot \underline{\underline{\partial v}}_0 - \underline{v}_0 \cdot \underline{\underline{\partial w}}_k - \partial \underline{w}_k/\partial\theta \\ 0 \end{pmatrix} \\
&= \begin{pmatrix} [\underline{w}_k, \underline{v}_0] - \partial \underline{w}_k/\partial\theta \\ 0 \end{pmatrix} \\
&= -\underline{\underline{J}} \cdot \underline{\partial} \begin{pmatrix} [S_k, H_0] + \partial S_k/\partial\theta \\ 0 \end{pmatrix}.
\end{aligned}$$

The only change that must be made in Eq. (5.59) in order to extend it to nonautonomous Hamiltonians is to replace $[S_k, H_0] \to [S_k, H_0] + \partial S_k/\partial\theta$. Both cases can be handled succinctly by using the symbol $(dS_k/d\theta)_{H_0}$, meaning the total (convective, extended Lie) derivative of S_k along an orbit of H_0.

$$\mathbf{D}S_k = (dS_k/d\theta)_{H_0} = H_k - K_k - \sum_{m=1}^{k-1} \left\{ \binom{k-1}{m} \mathcal{K}_m(K_{k-m}) + \binom{k-1}{m-1} [S_m, H_{k-m}] \right\}. \tag{5.62}$$

5.2.3 Synchrotron Models

We apply these procedures to Hamiltonian models arising in accelerator physics, specifically, storage rings, the principal feature of which is that the dynamics are periodic in the orbit parameter, θ. Not surprisingly, the terms of $O(\epsilon)$ will reproduce the shear and resonance models of the previous chapter, but using coordinates "distorted" by the transformation. Our first task is to find solutions to the homological equations, which possess the periodic boundary conditions of the accelerator.

Solving the Homological Equations As always, solving the homological equations is facilitated by working in an eigenbasis of the homological operator, which in this case is the vector field (or Lie operator) on extended phase space.[11]

$$\mathbf{D} \equiv \left(\frac{d}{d\theta}\right)_{H_0} = X_{H_0} + \frac{\partial}{\partial\theta} = \nu(\underline{I}) \cdot \frac{\partial}{\partial\underline{\varphi}} + \frac{\partial}{\partial\theta}.$$

[11] Recall that $X_{H_0} = -\underline{\partial} H_0 \cdot \underline{\underline{J}} \cdot \underline{\partial}$ is the vector field associated with H_0.

We assume that H_0 is in a normal form expressible in terms of action coordinates, that is, $\partial H_0/\partial \underline{\varphi} = \underline{0}$. In this case, eigenfunctions of $\mathbf{D}$ *obeying periodic boundary conditions* are of the form $f(\underline{I}) \exp i(n\theta + \underline{m} \cdot \underline{\varphi})$, where $f(\underline{I})$ is any real-valued function of the action coordinates, n is an integer, and $\underline{m}$ is an array of integers.

$$\begin{aligned} \mathbf{D}(f(\underline{I})e^{i(n\theta+\underline{m}\cdot\underline{\varphi})}) &= \left(\underline{\nu}(\underline{I}) \cdot \frac{\partial}{\partial \underline{\varphi}} + \frac{\partial}{\partial \theta}\right)(f(\underline{I})e^{i(n\theta+\underline{m}\cdot\underline{\varphi})}) \\ &= i(n + \underline{m} \cdot \underline{\nu}(\underline{I}))(f(\underline{I})e^{i(n\theta+\underline{m}\cdot\underline{\varphi})}). \end{aligned}$$

Notice that the null space of the homological operator — the set of (complex) observables annihilated by $\mathbf{D}$ — is spanned by those exponentials for which $n + \underline{m} \cdot \underline{\nu} = 0$. This is the resonance condition, and again it is the task of the Hamiltonian K to swallow these terms so that the homological equations (Eq. (5.62))can be solved.

Having found eigenfunctions of $\mathbf{D}$ we use them as a basis for expanding observables. Anticipating applications to the multipolar fields, we expect to be working with polynomials in x_1 and x_2. This suggests using products of $\sqrt{I_1}$ and $\sqrt{I_2}$ for the arbitrary functions $f(\underline{I})$ appearing above. Accordingly, we define our basis set as

$$\Phi_{\underline{e}\,\underline{m}}(\underline{\varphi}, \underline{I}) \equiv \prod_{k=1}^{2} I_k^{e_k/2} e^{i m_k \varphi_k} = I_1^{e_1/2} I_2^{e_2/2} e^{i\underline{m}\cdot\underline{\varphi}}, \tag{5.63}$$

indexed by a doublet of positive integers, $\underline{e} \equiv (e_1, e_2)$, and a doublet of integers, $\underline{m} \equiv (m_1, m_2)$. In order for such functions to arise from powers of x_1 and x_2 we impose the condition, $e_k \simeq m_k (\text{mod } 2)$. These may also be defined using the complex coordinates a and a^* as defined in the de Moivre representation of Eq. (2.9), but with I and φ the action-angle coordinates of H_0 (as in Eq. (3.100), for example).

$$\Phi_{\underline{e}\,\underline{m}} \equiv \prod_{k=1}^{2} i^{m_k} a_k^{(e_k - m_k)/2} a_k^{*(e_k + m_k)/2}. \tag{5.64}$$

Since the perturbation series is driven by brackets, we record for future reference the Poisson bracket,

$$\begin{aligned} [\Phi_{\underline{e}\,\underline{m}}, \Phi_{\underline{e}'\underline{m}'}] &= \frac{i}{2}(e_1' m_1 - e_1 m_1')\, \Phi_{(e_1+e_1'-2, e_2+e_2')\,\underline{m}+\underline{m}'} \\ &+ \frac{i}{2}(e_2' m_2 - e_2 m_2')\Phi_{(e_1+e_1', e_2+e_2'-2)\,\underline{m}+\underline{m}'}. \end{aligned} \tag{5.65}$$

We now formally expand all the relevant observables in this basis; for example,

$$S_k = \sum_{\underline{e},\underline{m}} S_{k\underline{e}\,\underline{m}}(\theta)\Phi_{\underline{e}\,\underline{m}}.$$

Because the vector field **D** is a linear operator and *does not mix the basis vectors,* we can solve the homological equations one "component" at a time. Rewrite Eqs. (5.62) symbolically as

$$\mathbf{D}S_k = \underline{\nu}(\underline{I}) \cdot \frac{\partial S_k}{\partial \underline{\varphi}} + \frac{\partial S_k}{\partial \theta} \equiv \xi_k - K_k,$$

$$\mathbf{D}(S_{k\underline{e}\underline{m}}(\theta)\Phi_{\underline{e}\underline{m}}) = \left(\left(\frac{\partial}{\partial \theta} + i\underline{m} \cdot \underline{\nu}\right) S_{k\underline{e}\underline{m}}(\theta)\right) \Phi_{\underline{e}\underline{m}} = (\xi_k - K_k)_{\underline{e}\underline{m}} \Phi_{\underline{e}\underline{m}}.$$

The solutions are obtained by integration,

$$S_{k\underline{e}\underline{m}}(\theta) = \oint_0^{2\pi} d\theta' G_{\underline{m}}(\theta - \theta')(\xi_k - K_k)_{\underline{e}\underline{m}}(\theta'),$$

where each $G_{\underline{m}}$ is a Green's function obeying periodic boundary conditions.

$$\begin{aligned}
\mathbf{D}\left(G_{\underline{m}}(\theta - \theta')\Phi_{\underline{e}\underline{m}}\right) &= \left(\sum_{n=-\infty}^{\infty} \delta(\theta - \theta' - 2\pi n)\right) \Phi_{\underline{e}\underline{m}} \\
&= \delta_{2\pi}(\theta - \theta')\Phi_{\underline{e}\underline{m}}, \\
G_{\underline{m}}(\tau) = \left(\frac{\partial}{\partial \tau} + i\underline{m} \cdot \underline{\nu}\right)^{-1} \delta_{2\pi}(\tau) &= \frac{1}{2\pi i} \sum_{n=-\infty}^{\infty} \frac{e^{in\tau}}{n + \underline{m} \cdot \underline{\nu}} \\
&= \frac{e^{-i\underline{m}\cdot\underline{\nu}\,\mathrm{saw}(\tau)}}{2i \sin(\pi \underline{m} \cdot \underline{\nu})}.
\end{aligned} \tag{5.66}$$

In the last line we have introduced the discontinuous, 2π periodic function, saw: $R \rightarrow (-\pi, \pi)$,

$$\begin{aligned}
\forall \tau \in (0, 2\pi): \quad & \mathrm{saw}(\tau) = \tau - \pi, \\
& \mathrm{saw}(0) = 0, \\
\forall n \in Z: \quad & \mathrm{saw}(\tau + 2\pi n) = \mathrm{saw}(\tau).
\end{aligned}$$

The middle line is added just for completeness. Using discontinuous, periodic functions in this way is a manner of working around the necessity of using an atlas of multiple charts.[12]

The generator of degree one is particularly important, and exceedingly simple.

$$S_{1\underline{e}\underline{m}}(\theta) = \oint d\theta' G_{\underline{m}}(\theta - \theta')(H_1 - K_1)_{\underline{e}\underline{m}}(\theta'). \tag{5.67}$$

[12] The "Green's function" measures the response at one point in the ring to a local perturbation at another. Forest [38] recently has written a derivation of it using map methods, without appeal to Hamiltonian flow equations.

It frequently happens that we can set $K_1 = 0$, which simplifies the integrand even further. (For example, recall the skew multipoles and order $4n + 2$ normal multipoles.)

Near a resonance, the situation is only a little more complicated. Suppose that for some integer n_o we have $n_o + \underline{m} \cdot \underline{\nu} \approx 0$. Because $\mathbf{D}$ annihilates the functions $e^{in_o\theta}\Phi_{\underline{e}\,\underline{m}}$, K_k must swallow the offending terms, so that

$$\oint d\theta(\xi_k - K_k)_{\underline{e}\,\underline{m}}(\theta)e^{-in_o\theta} = 0, \quad \text{for all } \underline{e}.$$

The Green's function can be modified as well by subtracting the contribution coming from the resonance term under consideration.

$$G_{\underline{m}}(\tau) = \frac{e^{-i\underline{m}\cdot\underline{\nu}\,\mathrm{saw}(\tau)}}{2i\sin(\pi\underline{m}\cdot\underline{\nu})} - \frac{1}{2\pi i}\frac{e^{in_o\,\mathrm{saw}(\tau)}}{n_o + \underline{m}\cdot\underline{\nu}}. \tag{5.68}$$

This makes it better behaved as the resonance is approached. In practice, one should swallow resonance terms when working in regions of phase space where $n_o + \underline{m} \cdot \underline{\nu}$ is "sufficiently close" to zero. The judgment required to make that decision is nonalgorithmic.

Even in the absence of nontrivial resonances, one must swallow the $\underline{m} = \mathbf{\underline{0}}$, $n_o = 0$ terms, that is, the shear terms. Applying Eq. (5.68) in this limit, we get the result

$$G_{\underline{0}}(\tau) = -\frac{1}{2\pi}\,\mathrm{saw}(\tau), \tag{5.69}$$

and correspondingly, because we want K_k to depend only on action coordinates — i.e., it is autonomous and independent of $\underline{\varphi}$ — it must be that

$$K_k = \langle \xi_k \rangle,$$

where the average is carried out over the angle coordinates and the orbit parameter. Of course, all these arguments run exactly parallel to those considered previously, with other perturbation procedures. Notice that $K_1 = \langle H_1 \rangle$, which reestablishes the averaging procedure of the last chapter, but using the transformed coordinates.

Second Order Quadrupole Tune Shifts In the previous chapter we used a simple averaging argument to explain the tune shifts caused by small quadrupole "error" distributions. We shall formally apply perturbation theory to this problem and extend the previous results to the second order in field strength. Starting from the Hamiltonian of Eq. (3.95) we construct the normal form Hamiltonian,

$$K = \underline{\nu} \cdot \underline{J} + \epsilon K_1 + \tfrac{1}{2}\epsilon^2 K_2,$$

according to the rules developed. First rewrite H_1 using the basis functions.

$$\begin{aligned} H_1 &= \sum_{\underline{e},\underline{m}} H_{1;\underline{e},\underline{m}}(\theta)\Phi_{\underline{e},\underline{m}} \\ &= \tfrac{1}{2}b_1(\theta)(L_{(2,0)(0,0)}\Phi_{(2,0)(0,0)} - L_{(0,2)(0,0)}\Phi_{(0,2)(0,0)} \\ &\quad + \mathrm{Re}[L_{(0,2)(0,2)}\Phi_{(0,2)(0,2)} - L_{(2,0)(2,0)}\Phi_{(2,0)(2,0)}]) \\ &\quad + a_1(\theta)(\mathrm{Re}[L_{(1,1)(1,1)}\Phi_{(1,1)(1,1)} - L_{(1,1)(1,-1)}\Phi_{(1,1)(1,-1)}]), \\ L_{\underline{e},\underline{m}}(\theta) &= \frac{B_oR}{|B\rho|}\sqrt{\beta_1^{e_1}(\theta)\beta_2^{e_2}(\theta)}\, e^{i\underline{m}\cdot\underline{\tilde{\psi}}(\epsilon)}. \end{aligned}$$

The term K_1 swallows the averaged H_1,

$$\begin{aligned} \epsilon K_{01} &= \langle \epsilon H_1\rangle = \langle \epsilon b_1/2(L_{(2,0)(0,0)}\Phi_{(2,0)(0,0)} - L_{(0,2)(0,0)}\Phi_{(0,2)(0,0)})\rangle \\ &\equiv \Delta\nu_1^{(1)}J_1 - \Delta\nu_2^{(1)}J_2, \\ \Delta\nu_k^{(1)} &= \frac{1}{2}\oint \frac{B_oR}{|B\rho|}\frac{d\theta}{2\pi}\epsilon b_1(\theta)\beta_k(\theta), \end{aligned}$$

in agreement with Eq. (3.100) but now understood to be applied to the *transformed* coordinates, $\underline{J}$. We obtain the second order term by setting $k = 2$ in Eq. (5.62).

$$K_2 = \langle [H_1, S_1] - \mathcal{K}_1(K_1)\rangle = \oint \frac{d\theta}{2\pi}([H_1, S_1] - \mathcal{K}_1(K_1))_{\underline{m}=\underline{0}}\,.$$

This is simplified by the vanishing of the averaged $\mathcal{K}_1(K_1)$.

$$\begin{aligned} \mathcal{K}_1(K_1)|_{\underline{m}=\underline{0}} &= [S_1, K_1]|_{\underline{m}=\underline{0}} \\ &= \sum_{\underline{e}} K_{1;\underline{e}\underline{0}}\,[S_1, \Phi_{\underline{e}\underline{0}}]\Big|_{\underline{m}=\underline{0}} \\ &= \sum_{\underline{e}\underline{e}'} K_{1;\underline{e}\underline{0}}S_{1;\underline{e}'\underline{0}}[\Phi_{\underline{e}'\underline{0}}, \Phi_{\underline{e}\underline{0}}] \\ &= 0. \end{aligned}$$

K_2 is then simply the averaged value of the Poisson bracket:

$$\begin{aligned} K_2 &= \langle [H_1, S_1]\rangle = \oint \frac{d\theta}{2\pi}\,[H_1, S_1]|_{\underline{m}=\underline{0}} \\ &= \sum_{\substack{\underline{e},\underline{m}\\ \underline{e}',\underline{m}'}} \left(\oint \frac{d\theta}{2\pi}H_{1;\underline{e}\underline{m}}(\theta)S_{1;\underline{e}'\underline{m}'}(\theta)\right)[\Phi_{\underline{e}\underline{m}}, \Phi_{\underline{e}'\underline{m}'}]\Big|_{\underline{m}+\underline{m}'=\underline{0}} \\ &= \sum_{\underline{e}\underline{m}}(-ig_{\underline{e}\underline{m}})[\Phi_{\underline{e}\underline{m}}, \Phi_{\underline{e}-\underline{m}}] = \sum_{\substack{m_1>0 \text{ or}\\ m_1=0 \text{ and } m_2>0}} \mathrm{Re}(g_{\underline{e}\underline{m}})(-2i[\Phi_{\underline{e}\underline{m}}, \Phi_{\underline{e}-\underline{m}}]), \end{aligned}$$

where the coefficients $g_{\underline{e}\,\underline{m}}$ are given by the functionals,

$$g_{\underline{e}\,\underline{m}} \equiv \int\int \frac{d\theta}{2\pi}\frac{d\theta'}{2\pi} H_{1;\underline{e}\,\underline{m}}(\theta) 2\pi G_{-\underline{m}}(\theta - \theta') H_{1;\underline{e}-\underline{m}}(\theta'),$$

and obey the identity $-ig_{\underline{e}-\underline{m}} = (-ig_{\underline{e}\,\underline{m}})^*$. The nonvanishing brackets are tabulated below.

$\underline{e}$	$\underline{m}$	$[\Phi_{\underline{e}\,\underline{m}}, \Phi_{\underline{e}-\underline{m}}]$
2 0	2 0	$4i\Phi_{(20)(00)}$
0 2	0 2	$4i\Phi_{(02)(00)}$
1 1	1 1	$i\Phi_{(02)(00)} + i\Phi_{(20)(00)}$
1 1	1 −1	$i\Phi_{(02)(00)} - i\Phi_{(20)(00)}$

Writing the final answer is now just a matter of putting together the pieces. We can do this succinctly in terms of the auxiliary functions,

$$c_{\underline{e}}(\theta) \equiv \begin{cases} b_1(\theta), \underline{e} = (2,0) \text{ or } (0,2) \\ a_1(\theta), \underline{e} = (1,1) \end{cases},$$

and functionals,

$$\begin{aligned} f_{\underline{e}\,\underline{m}} \equiv &-\frac{\pi/2}{\sin \pi \underline{m} \cdot \underline{\nu}} \int\int \left(\frac{B_o R}{|B\rho|}\right)^2 \frac{d\theta}{2\pi}\frac{d\theta'}{2\pi} \times c_{\underline{e}}(\theta)\, c_{\underline{e}'}(\theta') \\ &\times (\beta_1(\theta)\beta_1(\theta'))^{e_1/2}(\beta_2(\theta)\beta_2(\theta'))^{e_2/2} \\ &\times \cos \underline{m} \cdot (\underline{\tilde{\psi}}(\theta) - \underline{\tilde{\psi}}(\theta') + \underline{\nu}(\text{saw}(\theta - \theta'))), \end{aligned}$$

as follows.

$$\begin{aligned} K_2 = &(f_{(2,0)(2,0)} + f_{(1,1)(1,1)} - f_{(1,1)(1,-1)})J_1 \\ &+ (f_{(0,2)(0,2)} + f_{(1,1)(1,1)} + f_{(1,1)(1,-1)})J_2. \end{aligned} \tag{5.70}$$

The "(1, 1)" terms contain the second order contribution to the tune shift from skew quadrupoles. Because K is decoupled — there is no $J_1 J_2$ term — the tunes obtained from it are *the eigenfrequencies of the originally coupled dynamic,* and the transformed coordinates are the eigencoordinates, to second order in the "error" fields.

Tune Spread from Sextupoles The averaging procedure used in the previous chapter predicted first order tune spreads from normal multipoles for which the number of poles was a multiple of four: quadrupoles, octupoles, dodecapoles, and so forth. In particular, according to Eqs.(4.22) and (4.23), the quadratic terms in H_{avg}

come from normal octupoles. We now shall demonstrate that normal sextupoles also produce such terms when the calculation is continued to second order.

The perturbing Hamiltonian we shall use is that of the sextupolar model written in Eqs.(4.18) and (4.27). We express it here in terms of our basis set via substitution from Eqs. (3.100).

$$\begin{aligned}
\epsilon H_1 &= \frac{1}{3}\frac{B_oR}{|B\rho|}\,\epsilon b_2(\theta)(x_1^3 - 3x_1x_2^2) \\
&= -i/6\sqrt{2}\sum_{\underline{e},\underline{m}} c_{\underline{e},\underline{m}}\, L_{\underline{e},\underline{m}}(\theta)\Phi_{\underline{e},\underline{m}}, \\
L_{\underline{e},\underline{m}}(\theta) &= \frac{B_oR}{|B\rho|}\,\epsilon b_2(\theta)(\beta_1^{e_1}(\theta)\beta_2^{e_2}(\theta))^{1/2} e^{i\underline{m}\cdot(\underline{\psi}(\theta)-\underline{\nu}\theta)} \qquad (5.71) \\
c_{(3,0)(3,0)} &= -1, c_{(1,2)(1,0)} = -6, \\
c_{(3,0)(1,0)} &= c_{(1,2)(1,2)} = c_{(1,2)(1,-2)} = 3.
\end{aligned}$$

In the absence of resonances, $K_1 = \mathcal{K}(K_1) = 0$, and once again the lowest order nontrivial term in the new representation is the averaged bracket, $K_2 = \langle [H_1, S_1]\rangle$. Starting from Eq. (5.67), the calculation of this bracket proceeds via manipulations that parallel those that led to Eq. (5.70). The final result is the transformed Hamiltonian,

$$\begin{aligned}
K(\underline{J}) + O(\epsilon^4) &= \underline{\nu}\cdot\underline{J} - J_1^2(\tfrac{1}{8}f_{3,0;3,0;3,0} + \tfrac{3}{8}f_{3,0;3,0;1,0}) \\
&\quad - J_2^2(\tfrac{1}{2}f_{1,2;1,2;1,0} + \tfrac{1}{8}f_{1,2;1,2;1,2} + \tfrac{1}{8}f_{1,2;1,2;1,-2}) \\
&\quad + J_1J_2(f_{3,0;1,2;1,0} - \tfrac{1}{2}f_{1,2;1,2;1,2} + \tfrac{1}{2}f_{1,2;1,2;1,-2}), \qquad (5.72)
\end{aligned}$$

where the coefficients are superpositions of the following quadratic functionals of the sextupole distribution, $b_2(\theta)$:

$$\begin{aligned}
f_{\underline{e};\underline{e}';\underline{m}} &= \iint \frac{d\theta\, d\theta'}{(2\pi)^2}\,\mathrm{Re}\,[L_{\underline{e},-\underline{m}}(\theta)2\pi i G_{\underline{m}}(\theta-\theta')L_{\underline{e}'\underline{m}}(\theta')] \\
&\equiv \iint \left(\frac{B_oR}{|B\rho|}\right)^2 \frac{d\theta\, d\theta'}{(2\pi)^2}\,\epsilon b_2(\theta)\mathcal{G}_{\underline{e};\underline{e}';\underline{m}}(\theta,\theta')\epsilon b_2(\theta'), \\
\mathcal{G}_{\underline{e};\underline{e}';\underline{m}}(\theta,\theta') &= \frac{\pi}{\sin \pi\underline{m}\cdot\underline{\nu}}(\beta_1^{e_1}(\theta)\beta_1^{e_1'}(\theta')\beta_2^{e_2}(\theta)\beta_2^{e_2'}(\theta'))^{1/2} \\
&\quad \times \cos(|\underline{m}\cdot(\underline{\psi}(\theta) - \underline{\psi}(\theta'))| - \pi\underline{m}\cdot\underline{\nu}).
\end{aligned}$$

The range of integration is $(\theta, \theta') \in [0, 2\pi) \times [0, 2\pi)$ in order that $\underline{\psi}$ be used in the integrand rather than $\tilde{\underline{\psi}}$.[13]

As a special case, we apply this result to an individual, thin sextupole, restricting the calculation to the "horizontal" sector: $J_2 = 0$. The double integrals of Eq. (5.72) are evaluated trivially in the impulse approximation, expressed in Eq. (3.113) but with ϵb_2 replacing b_2. Since we have set $J_2 = 0$, we are only interested in the coefficient of J_1^2 and drop the subscript.

$$\begin{aligned} K_S(J) &= \nu J + \frac{\epsilon^2}{2!} K_{2;40;00} \\ &= \nu J - J^2\left(\tfrac{1}{8} f_{3,0;3,0;3,0} + \tfrac{3}{8} f_{3,0;3,0;1,0}\right) \\ &= \nu J - \frac{1}{32\pi}\lambda^2 \beta(\cot 3\pi\nu + 3\cot \pi\nu) J^2 \end{aligned} \tag{5.73}$$

The variable λ was defined in Eq. (3.117), but with $\epsilon K_2 L$ replacing $K_2 L$. A subscript S has been attached to K to emphasize that this is the "shear part" of the transformed Hamiltonian, a "normal form." To it we can add resonance terms so as to make it a "resonance normal form."

Coordinate Transformations The coordinates appearing in the tune shift expressions are the action coordinates of normal form Hamiltonians, *not* the original coordinates defined in Eq. (3.100). To make the distinction, we shall here symbolize these earlier coordinates as $\underline{z} = (\underline{\varphi}^\circ, \underline{I}^\circ)$. They are related via the "distortion functions" $\underline{F}$ and $\underline{G}$, introduced in Eq. (5.19) and expanded using Eqs.(5.56) and (5.55). To second order, for Hamiltonian systems, we combine Eq. (5.57) with Eq. (5.61) to get

$$\begin{aligned} \underline{F}_1 &= \underline{w}_1 = \underline{\underline{J}} \cdot \underline{\partial} S_1, \\ \underline{F}_2 &= \underline{w}_2 - \underline{\mathcal{L}}_1(\underline{F}_1) \\ &= \underline{\underline{J}} \cdot \underline{\partial} S_2 - [S_1, \underline{w}_1]. \end{aligned}$$

We develop this a little further, projecting the angle and amplitude components.

$$\begin{aligned} \underline{\varphi}^\circ &= \underline{\varphi} + \epsilon \underline{\varphi}^{(1)} + \frac{\epsilon^2}{2!}\underline{\varphi}^{(2)} + \cdots, \\ \underline{\varphi}^{(1)} &= \partial S_1/\partial \underline{I}, \\ \underline{\varphi}^{(2)} &= \partial S_2/\partial \underline{\varphi} - [S_1, \partial S_1/\partial \underline{I}], \\ \underline{I}^\circ &= \underline{I} + \epsilon \underline{I}^{(1)} + \frac{\epsilon^2}{2!}\underline{I}^{(2)} + \cdots, \\ \underline{I}^{(1)} &= -\partial S_1/\partial \underline{\varphi}, \\ \underline{I}^{(2)} &= -\partial S_2/\partial \underline{\varphi} + [S_1, \partial S_1/\partial \underline{\varphi}]. \end{aligned} \tag{5.74}$$

[13] This is done only to make contact with classic expressions widely known among accelerator physicists.

As a concrete instance, we continue working with the sextupolar perturbation of Eq. (5.71). Using Eq. (5.67) we see that if we again confine the calculation to the horizontal sector the only relevant components of S_1 are $S_{1;30;10}$, $S_{1;30;30}$, and their complex conjugates.

$$\begin{aligned} S_1 &= S_{1;30;10}\Phi_{(3,0)(1,0)} + S_{1;30;30}\Phi_{(3,0)(3,0)} + \text{c.c.} \\ &= 2I_1^{3/2}\,\mathrm{Re}(S_{1;30;10}e^{i\varphi_1} + S_{1;30;30}e^{i3\varphi_1}). \end{aligned}$$

Similarly, because S_2 is generated via Poisson brackets, Eq. (5.65) implies that the only nonvanishing components are $S_{2;40;00}$, $S_{2;40;20}$, $S_{2;40;40}$, and their complex conjugates. Applying these observations to Eqs. (5.74), we obtain

$$\begin{aligned} \varphi_1^{(1)} &= 3I_1^{1/2}\,\mathrm{Re}(S_{1;30;10}e^{i\varphi_1} + S_{1;30;30}e^{i3\varphi_1}), \\ \varphi_1^{(2)} &= 2I_1(S_{2;40;00} + \mathrm{Re}((2S_{2;40;20} + 3iS_{1;30;-10}S_{1;30;30} + \tfrac{3i}{2}S_{1;30;10}^2)e^{i2\varphi_1} \\ &\quad + (2S_{2;40;40} + 6iS_{1;30;10}S_{1;30;30})e^{i4\varphi_1} + (\tfrac{9i}{2}S_{1;30;30}^2)e^{i6\varphi_1})), \\ I_1^{(1)} &= I_1^{3/2}\mathrm{Im}(S_{1;30;10}(\theta)e^{i\varphi_1} + 3S_{1;30;30}e^{i3\varphi_1}) \\ I_1^{(2)} &= I_1^2(27|S_{1;30;30}|^2 + 3|S_{1;30;10}|^2 \\ &\quad + 2\,\mathrm{Re}((12S_{1;30;-10}S_{1;30;30} - iS_{2;40;20})e^{i2\varphi_1} \\ &\quad + (3S_{1;30;10}S_{1;30;30} - 2iS_{2;40;40})e^{i4\varphi_1})). \end{aligned} \tag{5.75}$$

In particular, using Eq. (5.71), Eq. (5.66), and Eq. (5.67), we evaluate these expressions in the case of a single, thin sextupole. The degree one components are

$$\begin{aligned} \epsilon S_{1;30;10}(\theta) &= -\frac{1}{4\sqrt{2}}\frac{1}{\sin\pi\nu}\frac{B_oR}{|B\rho|}\oint d\theta'\,(\epsilon b_2(\theta')\beta^{3/2}(\theta')e^{i\tilde{\psi}(\theta')})e^{-i\nu[\mathrm{saw}(\theta-\theta')]} \\ &\approx -\frac{1}{4\sqrt{2}}\frac{1}{\sin\pi\nu}\lambda\beta_s^{1/2}e^{i\tilde{\psi}_s}e^{-i\nu[\mathrm{saw}(\theta-\theta_s)]}, \\ \epsilon S_{1;30;30}(\theta) &\approx \frac{1}{12\sqrt{2}}\frac{1}{\sin 3\pi\nu}\lambda\beta_s^{1/2}e^{i3\tilde{\psi}_s}e^{-i3\nu[\mathrm{saw}(\theta-\theta_s)]}, \end{aligned} \tag{5.76}$$

where the subscript s means "evaluated at the sextupole location." Those of degree two are

$$\begin{aligned} \epsilon^2 S_{2;40;40}(\theta) &= \frac{i}{32}\frac{1}{\sin 4\pi\nu}\lambda^2\beta_s e^{i4\tilde{\psi}_s}(\cot 3\pi\nu - \cot\pi\nu)e^{-i4\nu\mathrm{saw}(\theta-\theta_s)}, \\ \epsilon^2 S_{2;40;20}(\theta) &= -\frac{i}{16}\frac{1}{\sin 2\pi\nu}\lambda^2\beta_s e^{i2\tilde{\psi}_s}(\cot 3\pi\nu + \cot\pi\nu)e^{-i2\nu\mathrm{saw}(\theta-\theta_s)}, \\ \epsilon^2 S_{2;40;00} &= \frac{1}{16\pi}\left(1 - \frac{1}{2\pi}\right)\mathrm{saw}(\theta-\theta_s)\beta_s\lambda^2[3\cot\pi\nu + \cot 3\pi\nu] \end{aligned} \tag{5.77}$$

It is left to the masochistically inclined reader to substitute these back into Eqs. (5.75).

Resonance Normal Form: the Hénon Map A characteristic of these manipulations — indeed, of *any* perturbation theoretic calculation — is that they become extremely burdensome beyond the first order; this is already evident in the second order of this very simple model. *They are meant to be carried out by computers,* and the algorithmic nature of the Lie algebraic methods — especially the one to be discussed in the next section — makes them admirably suited for either symbolic or numeric automatic computation. The one *nonalgorithmic* step is the decision to absorb a resonance into the "secular terms," $\underline{u}_k$ (or for Hamiltonian systems, K_k). We have mentioned that resonances manifest themselves in perturbation theories as "small denominators." Now, there are two things that are constructed: (a) a transformation that "distorts" the phase space chart in which one represents the dynamics and (b) the "normal form" representation of the vector field on the new chart. Small denominators that appear while building the transformation can be avoided by absorbing the most "important" resonances into the normal form, via the homological equations, so as to keep the solutions for $\underline{w}_k$ (or S_k), the generator of the transformation, within reasonable limits. Interpreting the word "important" is the nonalgorithmic step. It depends strongly on both the parameters of the problem and the phase space region of interest. The validity of Eq. (5.72), for example, depends on being far from prevailing first and second order sextupole resonances, which otherwise would have to be appended: (first order) $3\nu_1$, $\nu_1 \pm 2\nu_2$, and ν_1; (second order) $4\nu_1$, $4\nu_2$, $2\nu_1$, $2\nu_2$ and $2\nu_1 \pm 2\nu_2$. Within the influence of any of these the resonance term should be absorbed into the transformed Hamiltonian. If only one resonance term is absorbed, then the normal form becomes an integrable, isolated resonance model of the type considered in the previous chapter; we call this a "resonance normal form."

Consider, once again, the specific example of a ring whose only nonlinearity is an individual thin sextupole.[14] We have noted already that the Poincaré map of this system is equivalent to a Hénon map, as given in Eq. (3.116), a few orbits of which, generated at the tune $\nu = 0.29$, were displayed in Figure 2.13. More orbits, at other tunes, are shown in Figure 5.3 and compared to the results of applying the perturbation theoretic calculations of this section (or, for that matter, the next). In all but Figures 5.3(e), (f), and (h), the integrable separatrix of an isolated resonance normal form was calculated and then transformed back to the original phase space of Eq. (3.116) using Eqs. (5.75) (i.e., only to second order in the sextupole strength). These are shown as solid lines. Superposed are a few orbits, obtained using Eq. (3.116), in the same region of phase space. Because the motion is confined to the horizontal sector, the only resonance terms that can be absorbed through $O(\epsilon^2)$ are $m\nu + n = 0$, where $|m| \leqslant 4$. Figure 5.3(a) shows the effect of the integer resonance at the tune $\nu = 0.15$. Of course, $\nu = 0.15$ is close to the $1/5$ integer resonance as well, but that is a much weaker excitation, not appearing until third order in the sextupole strength. The companion Figure 5.3(b) shows a closeup in the neighborhood of the resonant orbit. Although many very high order resonances can be identified visually, the actual stability boundary falls exceedingly close to the calculated separatrix. (It is, essentially, the chaotic separatrix of an integer resonance produced by the map, a

[14]The results described in this section were first reported in Reference [71] and later repeated in [70].

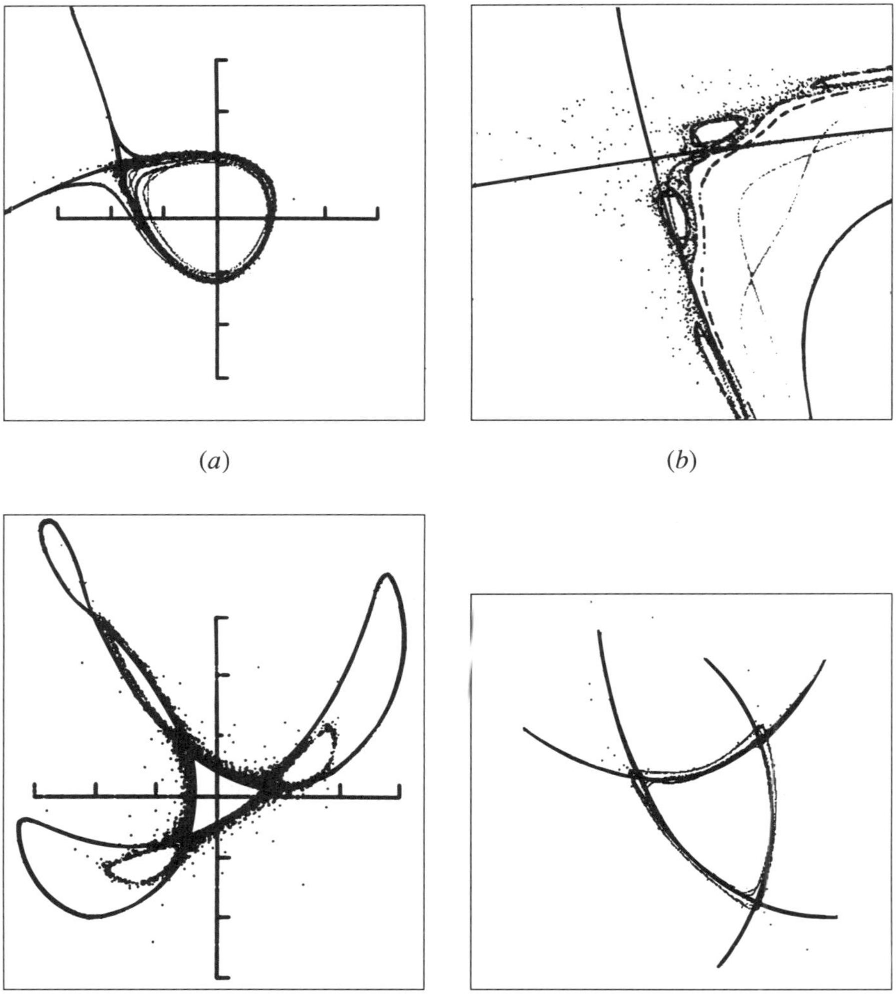

(a) (b) (c) (d)

FIGURE 5.3 Perturbation theory suggests that the outer stability limits of a Hamiltonian system are "seeded" by low order resonances. (The tic marks on the axes are separated by 0.5.)

separatrix which our first order isolated resonance model approximates very nicely.) Figure 5.3(c) and (d) show the stability boundaries a little below ($\nu = 0.32$) and a little above ($\nu = 0.37$) the third integer resonance. The triangular shaped separatrix of the corresponding isolated resonance model (see Figure 4.6), distorted by the transformation, is identified easily. At very large amplitude the correspondence with tracking is poor; so much so that the transformation is not even invertible. But at the important amplitudes, the boundary of the central stable region is predicted remarkably well in both figures, although a little better in Figure 5.3(c), which is closer to the resonance. Figures 5.3(e), (f), and (g) illustrate conditions in the vicinity

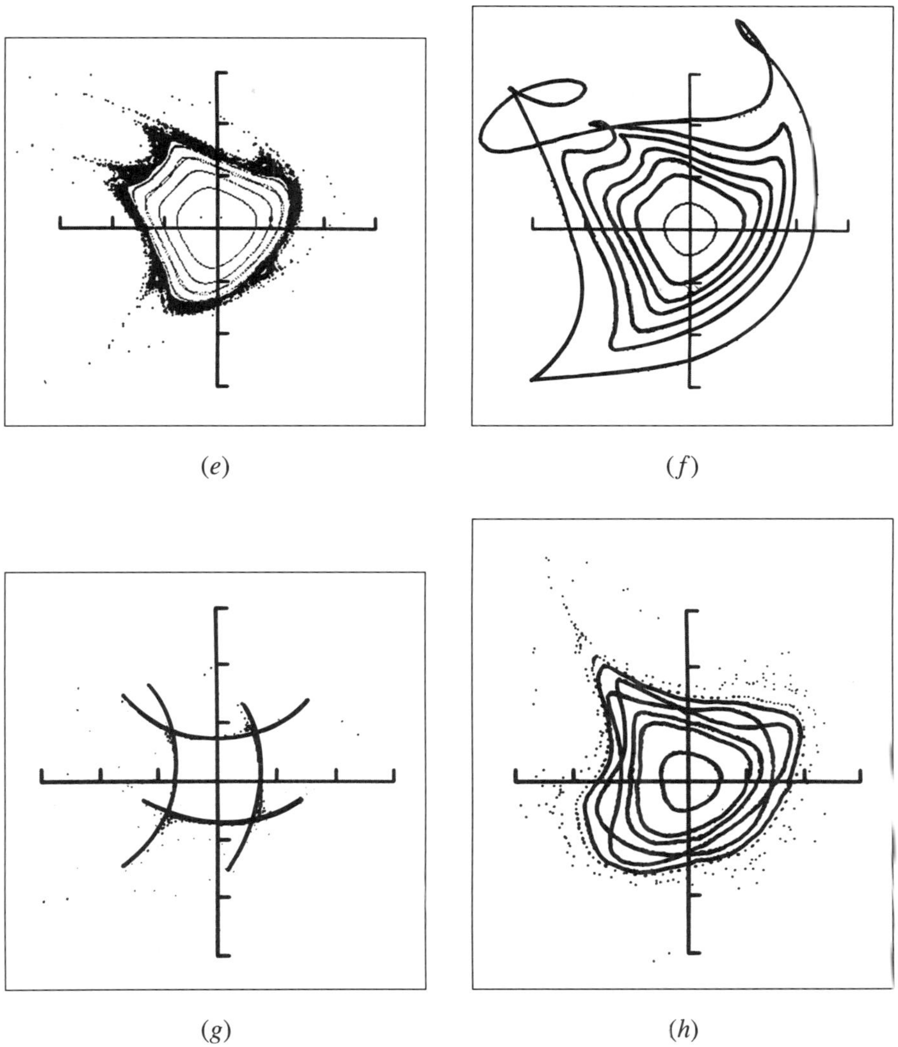

FIGURE 5.3 *(continued)*

of the fourth integer resonance. Just below resonance ((e): $\nu = 0.235$) there is no evidence of a fourth integer separatrix, and the perturbative calculation (f) is carried out using shear terms alone. On resonance, a pitchfork bifurcation occurs — much like the one sketched in Figure 4.9 — so that the separatrix is very much in evidence above resonance ((g): $\nu = 0.26$). The actual stability boundary, which is indeed the chaotic separatrix of a fourth integer resonance, and the separatrix of the fourth integer resonance model coincide so closely that they are visually inseparable on the scale of the picture.

Figure 5.3(h) shows the application of a two-resonance model, at the same tune as Figure 2.13 ($\nu = 0.29$). The dominant feature of Figure 2.13 was the very large system of seven islands *apparently* produced by a 2/7 resonance. This is puzzling: according to the perturbation procedure, seventh order resonances do not appear until terms of fifth degree in the sextupole strength, while the island chain seems to be a much stronger effect. In fact it is due to an *interference* between the 1/3 resonance, which appears at first order in the perturbation expansion, and the 1/4 resonance, which appears at second. This is confirmed by the perturbation theoretic calculation in which *both* resonances were absorbed, giving rise to a nonintegrable, two-resonance normal form possessing orbits shown in Figure 5.3(h) and obtained by numerical integration. The 2/7 resonance, *which was not explicitly put into the Hamiltonian,* nonetheless surfaces, a result of interference between the two lower order resonances.

These simple tests and others — at different values of the tune and with other multipoles (esp., octupoles [45]) — indicate that the "dynamic aperture" of these autonomous dynamics, that chaotic boundary layer separating stable orbits from those that very quickly go off to infinity, is built from a foundation provided by the separatrix of a low order resonance. In fact, the stability boundary is (frequently) the chaotic separatrix of a low order resonance, whose integrable, isolated resonance model is calculable via low order perturbation theory. *This is a statement about physics rather than any particular way of calculating the resonance model.* In light of all the complexity exhibited in this region of phase space, it is rather remarkable that such simple models produce such useful information.

5.3 EXPONENTIAL MAPS

Deprit's algorithm for constructing perturbative expansions was originally devised to solve problems in celestial mechanics and continues to be used in that field [23]. Despite similarities, accelerator physics — especially, the physics of high energy synchrotrons — possesses problems with properties significantly different from celestial mechanics. For example, it was necessary to enlarge the scope of the algorithm from autonomous dynamics to nonautonomous, periodic dynamics and to absorb the resulting resonances. One of the most important differences, however, is that synchrotrons, storage rings, linacs, beamlines, and the like are collections of discrete, compact, separated devices, sources of *highly localized* electromagnetic environments. Because of this, the chain of maps strung out from the Chapman-Kolmogorov equation, as done below Eq. (1.4), which was associated with the "time-ordered product" construction, is now more practical than formal. It describes the dominant mathematical modelling tool of the entire field: the tracking program, with "time" interpreted as the θ orbit parameter marking azimuthal (longitudinal) location along the ring or beamline. It is the Chapman-Kolmogorov equation in its discrete form, Eq. (1.4), that models the traversal of an entire ring; the continuous form, Eq. (1.9), is used, to construct the maps (or orbit segments) through individual devices. Adapting Deprit's algorithm to this class of problems is practical for simpler problems in which a flow field can be represented easily and the order of the calculation is sufficiently low that iterative evaluation of integrals does not present a bottleneck. For complex problems more easily modelled with maps than with flows — problems in which intractable complications arise from using the representation of a Hamiltonian (see Section 3.4.2) constructed from a global, smooth, closed curve fiducial rather than a collection of local fiducials — it is better to use a perturbation procedure that works directly with maps.[15] We glanced briefly at a basic procedure of this type in Section 5.1.1. Here we will study another, one that is built on the notion of a map being the exponential of a Lie operator, and therefore more suitable in problems in which the symplectic condition is satisfied, or at least violated only weakly. The algorithm was described originally by Forest, Berz, and Irwin [37]; the pioneering work in constructing multidimensional accelerator maps using Lie algebraic methods was done by Dragt [31, 30] and his collaborators.

5.3.1 Preconditioning

Let us indulge ourselves, at least temporarily, with the hypothesis that the mapping under consideration — which is a Poincaré, period-advance, or some other evolution map of our dynamic — can be written as a composite of exponential maps of the form

$$\begin{aligned} \underline{M} &= \mathcal{M}\underline{I}, \text{ where} \\ \mathcal{M} &= e^{\epsilon \mathbf{M}_1} e^{\epsilon^2 \mathbf{M}_2} e^{\epsilon^3 \mathbf{M}_3} \cdots \\ &= \prod_k^{(o)} e^{\epsilon^k \mathbf{M}_k}, \end{aligned} \tag{5.78}$$

where each $\mathbf{M}_k$ is a Lie derivative operator, i.e., a vector field. The superscript "(o)" serves to remind that this is an ordered product, with lower powers of ϵ appearing to the left. It is not important that $\mathcal{M}$ be *explicitly given* in this form, only that it *can be*. We shall leave open the question of whether this is a finite or infinite product, taking refuge in the excuse that this is an "intermediate," not "advanced," text to avoid concerning ourselves with difficult questions about convergence — or any other difficult questions, for that matter.

Now, in fact, we are not generally given such maps *a priori*. For one thing, they possess the property of being near-identity transformations in ϵ; that is, $\lim_{\epsilon \to 0} \mathcal{M} = I$. Furthermore, for computational purposes, it is important that the origin be a fixed point of $\underline{M}$; that is, we shall require that $\underline{M}(\underline{0}) = (\mathcal{M}\underline{I})_{(\underline{0})} = \underline{0}$. In general, we start with a map, say $\underline{z} \longmapsto \underline{M}^{\oplus}(\underline{z})$ obtained from a tracking program, numerical integration, or whatever, which does not possess these attributes and therefore must be preconditioned before entering the perturbation procedure itself.

The first step is to change to another map that treats the origin as a fixed point; alternatively, think of this as the same map represented on a different coordinate chart. If the physical system is a circular machine, then we look for a fixed point of $\underline{M}^{\oplus}$, which represents the closed orbit in the machine. Let this be $\underline{z}^*$, so that $\underline{M}^{\oplus}(\underline{z}^*) = \underline{z}^*$.

[15] This point of view has been stressed by Forest [36, 38].

Now define $\underline{M}^{\sim}$ according to

$$\underline{z}^* + \underline{M}^{\sim}(\underline{z}) = \underline{M}^{\oplus}(\underline{z}^* + \underline{z}), \tag{5.79}$$

which establishes a morphism between $\underline{M}^{\sim}$ and $\underline{M}^{\oplus}$ under a translation through $\underline{z}^*$. Clearly, $\underline{M}^{\sim}(\underline{0}) = \underline{0}$; we have accomplished the first of our objectives.

Equation (5.79) must be modified slightly for studying a transfer line, linac, spectrometer, or any device for which the concept of a closed orbit is meaningless. Rather, let $\underline{z}_a$ be the initial coordinates of the fiducial (design) orbit, and let $\underline{M}^{\oplus}(\underline{z}_a) = \underline{z}_b$. Equation (5.79) is replaced by

$$\underline{z}_b + \underline{M}^{\sim}(\underline{z}) = \underline{M}^{\ominus}(\underline{z}_a + \underline{z}).$$

However, the perturbation problem is different from that of circular machines; a goal of the normal form procedure for a spectrometer or electron microscope, for example, is to identify and evaluate aberration coefficients. We shall not examine these problems but accept in everything that follows that the analysis is being carried out for a circular machine.

Now, let us assume that in the limit $\epsilon \to 0$ the map $\underline{M}^{\sim}$ becomes linear.[16] In the spirit of perturbation theory, we expand in the order parameter, ϵ, and write the linear leading term explicitly.

$$\underline{M}^{\sim}(\underline{z}) = \underline{\underline{A}} \cdot \underline{z} + \epsilon \underline{M}_1^{\sim}(\underline{z}) + O(\epsilon^2). \tag{5.80}$$

We cannot express $\underline{M}^{\sim}$ as a composition of exponential maps (see Eq. (5.78)) since it is not a near-identity transformation in ϵ. We could fix this by factoring,

$$\underline{M}^{\sim}(\underline{z}) \equiv \underline{\underline{A}} \cdot \underline{M}(\underline{z}),$$

and working with $\underline{M}$. However, the algorithm is written in terms of compositions of maps, and this would not fit into its framework. Instead, the linear form is factored out by writing

$$\underline{M}^{\sim}(\underline{z}) \equiv \underline{M}(\underline{\underline{A}} \cdot \underline{z}), \quad \text{which means} \quad \underline{M}(\underline{z}) \equiv \underline{M}^{\sim}(\underline{\underline{A}}^{-1} \cdot \underline{z}). \tag{5.81}$$

It is the map $\underline{M}$ that finally possesses the right properties and which we shall assume could be expressed as the product of exponential maps written in Eq. (5.78).

5.3.2 Operator Products

We suppose that the map $\underline{M}^{\sim}$ is equivalent to another map, say, $\underline{N}^{\sim}$, its "normal form." The transformation connecting the two will be called $\underline{T}$, and as always, it

[16]This is a *nontrivial* assumption. Physically it means that any nonlinearity in the map — a magnetic multipole strength, for example — comes from the order parameter, which is *assumed to be "small."* It also means the hypothesis of the KAM Theorem is not satisfied.

is our task to determine $\underline{T}$ in such a way that $\underline{N}^{\sim}$ is "as simple as possible," with the linear map as our initial target. The functional relationship that expresses this equivalence is given by

$$\underline{M}^{\sim}(\underline{T}(\underline{z})) = \underline{T}(\underline{N}^{\sim}(\underline{z})), \quad \text{or}$$
$$\underline{M}(\underline{\underline{A}} \cdot \underline{T}(\underline{z})) = \underline{T}(\underline{N}(\underline{\underline{A}} \cdot \underline{z})). \tag{5.82}$$

All the functional relations we have written are collected and shown diagramatically in Figure 5.4(a). As usual, the construction of $\underline{T}$ will not be done in a single step but via a sequence of transformations, as shown in Figure 5.4(b). The objectives are: (a) $\underline{M}^{\sim}$ and $\underline{N}_k^{\sim}$ are $O(\epsilon^{k+1})$ equivalent, and (b) the first "non-normal" term in $\underline{N}_k^{\sim}$ is $O(\epsilon^{k+1})$. The initial target normal form is the linear, $\underline{\underline{A}} \cdot \underline{z}$. The procedure to be described, like those of the preceding sections, is recursive: it is applied to each rung of the ladder until $\underline{N}^{\sim}$ is constructed to the desired accuracy. Also like the others, it will fail to reach its target but require that shear and resonant terms be absorbed into $\underline{N}^{\sim}$, as topology demands it must.

We now recast Eq. (5.82) into the language of exponential maps. To this end, we assume all the maps appearing in Eq. (5.82) can be expressed as exponentials of Lie operators acting on the identity map. (See the discussion above Eq. (2.64).) In particular, let $\underline{\underline{W}}$ be a logarithm of the nonsingular matrix $\underline{\underline{A}}$, so that $\underline{\underline{A}} = \exp \underline{\underline{W}}$. Let $\mathbf{W}$ be the linear vector field (Lie operator) associated with $\underline{\underline{W}}$.

$$\mathbf{W} \equiv \underline{z} \cdot \underline{\underline{W}}^T \cdot \frac{\partial}{\partial \underline{z}}.$$

Finally, recall from Eq. (2.64) that exponential Lie operators obey a functional absorption law that allows us to write

$$\underline{M}(\underline{\underline{A}} \cdot \underline{z}) = \underline{M}(e^{\underline{\underline{W}}} \cdot \underline{z}) = \underline{M}((e^{\mathbf{W}}\underline{I})_{(\underline{z})}) = (e^{\mathbf{W}}\underline{M})_{(\underline{z})}.$$

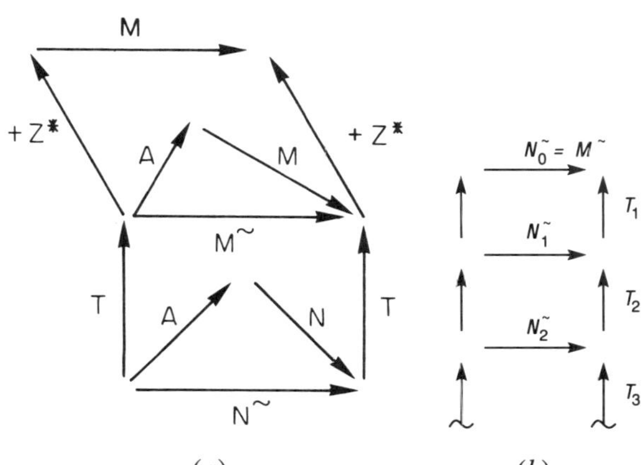

FIGURE 5.4 Diagrammatic representation of a normal form transformation.

Now, using Eq. (2.64) repeatedly, as in Eq. (2.65), and making use of the obvious composition identity, $\underline{I} \circ \underline{f} = \underline{f}$, the left side of Eq. (5.82) becomes

$$\begin{aligned}\underline{M}(\underline{A} \cdot \underline{T}(\underline{z})) &= (e^{\mathbf{W}}\underline{M})_{(\underline{T}(\underline{z}))} \\ &= ((e^{\mathbf{W}}\underline{M}) \circ \underline{T})_{(\underline{z})} \\ &= ((e^{\mathbf{W}}\mathcal{M}\underline{I}) \circ (e^{\mathbf{T}}\underline{I}))_{(\underline{z})} \\ &= (\underline{I} \circ (\mathcal{M}\underline{I}) \circ (e^{\mathbf{W}}\underline{I}) \circ (e^{\mathbf{T}}\underline{I}))_{(\underline{z})} \\ &= (e^{\mathbf{T}}e^{\mathbf{W}}\mathcal{M}\underline{I})_{(\underline{z})},\end{aligned}$$

and the right side becomes

$$\begin{aligned}\underline{T}(\underline{N}(\underline{A} \cdot \underline{z})) &= (\underline{T} \circ (e^{\mathbf{W}}\underline{N}))_{(\underline{z})} \\ &= ((e^{\mathbf{T}}\underline{I}) \circ (e^{\mathbf{W}}\mathcal{N}\underline{I}))_{(\underline{z})} \\ &= ((e^{\mathbf{T}}\underline{I}) \circ \underline{I} \circ (\mathcal{N}\underline{I}) \circ (e^{\mathbf{W}}\underline{I}))_{(\underline{z})} \\ &= (\underline{I} \circ (e^{\mathbf{T}}\underline{I}) \circ (\mathcal{N}\underline{I}) \circ (e^{\mathbf{W}}\underline{I}))_{(\underline{z})} \\ &= (e^{\mathbf{W}}\mathcal{N}e^{\mathbf{T}}\underline{I})_{(\underline{z})}.\end{aligned}$$

Equating these two gives us the following:

$$e^{\mathbf{T}}e^{\mathbf{W}}\mathcal{M} = e^{\mathbf{W}}\mathcal{N}e^{\mathbf{T}}.$$

Clearly, if $\mathcal{M}$ is a product of exponential maps, then so is $\mathcal{N}$.

$$\mathcal{N} = e^{-\mathbf{W}}e^{\mathbf{T}}e^{\mathbf{W}}\mathcal{M}e^{-\mathbf{T}}. \tag{5.83}$$

5.3.3 Iterative Procedure

Now, we shall develop $\underline{T}$ perturbatively as a product of exponential maps whose arguments are ordered by ϵ. The ladder diagram of Figure 5.4(b) serves to orient this procedure. We think of Eq. (5.83) as applying to each rung of the ladder, starting from $\mathcal{N}_0 \equiv \mathcal{M}$, and iteratively move down the ladder rung by rung using the by now familiar strategy of eliminating one power of ϵ at a time via the ansatz,

$$\underline{T}_k = e^{\epsilon^k \mathbf{T}_k}\underline{I} = \underline{I} + O(\epsilon^k).$$

Accordingly, Eq. (5.83) is replaced by the sequence of equations,

$$\mathcal{N}_k = e^{-\mathbf{W}}e^{\epsilon^k \mathbf{T}_k}e^{\mathbf{W}}\mathcal{N}_{k-1}e^{-\epsilon^k \mathbf{T}_k}.$$

Terms of O(ε) We start from $\mathcal{N}_0 = \mathcal{M}$ itself, noting that $\mathcal{N}_0 = 1 + O(\epsilon)$, and construct $\mathcal{N}_1$. Adopt the notation $(\mathcal{N})_{.n}$ to indicate the coefficient of ϵ^n in an expansion of $\mathcal{N}$.

$$\begin{aligned}\mathcal{N}_1 &= e^{-\mathbf{W}} e^{\epsilon \mathbf{T}_1} e^{\mathbf{W}} \mathcal{N}_0 e^{-\epsilon \mathbf{T}_1} \\ &= e^{-\mathbf{W}}(1+\epsilon \mathbf{T}_1) e^{\mathbf{W}}(1+\epsilon(\mathcal{N}_0)_{.1})(1-\epsilon \mathbf{T}_1) + O(\epsilon^2) \\ &= 1+\epsilon(e^{-\mathbf{W}} \mathbf{T}_1 e^{\mathbf{W}} + (\mathcal{N}_0)_{.1} - \mathbf{T}_1) + O(\epsilon^2) \\ &\equiv 1+\epsilon \mathbf{N}_1 + O(\epsilon^2) \\ &= e^{\epsilon \mathbf{N}_1} + O(\epsilon^2).\end{aligned} \tag{5.84}$$

We can rewrite the homological equation for $\mathbf{N}_1$ into a form more familiar from Section 5.1.2.

$$[\mathbf{T}_1, e^{\mathbf{W}}] = e^{\mathbf{W}}(\mathbf{N}_1 - (\mathcal{N}_0)_{.1}). \tag{5.85}$$

This is virtually identical in structure to Eq. (5.11), as it should be. As with Eq. (5.11) we would like to zero $\mathbf{N}_1$ and then solve for $\mathbf{T}_1$. As always, shear and resonance terms prevent us from doing that and we must absorb the offending terms from $(\mathcal{N}_0)_{.1}$ into $\mathbf{N}_1$ before proceeding with the solution.

Before doing this we need to construct $(\mathcal{N}_0)_{.1}$ itself. In accordance with Eq. (5.80) and Eq. (5.81),

$$\underline{N}_0(\underline{z}) = \underline{M}(\underline{z}) = \underline{z} + \epsilon \underline{M}_1(\underline{z}) + O(\epsilon^2). \tag{5.86}$$

Now we seek to find a Lie operator, say, $(\mathcal{N}_0)_{.1} = \underline{f}_{-1} \cdot \underline{\partial}$, such that

$$\begin{aligned}\underline{M} &= e^{\epsilon(\mathcal{N}_0)_{.1}} \underline{I} + O(\epsilon^2) \\ &= \underline{I} + \epsilon(\mathcal{N}_0)_{.1} \underline{I} + O(\epsilon^2).\end{aligned}$$

Combining this with Eq. (5.86) and Eq. (5.81) provides the answer:

$$(\mathcal{N}_0)_{.1} \underline{I} = \underline{M}_1, \quad \text{or} \quad \underline{f}_{-1}(\underline{z}) = \underline{M}_1(\underline{z}) \quad \text{and} \quad (\mathcal{N}_0)_{.1} = \underline{M}_1(\underline{z}) \cdot \frac{\partial}{\partial \underline{z}}. \tag{5.87}$$

Assuming that we have satisfied Eq. (5.85) with $\mathbf{N}_1$ possessing a minimal number of terms, we return to Eq. (5.84) to calculate $\mathcal{N}_1$ accurately. By this construction,

$$e^{-\epsilon \mathbf{N}_1} \mathcal{N}_1 = 1 + O(\epsilon^2), \tag{5.88}$$

as required.

Solving the Homological Equation As always, the calculations are expedited by using normal form (eigen) coordinates of the linear vector field, **W**. We shall consider only the case (a) that the origin is an elliptic fixed point, which implies that the eigenvalues of $\mathbf{W}$ ($e^{\mathbf{W}}$) are on the imaginary axis (unit circle), and (b) that $\underline{\underline{W}}$ is diagonalizable; our arguments can be extended to the more general case. Under these assumptions, the linear part of the map arises from a Hamiltonian function, which can be written in action-angle representation as in Eq. (3.103) (a first degree polynomial in the action coordinates) and which can be transferred to the complex normal form (deMoivre) chart using Eq. (2.9).

$$H = \underline{\nu} \cdot \underline{I} = \sum_k \nu_k a_k^* a_k.$$

We use Eq. (3.56) to write the vector field produced by this Hamiltonian.

$$X_H = i \sum_k \nu_k \left(a_k^* \frac{\partial}{\partial a_k^*} - a_k \frac{\partial}{\partial a_k} \right) = \underline{\nu} \cdot \frac{\partial}{\partial \underline{\varphi}}.$$

With regard to the last expression, see Problem 2.4. Monomials in these coordinates are clearly eigenfunctions of X_H.

$$\Psi_{\underline{m}\,\underline{n}} \equiv \prod_j a_j^{m_j} a_j^{*n_j},$$

$$X_H \Psi_{\underline{m}\,\underline{n}} = X_H \prod_j a_j^{m_j} a_j^{*n_j} = \left(i \sum_k \nu_k (n_k - m_k) \right) \prod_j a_j^{m_j} a_j^{*n_j}$$
$$= i\underline{\nu} \cdot (\underline{n} - \underline{m}) \Psi_{\underline{m}\,\underline{n}}.$$

Apart from a trivial multiplicative constant, these are essentially the same basis states used earlier; for example, compare this to Eqs.(5.63) and (5.64). Anticipating applications to the transverse dynamics of storage rings, we assume the linear part of our map is an evolution through 2π in the orbit parameter, θ,

$$e^{\mathbf{W}} = e^{2\pi X_H} = e^{-2\pi \underline{\nu} \cdot :\underline{I}:} = e^{2\pi \underline{\nu} \cdot \partial / \partial \underline{\varphi}},$$

from which it follows that the $\Psi_{\underline{m}\,\underline{n}}$ are the eigenfunctions of $e^{\mathbf{W}}$ that we desire. We have written this using both Dragt notation, familiar to accelerator physicists, and the mathematicians' Hamiltonian vector field notation. Recall, from the discussion below Eq. (3.11), that the correspondence is $X_f = -{:}f{:}$.

$$e^{\mathbf{W}} \Psi_{\underline{m}\,\underline{n}} = \exp[i 2\pi \underline{\nu} \cdot (\underline{n} - \underline{m})]\, \Psi_{\underline{m}\,\underline{n}}.$$

We can already sense the resonance condition peeking through this equation.

Our goal is to solve the homological equation, Eq. (5.85), employing the $\Psi_{\underline{m}\,\underline{n}}$ as a set of basis functions. It will be convenient to premultiply both sides of the equation by $e^{-\mathbf{W}}$ and rewrite it as follows:

$$e^{-\mathbf{W}} \mathbf{T}_1 e^{\mathbf{W}} - \mathbf{T}_1 = \mathbf{N}_1 - (\mathcal{N}_0)_{.1}. \tag{5.89}$$

If we now define "matrix elements" of $\mathbf{T}$ — temporarily dropping the subscript "1" for convenience — according to

$$\mathbf{T}\Psi_{\underline{m}\,\underline{n}} \equiv \sum_{\underline{m}',\underline{n}'} \mathbf{T}_{\underline{m}'\underline{n}';\underline{m}\,\underline{n}} \Psi_{\underline{m}'\underline{n}'},$$

then the matrix elements of the left side of Eq. (5.89) become

$$(e^{i2\pi\underline{\nu}\cdot(\underline{m}'-\underline{n}'-\underline{m}+\underline{n})} - 1)\mathbf{T}_{\underline{m}'\underline{n}';\underline{m}\,\underline{n}}. \tag{5.90}$$

We recognize once again that the right side of Eq. (5.89) must vanish whenever the indices obey a resonance condition with $\underline{\nu}$; the shear terms correspond to the diagonal elements: $\underline{n} - \underline{m} = \underline{n}' - \underline{m}'$. (Of course, if the eigenvalues of $\mathbf{W}$ are not pure imaginary, then resonances do not occur, and one searches for attractors instead.)

Let us consider the implications for the right side of Eq. (5.89): which terms need to be absorbed into $\mathbf{N}_1$ to match the zeros of the left side? We shall let $\mathbf{R} = \mathbf{N}_1 - (\mathcal{N}_0)_1$ symbolize whatever vector field is on the right. To keep matters simple we provisionally accept that we are studying a Hamiltonian system (although the perturbation theory does not require this) and that $\mathbf{R}$ can be written as a Hamiltonian vector field. Furthermore, we expand $\mathbf{R}$ in the vector field basis generated by the $\Psi_{\underline{m}\,\underline{n}}$. That is, define (see Eq. (3.56))

$$X_{\underline{k}\,\underline{j}} \equiv X_{\Psi_{\underline{k}\,\underline{j}}} = -{:}\Psi_{\underline{k}\,\underline{j}}{:} = i\sum_l \left(\frac{\partial \Psi_{\underline{k}\,\underline{j}}}{\partial a_l}\frac{\partial}{\partial a_l^*} - \frac{\partial \Psi_{\underline{k}\,\underline{j}}}{\partial a_l^*}\frac{\partial}{\partial a_l} \right), \tag{5.91}$$

and accept, subject to later verification, that

$$\mathbf{R} = \sum_{\underline{k}\,\underline{j}} R_{\underline{k}\,\underline{j}} X_{\underline{k}\,\underline{j}} = -\sum_{\underline{k}\,\underline{j}} R_{\underline{k}\,\underline{j}} {:}\Psi_{\underline{k}\,\underline{j}}{:},$$

for some set of (complex) coefficients, $R_{\underline{k}\,\underline{j}}$. Computing the action of $\mathbf{R}$ on $\Psi_{\underline{m}\,\underline{n}}$ is made easier by first computing the action of $X_{\underline{k}\,\underline{j}}$, using Eq. (5.91).

$$X_{\underline{k}\,\underline{j}}\Psi_{\underline{m}\,\underline{n}} = i\sum_l (k_l n_l - j_l m_l)\Psi_{\underline{k}+\underline{m}-\underline{e}_l,\underline{j}+\underline{n}-\underline{e}_l}.$$

Here we have used the obvious identity, $\Psi_{\underline{k}\,\underline{j}}\Psi_{\underline{m}\,\underline{n}} = \Psi_{\underline{k}+\underline{m},\underline{j}+\underline{n}}$ and defined $\underline{e}_l$ to be the array with components $(\underline{e}_l)_i = \delta_{il}$. We identify $\underline{m}' = \underline{k} + \underline{m} - \underline{e}_l$ and $\underline{n}' = \underline{j} + \underline{n} - \underline{e}_l$, and (because I am running out of integer variable symbols) we shall let the tunes $\underline{\nu}$ obey a resonance condition, $\underline{r}\cdot\underline{\nu} + t = 0$. Then, according to Eq. (5.90), matrix elements of $\mathbf{R}$ that must be zeroed correspond to setting

$$(\underline{m}' - \underline{n}') - (\underline{m} - \underline{n}) = (\underline{k} + \underline{m} - \underline{e}_l) - (\underline{j} + \underline{n} - \underline{e}_l) - (\underline{m} - \underline{n}) = \underline{k} - \underline{j} = \pm\underline{r}.$$

That is, the coefficients $R_{\underline{k}\,\underline{j}}$ which must be set to zero are all those for which $\underline{k} - \underline{j} = \pm\underline{r}$.

The shear terms are obtained by setting $\underline{r} = \underline{0}$ and $t = 0$. These will comprise the subspace of terms spanned by $X_{\underline{p}\underline{p}}$, for integer arrays, $\underline{p}$. Correspondingly, these are generated by Hamiltonian terms of the form

$$\Psi_{\underline{p}\underline{p}} = \prod_j (a_j a_j^*)^{p_j} = \prod_j I_j^{p_j}. \tag{5.92}$$

In total agreement with everything that has gone before — averaging, Deprit's algorithm, and so forth — these are integer powers of the action coordinates alone. The operators themselves are written using Eq. (5.91).

$$\begin{aligned} X_{\underline{p}\underline{p}} &= i \sum_l p_l \left(\Psi_{\underline{p}-\underline{e}_l \underline{p}} \frac{\partial}{\partial a_l^*} - \Psi_{\underline{p}\underline{p}-\underline{e}_l} \frac{\partial}{\partial a_l} \right) \\ &= \sum_l p_l \left(\prod_j (a_j a_j^*)^{p_j} \right) \frac{1}{a_l a_l^*} i \left(a_l^* \frac{\partial}{\partial a_l^*} - a_l \frac{\partial}{\partial a_l} \right) \\ &= \underline{\nu}(\underline{I}) \cdot \frac{\partial}{\partial \underline{\varphi}}. \end{aligned}$$

According to Eq. (5.87), which serves as a template for what is done at all the higher orders of the algorithm, these will arise from map terms, expressed as polynomials in $\underline{a}$ and $\underline{a}^*$, in which the exponent of one and only one of the a_l differs from that of the corresponding a_l^* by one, all others being the same. That is, they arise from map terms that could be written as

$$\left(\prod_j (a_j a_j^*)^{p_j} \right) \frac{1}{a_l} \frac{\partial}{\partial a_l^*} \quad \text{or} \quad \left(\prod_j (a_j a_j^*)^{p_j} \right) \frac{1}{a_l^*} \frac{\partial}{\partial a_l}, \tag{5.93}$$

for some $\underline{p}$ and l. These are the terms from the map that, at each order of the algorithm, must be absorbed by the Lie operator of the nonresonant normal form so that homological equations, like Eq. (5.85), can be solved. One can, in fact, see that this must be so without reference to Hamiltonians at all, directly from Eq. (5.90). This is left as an exercise for the reader.

The nontrivial resonant terms will be generated by Hamiltonians of the form

$$R_{\underline{p},\underline{p}+\underline{r}} \Psi_{\underline{p},\underline{p}+\underline{r}} + R_{\underline{p}+\underline{r},\underline{p}} \Psi_{\underline{p}+\underline{r},\underline{p}}$$

for integer arrays, $\underline{p}$. These may be collected into terms of the form

$$\left(\prod_j I_j^{p_j + r_j/2} \right) \cos(\underline{r} \cdot \underline{\varphi}) \quad \text{and} \quad \left(\prod_j I_j^{p_j + r_j/2} \right) \sin(\underline{r} \cdot \underline{\varphi}),$$

which will, as before, give rise to resonant normal form (effective) Hamiltonians.

IMPORTANT: Before moving on, let us emphasize: the $\underline{a}$ and $\underline{a}^*$ are really *defined* to be the eigencoordinates (or normal form coordinates) of the linear part of the map. We do not need to assume a correspondence with "horizontal" and "vertical" directions; *there is no need to restrict ourselves to linear dynamics that are decoupled along these sectors.* Rather, in analogy with the derivation of Eq. (2.8), let $\underline{e}_k^{(\pm)}$ be defined according to

$$\underline{\underline{W}} \cdot \underline{e}_k^{(\pm)} = \pm i\nu_k \underline{e}_k^{(\pm)},$$

and let the dual vectors be $\underline{u}_k^{(\pm)}$. Then linear normal form coordinates are defined as

$$a_k = \underline{u}_k^{(-)\dagger} \cdot \underline{z} \quad \text{and} \quad a_k^* = \underline{u}_k^{(+)\dagger} \cdot \underline{z}.$$

Normalization can be imposed by extra conditions. For example, the condition $\underline{\underline{M}}\,\underline{\underline{J}}\,\underline{\underline{M}}^T = -i\underline{\underline{J}}$, which appears below Eq. (3.55). (See Problem 5.16.)

Example: Octupole Detuning, from a Hamiltonian To illustrate how some of this works, we shall examine once again the first order detuning produced by an octupole. We model passage through the octupole using Eqs.(3.104) and (3.110); that is, we take it to be governed by the Hamiltonian

$$H = \frac{B_o R}{|B\rho|}\frac{b_3}{4}(x_1^4 - 6x_1^2x_2^2 + x_2^4).$$

In this example we shall confine our attention to the horizontal plane, set $x_2 = 0$,[17] and suppress the subscripts. The map $\mathcal{M}$ is the orbit evolution operator through the octupole,

$$\mathcal{M} = e^{\Delta\theta X_H} = e^{(L/R)X_H} = e^{\epsilon \mathbf{M}_1},$$

which identifies

$$\epsilon \mathbf{M}_1 = \epsilon(\mathcal{N}_0)_{.1} = -\frac{B_o L}{|B\rho|}\frac{b_3}{4}{:}x^4{:}.$$

As always, ϵ should be set so as to carry the units and scale of the problem, so that all dynamical coordinates are of order unity; it is the order parameter. This expression is exact provided that b_3 is constant along the length of the octupole; it is not a thin element, impulse, or kick approximation.

The action-angle representation of a transverse Hamiltonian producing the linear part of the map, Eq. (3.103), was obtained in Chapter 2 using Snowdon's transformation, Eq. (3.100). We employ the transformation here to express x on the local

[17] Recall that there are then no forces to move the orbit out of the horizontal plane.

"de Moivre chart."

$$\begin{aligned} x &= \sqrt{\beta/2}\,(a^* e^{i\tilde{\psi}} + a e^{-i\tilde{\psi}}), \\ \alpha x + \beta p &= i\sqrt{\beta/2}\,(a^* e^{i\tilde{\psi}} - a e^{-i\tilde{\psi}}). \end{aligned} \tag{5.94}$$

If we *now* employ a thin element approximation, because of the arbitrariness of choosing the zero of phase, $\tilde{\psi} = \psi - \nu\theta$ has no physical meaning in this example and could be dropped; if there were more than one octupole — or if the octupole were extraordinarily long — the phase differences between them would enter into calculations of resonance terms. Nonetheless, we shall retain the factor if only to show how it gets carried through the next step, which is to expand x^4.

$$\begin{aligned} x^4 &= (\beta/2)^2(6(a^*a)^2 + 4(a^{*3}ae^{i2\tilde{\psi}} + a^3a^*e^{-i2\tilde{\psi}}) + (a^{*4}e^{i4\tilde{\psi}} + a^4e^{-i4\tilde{\psi}})) \\ &= (\beta/2)^2(6\Psi_{22} + 4(e^{i2\tilde{\psi}}\Psi_{13} + e^{-i2\tilde{\psi}}\Psi_{31}) + (e^{i4\tilde{\psi}}\Psi_{04} + e^{-i4\tilde{\psi}}\Psi_{40})). \end{aligned} \tag{5.95}$$

From our earlier discussion we know that the shear term will come from Ψ_{22}, which term therefore must be absorbed into $\mathbf{N}_1$.

$$\epsilon\mathbf{N}_1 = -\frac{B_oL}{|B\rho|}\frac{b_3}{4}(\beta/2)^2 6{:}(a^*a)^2{:} \equiv -2\pi g{:}I^2{:} = 4\pi g I\frac{\partial}{\partial\varphi} \tag{5.96}$$

The complete normal form map, to first order, is then given by the following (see Eq. (5.83), and the one above it):

$$e^{\mathbf{W}}\mathcal{N}_1 + O(\epsilon^2) = e^{\mathbf{W}}e^{\epsilon\mathbf{N}_1} = e^{-2\pi\nu{:}I{:}}e^{-2\pi g{:}I^2{:}} = e^{-2\pi(\nu{:}I{:}+g{:}I^2{:})}.$$

This normal form is the 2π orbit-evolution map of an autonomous Hamiltonian — sometimes called the "effective Hamiltonian" or "pseudo-Hamiltonian" — $H = \nu I + gI^2$. The first order octupole detuning thus obtained agrees with our previous results, as written in Eq. (4.22), for example.

As always, we have the freedom to absorb even more terms into $\mathbf{N}_1$, thereby making the normal form more complicated and the transformation simpler. We see from the terms appearing in Eq. (5.95) — and from our discussion above as to how these enter into resonance driving terms — that, at the minimum, isolated resonance models should be employed near fourth- and half-integer tunes. This cancels the singularity that would otherwise arise from the "small denominator" implicit in Eqs.(5.89) and (5.90). It is left to the reader to write the resonance normal forms so obtained.

Example: Octupole Detuning, from a map We repeat the previous calculation but without direct reference to an underlying Hamiltonian. Recall from Eq. (3.114) that passage through a thin octupole produces a cubic kick,

$$x \longmapsto x, p \longmapsto p + kx^3, \tag{5.97}$$

where k is proportional to the integrated third derivative of the magnetic field. Using Eq. (5.94), with the obnoxious and irrelevant $\tilde{\psi}$ set to zero, we see that $\Delta x = 0$ and

$\Delta p = kx^3$ translate to

$$\begin{aligned}\Delta a &= -\Delta a^*, \quad \text{and}\\ 2\Delta a &= i\beta\Delta p/\sqrt{\beta/2}\\ &= ik\sqrt{2\beta}x^3\\ &= \frac{i}{2}k\beta^2(a+a^*)^3.\end{aligned}$$

This, from the octupole kick alone; the full map comprises the kick followed by traversal of a ring with linear tune ν. Since a and a^* are the irreducible eigencoordinates of the local one-turn map, this is written

$$\underline{M}^{\sim}: \begin{pmatrix} a \\ a^* \end{pmatrix} \mapsto \begin{pmatrix} e^{-i2\pi\nu} & 0 \\ 0 & e^{i2\pi\nu} \end{pmatrix} \begin{pmatrix} a + i\dfrac{k\beta^2}{4}(a+a^*)^3 \\ a^* - i\dfrac{k\beta^2}{4}(a+a^*)^3 \end{pmatrix}.$$

The map $\underline{M}$ that is put into the iterative algorithm is obtained from this using Eq. (5.81).

$$\begin{aligned}\underline{M}: \begin{pmatrix} a \\ a^* \end{pmatrix} &\mapsto \begin{pmatrix} e^{-i2\pi\nu} & 0 \\ 0 & e^{i2\pi\nu} \end{pmatrix} \begin{pmatrix} e^{i2\pi\nu}a + i\dfrac{k\beta^2}{4}(e^{i2\pi\nu}a + e^{-i2\pi\nu}a^*)^3 \\ e^{-i2\pi\nu}a^* - i\dfrac{k\beta^2}{4}(e^{i2\pi\nu}a + e^{-i2\pi\nu}a^*)^3 \end{pmatrix}\\ &= \begin{pmatrix} a + i\dfrac{k\beta^2}{4}\left(e^{i4\pi\nu}a^3 + 3a^2a^* + 3e^{-i4\pi\nu}aa^{*2} + e^{-i8\pi\nu}a^{*3}\right) \\ a^* - i\dfrac{k\beta^2}{4}\left(e^{-i4\pi\nu}a^{*3} + 3aa^{*2} + 3e^{i4\pi\nu}a^2a^* + e^{i8\pi\nu}a^3\right) \end{pmatrix}.\end{aligned}$$

From this we extract $(\mathcal{N}_0)_{.1}$.

$$\begin{aligned}(\mathcal{N}_0)_{.1} &= i\frac{k\beta^2}{4}(e^{i4\pi\nu}a^3 + 3a^2a^* + 3e^{-i4\pi\nu}aa^{*2} + e^{-i8\pi\nu}a^{*3})\frac{\partial}{\partial a}\\ &\quad - i\frac{k\beta^2}{4}(e^{-i4\pi\nu}a^{*3} + 3aa^{*2} + 3e^{i4\pi\nu}a^2a^* + e^{i8\pi\nu}a^3)\frac{\partial}{\partial a^*}\end{aligned} \tag{5.98}$$

The minimal nonresonant normal form $\mathbf{N}_1$ that must be filtered from this in order to solve Eq. (5.85) is

$$\begin{aligned}\mathbf{N}_1 &= 3i\frac{k\beta^2}{4}\left(a^2a^*\frac{\partial}{\partial a} - aa^{*2}\frac{\partial}{\partial a^*}\right)\\ &= 3\frac{k\beta^2}{4}aa^* \cdot i\left(a\frac{\partial}{\partial a} - a^*\frac{\partial}{\partial a^*}\right)\\ &= -3\frac{k\beta^2}{4}I\frac{\partial}{\partial\varphi}.\end{aligned} \tag{5.99}$$

It is left to the reader to verify that this is equivalent to our previous results by evaluating k. We note in passing that $\mathbf{N}_1 = -\frac{3}{8}k\beta^2 X_{22}$.

Having absorbed shear terms into the normal form, we are ready to solve the homological equation, either in the form of Eq. (5.85) or that of Eq. (5.89). The structure of Eq. (5.85) is identical to that of Eq. (5.17) and we solve it in basically the same way: by solving a linear equation to determine a set of coefficients. However, we are here sufficiently enlightened to be working in the eigencoordinates of the linear vector field, thereby diagonalizing the system of equations which must be solved. It makes little difference, but it is slightly easier to work with Eq. (5.89), which we write here in functional, rather than operator, form by acting on the identity map.

$$\underline{\underline{A}} \cdot \underline{\underline{T}}(\underline{\underline{A}}^{-1} \cdot \underline{z}) - \underline{\underline{T}}(\underline{z}) = \underline{\underline{R}}(\underline{z})$$

The right side is symbolized by $\underline{R}$. We explicitly write this in component form, using the fact that $\underline{z} = (a, a*)$, the eigencoordinates of $\underline{\underline{A}}$.

$$\begin{pmatrix} e^{-i2\pi\nu}T_a\left(e^{i2\pi\nu}a, e^{-i2\pi\nu}a^*\right) - T_a(a, a^*) \\ e^{i2\pi\nu}T_{a^*}\left(e^{i2\pi\nu}a, e^{-i2\pi\nu}a^*\right) - T_{a^*}(a, a^*) \end{pmatrix} = \begin{pmatrix} R_a(a, a^*) \\ R_{a^*}(a, a^*) \end{pmatrix} \tag{5.100}$$

The operator $\mathbf{R} = R_a\partial/\partial a + R_{a^*}\partial/\partial a^*$ comprises everything in $-(\mathcal{N}_0)_{.1}$ that was *not* absorbed by $\mathbf{N}_1$. Functional linearity of Eq. (5.100) combined with the happy circumstance of using the eigencoordinate chart of $\underline{\underline{A}}$ offer us the luxury of handling these terms one at a time. For example, suppose that T_a contains the term ca^3, for some constant c. Plugging this into Eq. (5.100) and using Eq. (5.98) to provide the a^3 term of R_a, yields

$$ce^{-i2\pi\nu}(e^{i2\pi\nu}a)^3 - ca^3 = c(e^{i4\pi\nu} - 1)a^3 = -i\frac{k\beta^2}{4}e^{i4\pi\nu}a^3.$$

From this we get that the coefficient of a^3 in T_a is $c = -i(k\beta^2/4)/(1 - e^{-i4\pi\nu})$. Working out the remaining terms, and reverting to operator notation, the final result is written as follows:

$$\begin{aligned}\mathbf{T}_1 &= T_a\frac{\partial}{\partial a} + T_{a^*}\frac{\partial}{\partial a^*} \\ &= -i\frac{k\beta^2}{4}\left(\frac{a^3}{1 - e^{-i4\pi\nu}} + 3\frac{aa^{*2}}{1 - e^{i4\pi\nu}} + \frac{a^{*3}}{1 - e^{i8\pi\nu}}\right)\frac{\partial}{\partial a} \\ &\quad + i\frac{k\beta^2}{4}\left(\frac{a^{*3}}{1 - e^{i4\pi\nu}} + 3\frac{a^2a^*}{1 - e^{-i4\pi\nu}} + \frac{a^3}{1 - e{-i8\pi\nu}}\right)\frac{\partial}{\partial a^*}\end{aligned}$$

The shear terms, which could have given trouble, because of "small denominators," were avoided by absorption into $\mathbf{N}_1$. (See what happens when you plug a^2a^* for T_a in Eq. (5.100).) If any other denominator is, in fact, small — because the tune, ν, is too close to a half-integer or quarter-integer resonance — then these also should be absorbed before solving the homological equation. This would result in $\mathbf{N}_1$ becoming a resonance normal form, rather than the nonresonant, shear normal form.

Through $O(\epsilon^3)$ Proceeding to the second order, we write

$$\mathcal{N}_2 = e^{-\mathbf{W}} e^{\epsilon^2 \mathbf{T}_2} e^{\mathbf{W}} \mathcal{N}_1 e^{-\epsilon^2 \mathbf{T}_2}.$$

There is a problem with this as written: we now want all the ϵ-dependent factors appearing on the right to be $1 + O(\epsilon^2)$, but $\mathcal{N}_1$ is $1 + O(\epsilon)$. This prevents us from expressing the right-hand side as $1 + O(\epsilon^2)$, which is needed to identify it as $e^{\epsilon^2 \mathbf{N}_2} + O(\epsilon^3)$. To overcome this hurdle we multiply on the left by $e^{-\epsilon \mathbf{N}_1}$, use Eq. (5.88), and the fact that the commutator

$$[e^{-\mathbf{W}} e^{\epsilon^2 \mathbf{T}_2} e^{\mathbf{W}}, e^{-\epsilon \mathbf{N}_1}] = [1 + O(\epsilon^2), 1 + O(\epsilon)] = O(\epsilon^3),$$

to get

$$\begin{aligned} e^{-\epsilon \mathbf{N}_1} \mathcal{N}_2 &= e^{-\mathbf{W}} e^{\epsilon^2 \mathbf{T}_2} e^{\mathbf{W}} (e^{-\epsilon \mathbf{N}_1} \mathcal{N}_1) e^{-\epsilon^2 \mathbf{T}_2} + O(\epsilon^3) \\ &= 1 + O(\epsilon^2) \\ &\equiv e^{\epsilon^2 \mathbf{N}_2} + O(\epsilon^3). \end{aligned} \tag{5.101}$$

If we now expand this through second order, using

$$e^{-\epsilon \mathbf{N}_1} \mathcal{N}_1 = 1 + (e^{-\epsilon \mathbf{N}_1} \mathcal{N}_1)_{.2} \epsilon^2 + O(\epsilon^3),$$

we obtain an equation exactly like Eq. (5.85) but with all subscripts raised by one and $(e^{-\epsilon \mathbf{N}_1} \mathcal{N}_1)_{.2}$ replacing $(\mathcal{N}_0)_{.1}$.

$$[\mathbf{T}_2, e^{\mathbf{W}}] = e^{\mathbf{W}} (\mathbf{N}_2 - (e^{-\epsilon \mathbf{N}_1} \mathcal{N}_1)_{.2}).$$

Proceeding immediately to third order,

$$\mathcal{N}_3 = e^{-\mathbf{W}} e^{\epsilon^3 \mathbf{T}_3} e^{\mathbf{W}} \mathcal{N}_2 e^{-\epsilon^3 \mathbf{T}_3}. \tag{5.102}$$

Similar to the second order, we must use Eq. (5.101),

$$\begin{gathered} e^{-\epsilon^2 \mathbf{N}_2} e^{-\epsilon \mathbf{N}_1} \mathcal{N}_2 = 1 + O(\epsilon^3), \quad \text{and} \\ [e^{-\mathbf{W}} e^{\epsilon^3 \mathbf{T}_3} e^{\mathbf{W}}, e^{-\epsilon^2 \mathbf{N}_2} e^{-\epsilon \mathbf{N}_1}] = [1 + O(\epsilon^3), 1 + O(\epsilon)] = O(\epsilon^4), \end{gathered}$$

to rewrite this as

$$\begin{aligned} e^{-\epsilon^2 \mathbf{N}_2} e^{-\epsilon \mathbf{N}_1} \mathcal{N}_3 &= e^{-\mathbf{W}} e^{\epsilon^3 \mathbf{T}_3} e^{\mathbf{W}} (e^{-\epsilon^2 \mathbf{N}_2} e^{-\epsilon \mathbf{N}_1} \mathcal{N}_2) e^{-\epsilon^3 \mathbf{T}_3} + O(\epsilon^4) \\ &= 1 + O(\epsilon^3) \\ &\equiv e^{\epsilon^3 \mathbf{N}_3} + O(\epsilon^4). \end{aligned}$$

Expanding through third order, the characteristic equation relating $\mathbf{T}_3$ and $\mathbf{N}_3$ is written similarly to the lower order ones.

$$[\mathbf{T}_3, e^{\mathbf{W}}] = e^{\mathbf{W}}(\mathbf{N}_3 - (e^{-\epsilon^2\mathbf{N}_2}e^{-\epsilon\mathbf{N}_1}\mathcal{N}_2)_{.3}).$$

Once this is solved, Eq. (5.102) is used to calculate $\mathcal{N}_3$ exactly, which completes consideration of the third order.

Homological Equations The pattern is now evident. At the $(k+1)^{\text{th}}$ order we shall have to solve the homological equation

$$[\mathbf{T}_{k+1}, e^{\mathbf{W}}] = e^{\mathbf{W}}\left(\mathbf{N}_{k+1} - \left(\prod_{j=1}^{k}{}^{(r)} e^{-\epsilon^j\mathbf{N}_j}\mathcal{N}_k\right)_{.k+1}\right), \tag{5.103}$$

for $\mathbf{T}_{k+1}$ and $\mathbf{N}_{k+1}$, bearing in mind the objective of keeping $\mathbf{N}_{k+1}$ minimal. The "(r)" superscript indicates "reverse order." After this, we return to solve for $\mathcal{N}_{k+1}$ using

$$\mathcal{N}_{k+1} = e^{-\mathbf{W}}e^{\epsilon^{k+1}\mathbf{T}_{k+1}}e^{\mathbf{W}}\mathcal{N}_k e^{-\epsilon^{k+1}\mathbf{T}_{k+1}}.$$

The homological equation assures that

$$\mathcal{N}_k = \prod_{j=1}^{k}{}^{(o)} e^{\epsilon^j\mathbf{N}_j} + \mathcal{O}(\epsilon^{k+1}), \quad \text{for all } k.$$

We are left with the problem of solving the sequence of homological equations for the $\mathbf{T}_k$ and $\mathbf{N}_k$. As with the other forms of perturbation theory we have discussed, we ideally would like each $\mathbf{N}_k$ to vanish, but are forced to let it absorb shear and resonance terms. We adopt the notation [37]

$$\mathcal{N}_k^{\flat} \equiv \left(\prod_{j=1}^{k}{}^{(r)} e^{-\epsilon^j\mathbf{N}_j}\mathcal{N}_k\right),$$

and, as before, premultiply by $e^{-\mathbf{W}}$ to rewrite the homological equations in the following form:

$$e^{-\mathbf{W}}\mathbf{T}_{k+1}e^{\mathbf{W}} - \mathbf{T}_{k+1} = (\mathbf{N}_{k+1} - (\mathcal{N}_k^{\flat})_{.k+1}). \tag{5.104}$$

As is typical of the methods we have seen, these equations all have the same structure and are solved precisely as was the first order homological equation. At each step the minimum shear and resonance terms are absorbed into the $\mathbf{N}_k$. If only shear terms are absorbed, each $\mathbf{N}_k$ will contain only vector fields generated by terms like those in Eq. (5.92). We then would obtain the normal form operator,

$$\mathcal{N} = e^{\epsilon\mathbf{N}_1}e^{\epsilon^2\mathbf{N}_2}e^{\epsilon^3\mathbf{N}_3}\cdots,$$

in which each $\mathbf{N}_k$ depends on the action coordinates alone; that is, each $\mathbf{N}_k$ is independent of the angle coordinates. Since the action coordinates are in involution — they all commute with each other — we can combine the $\mathbf{N}_k$'s into the argument of a single exponential.

$$\mathcal{N} = \exp\left[\sum_k \epsilon^k \mathbf{N}_k(\underline{J})\right].$$

But this is precisely the form of the orbit-evolution operator associated with an effective autonomous vector field. Let $H(\underline{J})$ be the representation of a shear, normal form Hamiltonian in its proper action coordinates. The orbits of this model are given by Eq. (2.63).

$$\begin{aligned}\underline{z}(\theta) = \underline{z}_\theta(\underline{z}(0)) &= \left(e^{\theta X_H}\underline{I}\right)_{(\underline{z}(0))} \\ &= \left(e^{-\theta:H(\underline{J}):}\underline{I}\right)_{(\underline{z}(0))} \\ &= \left(e^{\theta\underline{\nu}(\underline{J})\cdot\partial/\partial\underline{\varphi}}\underline{I}\right)_{(\underline{z}(0))}.\end{aligned}$$

Here we have used $\underline{\nu}(\underline{J}) = \partial H/\partial \underline{J}$. All we need is to identify

$$-:H(\underline{J}): = \frac{1}{2\pi}\sum_k \epsilon^k \mathbf{N}_k(\underline{J}) = \underline{\nu}(\underline{J})\cdot\frac{\partial}{\partial\underline{\varphi}}.$$

The linear part, arising from $e^{\mathbf{W}}$, is missing from this expression, but that is added trivially: recall that the full normal form operator is $e^{\mathbf{W}}\mathcal{N}$.

As a formal closing comment, we note that in Section 3.5 is introduced a linear operator, $\overline{\mathbf{W}}$, on the space of Hamiltonian vector fields such that

$$e^{-\mathbf{W}}\mathbf{T}_{k+1}e^{\mathbf{W}} - \mathbf{T}_{k+1} = \left(e^{-\overline{\mathbf{W}}} - 1\right)\mathbf{T}_{k+1}.$$

(See Lemma 3.4.) The entire issue of resonances and shear terms therefore boils down to understanding the null space, or kernel, of the operator, $\exp[-\overline{\mathbf{W}}] - 1$.

PROBLEMS

5.1 Write the group of scale transformations on extended phase space for the dynamical system, Eq. (5.1). Show that μ is an inessential parameter by finding a scale that eliminates it. How is it possible to have a bifurcation in the original system and none in the transformed system?

5.2 Calculate the coefficient of $(w/\sqrt{\mu})^6$ in Eq. (5.6).

5.3 Write the first degree coordinate transformation corresponding to Eq. (5.50). Continue the calculation of $\mathbf{W}$ to second degree.

5.4 Write the normal form Hamiltonian of the pendulum to third order in the action coordinate.

5.5 Confirm the definition of $G_{\underline{m}}$ given in Eq. (5.66).

5.6 Confirm Eq. (5.69) both (a) by taking the limit of Eq. (5.68) and (b) by applying the fundamental theorem of calculus to periodic functions with zero average.

5.7 Confirm Eq. (5.73) and extend it to four dimensions by including J_2^2 and J_1J_2 terms from Eq. (5.72).

5.8 Solve Eq. (5.17) again, but using the eigencoordinates of $\mathbf{V}^{(1)}$.

5.9 Write the isolated resonance models used to generate the separatrices in Figure 5.3. In particular, calculate the coupling constants.

5.10 Finish deriving the sextupole-induced tune spread written in Eq. (5.72).

5.11 Confirm the condition in Eq. (5.93) for nonresonant normal forms without reference to underlying Hamiltonians using Eq. (5.90) directly. What are the corresponding terms for resonant normal forms?

5.12 Confirm that the first order octupole detuning terms predicted by Eq. (4.22), Eq. (5.96), and Eq. (5.99) are equivalent. Obtain the same result using Deprit's algorithm.

5.13 Calculate the first order resonance driving terms for the fourth-integer resonance arising from an octupole distribution using (a) averaging, (b) Deprit's algorithm, and (c) the Forest, Berz, Irwin map procedure and show that they all give the same result.

5.14 Calculate the second order tune spread arising from sextupoles using the Forest, Berz, Irwin procedure and compare to Eq. (5.73). (Hint: see Forest, et al. [38].)

5.15 Extend the calculation of Eq. (5.99) to four-dimensional phase space.

5.16 Let the (square) matrix $\underline{\underline{M}}$ be the Jacobian of a Poincaré map evaluated at a fixed point; that is, $\underline{\underline{M}}$ represents the tangent map at the fixed point. Let the columns of the (square) matrix $\underline{\underline{B}}$ consist of the eigenvectors of $\underline{\underline{M}}$, and let $\underline{\underline{M}} \cdot \underline{\underline{B}} = \underline{\underline{B}} \cdot \underline{\underline{\Lambda}}$. (1) Describe the properties of $\underline{\underline{\Lambda}}$. If the coordinates of the original map are $\underline{z}$, define the (possibly complex) coordinates $\underline{a} \equiv \underline{\underline{B}}^{-1} \cdot \underline{z}$. (2) Show that under the tangent map, $a_k \longmapsto \lambda_k a_k$, and relate the constants λ_k to the matrix elements of $\underline{\underline{\Lambda}}$. (3) Show that each $|a_k|^2$ is an invariant when the fixed point is elliptic. (4) If the Poincaré map is symplectic, prove that $\underline{\underline{B}}^T \cdot \underline{\underline{J}} \cdot \underline{\underline{B}} \cdot \underline{\underline{J}}$ is diagonal. (Note: components of $\underline{a}$ are the linear normal form coordinates of the map. In the case of an elliptic fixed point of a symplectic map, $I_k = |a_k|^2$ are action coordinates labelling KAM tori of the tangent map, *provided that* $\underline{\underline{B}}$ is normalized so as to make the non-zero elements of the diagonal matrix in (4) have magnitude one. This obviously requires that $|\det\underline{\underline{B}}| = 1$.)

CHAPTER 6

Epilogue: What Next?

Every other man spoke a language entirely his own, which he had figured out by private thinking; he had his own ideas and peculiar ways. If you wanted to talk about a glass of water, you had to start back with God creating the heavens and the earth; the apple; Abraham; Moses and Jesus; Rome; the Middle Ages; gunpowder; the Revolution; back to Newton; up to Einstein; then war and Lenin and Hitler. After reviewing this and getting it all straight again you could proceed to talk .... "I'm fainting, please get me a little water." You were lucky even then to make yourself understood ....

— Saul Bellow
Seize the Day

If anyone thinks that he knows something, he has not yet known as he ought to know.

— St. Paul
1 Corinthians 8, 2

The academic lectures that were the starting point of this book grew from a growing, personal sense of frustration at how frequently classical mechanics is taught as though it were a closed, dead subject. In them I tried, in as simple a manner as possible,

(a) to define a vector field over a manifold as the appropriate geometric object associated with most of the "equations of motion" of dynamics.

(b) to establish the vector field as a Lie derivative operator acting on real- and complex-valued functions over phase space. It is in this way that Lie algebras and Lie groups enter into physics at a very fundamental level: *it is almost impossible even to begin to study dynamics carefully without introducing them.*

(c) to state the Liouville(-Arnold) theorem, the "fundamental theorem" of Hamiltonian dynamics. Indeed, I cannot remember the concept of "torus" ever making an appearance in any classical mechanics course that I took.

(d) to stress that resonances are geometrically characterized as the dimensional collapse of tori; that is, they correspond to tori of lower than maximal dimension.

(e) to describe the topology of the separatrices of multi-dimensional resonances.

(f) to develop usable perturbation methods, especially those based on the Lie derivative.

(g) to point out that the "small denominators" appearing in homological equations are the effect, not the cause, of resonances.[1] Fixed points, periodic orbits, resonances, and their separatrices belong to a class of structures, the "recurrent sets" of dynamics, which are the topological obstructions that keep all dynamics from being equivalent.

(h) to suggest connections between the separatrices of low order resonances and the stability boundary of maps.

Students should be exposed to these basic concepts — manifolds; vector fields; tangent bundles (and their generalization, fibre bundles); differential forms; Lie groups and algebras — as early as possible and within the natural, familiar setting that spawned them, classical mechanics.[2] Their introduction should not be put off until the study of general relativity or quantum gauge theories, when other conceptual challenges can confuse the issue. Rather, students who approach such advanced subjects — and even the ones who never do — should already feel as comfortable with these basic mathematical objects and their manipulation as with complex variables or line integrals. After all, when string theorists talk about homology groups, they are dealing with ideas rooted in such fundamental issues as the definition of action-angle coordinates. No one should consider the elements of differential topology esoteric or belonging only to the toolkits of quantum cosmologists and string theorists.[3] *These ideas are too important to be surrendered to the experts,* if for no other reason than experts have little inclination to get into the trenches and solve *your* particular problems.

Those who have the privilege and responsibility of guiding students should consider a different approach to teaching classical mechanics, one that does not assume it stopped developing shortly after the Napoleonic Wars, that makes connections with modern, advanced subjects, and that emphasizes real applications. If this book has failed to suggest a way of doing this, the fault is mine, but the goal itself is still valid.

****

[1]This effect is universal, appearing even in the perturbation theories of nonlinear waves. If one tries to generate an expansion about "free," asymptotically plane-wave states, the small-demonimators which occur are connected to the existence of soliton modes of excitation. Corresponding phenomena occur in quantum scattering theory as well. (See, for example, Equations (5.48) and (5.49).)

[2]The concept "manifold" probably was appropriated from complex analysis.

[3]Indeed, given sufficient early exposure to the basics, many more may feel an inclination to actually read some quantum cosmology or string theory.

There is much more that could be added to this book if time and space were unlimited. Among other things, I should (a) include material on polarization, a topic which would have fit in beautifully with our emphasis on the Lie algebraic structure of dynamics, (b) discuss better the physics of the beam-beam interaction and its comparison to "space-charge" forces, (c) add one more chapter on nonintegrable two-resonance models (or more generally, multi-resonance models), and (d) revise two or three sections so as to emphasize the map techniques a little more and the flow techniques a little less.[4] There is indeed much left to be done, but time is up and deadlines loom. Before ending, allow me to introduce very briefly a few other topics, if only that we be aware of how much is missing and to suggest some paths for the reader's further exploration.

KAM theorem The fundamental goal of Hamiltonian perturbation theory is to find coordinate charts on which a given Hamiltonian is at least locally representable by a shearing Hamiltonian, as in Eq. (3.77), the most benign form of twist map. The Liouville-Arnold theorem tells us that such charts exist if there is a complete commuting set of observables — i.e., N invariant observables in involution. But this is essentially never the case: almost all dynamical systems are not integrable, a fact that, a short time ago, was not given the attention it deserved. Nonetheless, if the dynamic is "near-integrable," most orbits will still exist on invariant tori and generally act *as though* they were governed by an integrable vector field. The proof of this remarkable statement was first accomplished by Kolmogorov [57], refined and published by Arnold [4], and further developed by Moser [73]. It is thus known as the KAM theorem, and the invariant tori which it describes are called KAM tori.[5]

Suppose that the autonomous (and most likely nonintegrable) Hamiltonian H is obtained from a shearing Hamiltonian H_0 via the addition of a "small" perturbation.

$$H(\underline{\varphi}^\star, \underline{I}^\star) = H_0(\underline{I}^\star) + \epsilon H_1(\underline{\varphi}^\star, \underline{I}^\star).$$

Here, ϵ is an order parameter which gives the relative scale between the two terms, and $\underline{\varphi}^\star$ and $\underline{I}^\star$ are action-angle coordinates on a chart appropriate for the tori of H_0. Under what conditions will H possess invariant N-dimensional tori of its own in the vicinity of $\epsilon \approx 0$? The KAM theorem, which partially answers this question, is not easy to state, but the gist of it is this: when H_0 is itself a *nonlinear* dynamical system (of a certain kind), most invariant tori of H_0 whose winding numbers (tunes) are "sufficiently off resonance" will survive small perturbations. They will be deformed into invariant tori of H.

[4]In any case, I hope I have succeeded in establishing connections between these alternative ways of viewing dynamics, and especially of developing perturbation theories.

[5]The notation and statement of the theorem which we shall use here have been combined from Arnold [4] and Thirring [88].

Let $\underline{\omega}(\underline{I}^\star) = \partial H_0(\underline{I}^\star)/\partial \underline{I}^\star$ be the frequencies associated with the invariant torus of H_0 labeled by $\underline{I}^\star$, and suppose that H_0 is "sufficiently" nonlinear so that

$$\text{either} \quad |\partial \underline{\omega}/\partial \underline{I}^\star| \neq 0 \quad \text{or} \quad \begin{vmatrix} \partial \underline{\omega}/\partial \underline{I}^\star & \underline{\omega} \\ \underline{\omega}^T & 0 \end{vmatrix} \neq 0. \tag{6.1}$$

The torus is said to be "off-resonance" if there exist real positive γ and η such that for all N-tuples of integers, $\underline{m}$,

$$|\underline{m} \cdot \underline{\omega}(\underline{I}^\star)| \geqslant \gamma \|\underline{m}\|^{-\eta}, \tag{6.2}$$

$$\|\underline{m}\| \equiv \sum_{k=1}^{N} |m_k|.$$

The numbers γ and η are not completely arbitrary: among other things, η is constrained by the number of degrees of freedom and γ depends on ϵ. There are numerous such details which we will not go into here. The key point is the meaning of "off-resonance" conveyed by the inequality. The first assertion of the theorem is that as $\epsilon \longrightarrow 0$ the phase space volume occupied by H_0 tori whose $\underline{\omega}(\underline{I}^\star)$ do *not* obey such a condition becomes arbitrarily small; "essentially all" H_0 tori will be off-resonance for small ϵ. The second is that those tori for which Eq. (6.2) *is* satisfied will be preserved, in the sense that

(a) there exists an invariant torus of H and a local $(\underline{\varphi}, \underline{I})$ action-angle chart such that:

$$\dot{\underline{\varphi}} = \underline{\omega}(\underline{I}^\star) \equiv \underline{\omega}^o \quad \text{and} \quad \dot{\underline{I}} = \underline{0};$$

(b) the connection between the two charts is a near-identity transformation:

$$\underline{I}^\star = \underline{I} + \underline{v}(\underline{\varphi}, \underline{I}, \epsilon, \underline{\omega}^o),$$

$$\underline{\varphi}^\star = \underline{\varphi} + \underline{u}(\underline{\varphi}, \underline{I}, \epsilon, \underline{\omega}^o);$$

(c) $\underline{u}$ and $\underline{v}$ are smooth functions of their arguments possibly excluding $\underline{\omega}^o$, and are zero-average, periodic functions of $\underline{\varphi}$; and

(d) $\underline{u}$ and $\underline{v}$ vanish as ϵ approaches zero: $\lim_{\epsilon\to 0} \underline{u} = \lim_{\epsilon\to 0} \underline{v} = \underline{0}$.

This collection of assertions comprise a (rather loose) statement of the theorem; the invariant tori whose existence is thus assured are called the KAM tori of H. The proof is at least partially constructive and includes a procedure for estimating — usually badly underestimating — the volume of phase space containing KAM tori [4]. Obviously, for autonomous systems up to two degrees of freedom, the existence of

KAM tori is enough to insure stability of orbits bounded by them; for nonautonomous systems with two degrees of freedom (the so-called $2\frac{1}{2}$ degree of freedom problems) or for phase spaces of larger dimensions, this is no longer the case. This was the original observation for the possibility of "Arnold diffusion." In general, the word "diffusion" is much overused and little understood these days — a tendency we shall avoid by not discussing it further.

Generically, tori too close to resonance, and in particular, the resonant tori themselves, will not survive perturbations intact. A subset of resonant orbits will survive but split into hyperbolic and elliptic families. The hyperbolic ones will become attached to separatrices — almost always chaotic, almost never integrable — a structure that Arnold has whimsically called "whiskered tori." (See, for example, Figure 4.16). Whether the local, macroscopic behavior of the dynamic can be associated for practical purposes, with an underlying isolated resonance model, or possibly a two- or three-resonance model, must be examined on a case-by-case basis.

The KAM theorem is tied up with the question of ergodicity in classical statistical mechanics. Recall that the ergodic hypothesis was required in order to equate time averages with ensemble averages: it was necessary that a system sample all the phase space available to it within the constraint of energy conservation. If it possessed dynamical invariants other than energy, however, this could not be the case. The question then is, are *nonintegrable* systems ergodic? The KAM theorem says no, not generally: even in the absence of a complete set of invariant observables, measureable regions of phase space will contain orbits confined to N-dimensional tori, just as though the invariants were still there.

Heuristically, the theorem assures us that most invariant tori of a shearing Hamiltonian will not be destroyed by a perturbation *provided that* there exist sufficiently large nonlinearities to detune resonances; that is the reason for the condition in Eq. (6.1). But suppose that this condition is not satisfied, as is the case, for example, with the harmonic oscillator. Is it possible for a perturbation to be so destabilizing that it could tear apart all the tori even for arbitrarily small ϵ? The answer is yes: one example, originally studied by Contopolous and others, can be found in Lichtenberg and Lieberman [59, p.164]. (The original work to be found in Lunsford and Ford, *J. Math. Phys,* **13**, 700 (1972) and G. Contopolous, *Cel. Mech.* **17**, 167 (1978).)

In its original form the KAM theorem treats autonomous Hamiltonians. Our problems are not of this type; we are concerned with, at best, periodic Hamiltonians and, more importantly, their maps. Some extensions of the original theorem exist; for a brief discussion, see Appendix 8 of Arnold [5].

In many ways the KAM theorem possesses sociological similarities to Gödel's famous theorem in logic. (a) Both are widely known and talked about, yet many people are rather vague on what the theorems actually state, and very few have actually read the proofs, much less validated them.[6] (b) Each has been called, by different mathematicians, the most important theorem of the twentieth century. (c) Neither is very useful for practical calculations: almost by definition, Gödel's theorem itself

[6] I regret that the present discussion does not improve that situation.

provides no hint on how to recognize undecidable propositions[7] and the stable phase space estimated by the KAM theorem is typically too conservative to be of value.[8]

Other topics in Hamiltonian dynamics There are a number of additional issues and results in Hamiltonian dynamics which we should have touched.

(a) Having dealt with single resonance models, the next logical step would be to consider what occurs when two resonances interfere with each other. This is the simplest step into the realm of nonintegrable dynamics. To be specific, suppose that both H_a and H_b are isolated resonance terms; that is, each is a product of a function of action coordinates with a cosine function whose argument is linear in the angle coordinates. Consider the Hamiltonians

$$H_1 = H_0 + H_a, \quad H_2 = H_0 + H_b, \quad \text{and} \quad H_3 = H_0 + H_a + H_b,$$

where H_0 is a shear Hamiltonian, a function of action coordinates only. We assume that H_a and H_b are of comparable strength. When the separatrices associated with the integrable Hamiltonians, H_1 and H_2, are widely separated in phase space, their "interference" in H_3 can be rather benign. One example of what can happen was shown in Figure 5.3(h), where third- and fourth-integer resonances interfered to produce a (nonintegrable) seventh-integer resonance. In fact, the strongest separatrix that will appear between the $m_a\nu + n_a = 0$ and $m_b\nu + n_b = 0$ resonances is generally driven by the $(m_a + m_b)\nu + (n_a + n_b) = 0$ resonance. This chaotic region associated with such resonance interference becomes larger as the separatrices of H_1 and H_2 approach each other and begin to overlap. If additional "overlapping resonances" be introduced, the chaotic region of phase space can become disastrously large. This has application in accelerator physics via the mechanism of "synchrobetatron" coupling, a cursory, inadequate treatment of which can be found in Appendix D.

(b) Although we have done so, the adiabatic theorem should not be neglected in any treatment of action-angle coordinates. Suppose that $\{H_\epsilon(\underline{z}) \mid \epsilon \in [0, 1]\}$ is a nondenumerably indexed set of integrable, autonomous Hamiltonians, depending continuously on ϵ. Consider the Hamiltonian

$$H(\underline{z}, t) = \begin{cases} H_0(\underline{z}), & t < 0 \\ H_{t/T}(\underline{z}), & 0 \leqslant t \leqslant T \\ H_1(\underline{z}) & T < t \end{cases}$$

for some T. Orbits of H change from those of H_0 to those of H_1 in the interval $t \in [0, T]$. If T is large enough, orbits on an invariant torus of H_0 will go over into orbits on a single invariant torus of H_1. Thus, this adiabatic process really maps tori

[7] That is done using the techniques of model theory.

[8] A colleague of mine once joked that the most useful application of the theorem had been as part of incantations — usually involving the mystic phrase "last KAM torus" — for separating funding agencies from their money. Of course, this was merely a humorous jest.

onto tori. The adiabatic theorem states that, under certain conditions, an invariant torus of H_0 will go over into that invariant torus of H_1 which possesses the same values of the action coordinates. That is, action coordinates are invariant under this sort of adiabatic change in the dynamical system.

The "certain conditions" that enter the hypothesis of this theorem require that the frequencies of the torus do not pass through a resonance. Geometrically, going from the initial to the final torus should not require passage through a separatrix, which would change the value of the action coordinate. (For example, consider what happens to a torus just inside the separatrix of a fourth-integer resonance when the separatrix is squeezed.)

(c) I have treated differential forms rather lightly in a deliberate attempt to keep the mathematics at the undergraduate level. The calculus of differential forms and their integration over manifolds should be part of the core graduate level physics curriculum, and Hamiltonian dynamics is a natural setting in which to introduce it. For example, Darboux's theorem (Theorem 3.5) is best stated in terms of two-forms. Their introduction would simplify the proof of the Liouville-Arnold theorem (Theorem 3.7) or the rather agonizing calculation in Section 3.2.2 demonstrating that canonical transformations are symplectic. Indeed, this last can be accomplished in a few lines, once the machinery of exterior derivatives is understood:

$$\begin{aligned}
dS &= (\underline{p} \cdot d\underline{x} - H\,dt) - (\underline{p}' \cdot d\underline{x}' - H'dt');\\
ddS &= (d\underline{p} \wedge d\underline{x} - dH \wedge dt) - (d\underline{p}' \wedge d\underline{x}' - dH' \wedge dt');\\
\text{but} \quad ddS &= 0, \quad \text{and} \quad (dH \wedge dt)_{(X_f,X_g)} = (dH \wedge dt')_{(X_f,X_g)} = 0;\\
\text{therefore} \quad & (d\underline{p} \wedge d\underline{x})_{(X_f,X_g)} = (d\underline{p}' \wedge d\underline{x}')_{(X_f,X_g)}.
\end{aligned}$$

That these lines replace essentially that entire section indicates the power of the method.

Detailed treatment of action coordinates is also simplified by employing integration of differential forms over manifolds. For example, the fact that the action coordinate is invariant under continuous deformation of the integration curve is an application of Stokes' theorem. Further consideration of this theorem could lead to a fruitful introduction to the connections between homology and cohomology, something worth doing if only to keep the student from being intimated by the very words.

A quick overview of the most important properties of differential forms can be found in Arnold [5]; the classic text by Flanders [35] contains a longer, very readable introduction to this important subject.

(d) In stressing Lie operators, we have given short shrift to methods which use generating functions. A good body of work in accelerator physics falls into this category. Noteworthy is the use of Hamilton-Jacobi theory — another topic we've omitted — to construct invariant tori numerically, using Fourier expansion in the angle coordinates, with coefficients depending on the action coordinates [90, 39]. Developing this further, symplectic one-turn maps have been produced by fitting the output of tracking programs via such expansions [12, 91].[9] The map is defined to be a transformation from the "old" coordinates $(\underline{\varphi}, \underline{J})$ to the "new" coordinates $(\underline{\varphi}', \underline{J}')$, which we formally write

$$\underline{\varphi}' = \underline{\varphi} + \underline{\Theta}(\underline{\varphi}, \underline{J}),$$
$$\underline{J}' = \underline{J} + \underline{R}(\underline{\varphi}, \underline{J}),$$

and which we assume is available via, for example, a tracking program. The fitted map is obtained by expressing the functions $\underline{\Theta}$ and $\underline{R}$ as a Fourier series in the angles and using a spline representation for the actions. However, in long term tracking applications, we require an exactly symplectic map (to computer accuracy). This can be achieved using a canonical transformation provided we find a generating function that reproduces $\underline{\Theta}$ and $\underline{R}$. Consider Fourier expanding the generating function in the angle coordinates.

$$S(\underline{\varphi}', \underline{J}) = \underline{\varphi}' \cdot \underline{J} + G(\underline{\varphi}', \underline{J})$$
$$G\left(\underline{\varphi}', \underline{J}\right) = \sum_{\underline{m}} g_{\underline{m}}(\underline{J})\, e^{i\underline{m}\cdot\underline{\varphi}'}.$$

The resulting transformation is given implicitly by the equations:

$$\underline{\varphi} = \underline{\varphi}' + \partial G\left(\underline{\varphi}', \underline{J}\right) / \partial \underline{J}, \tag{6.3}$$
$$\underline{J}' = \underline{J} + \partial G\left(\underline{\varphi}', \underline{J}\right) / \partial \underline{\varphi}'. \tag{6.4}$$

The Fourier coefficients are related to $\underline{R}$ by using one integration by parts.

$$\begin{aligned} g_{\underline{m}}(\underline{J}) &= \frac{1}{(2\pi)^n} \int_0^{2\pi} d\underline{\varphi}'\, e^{-i\underline{m}\cdot\underline{\varphi}'}\, G(\underline{\varphi}', \underline{J}) \\ &= \frac{1}{(2\pi)^n\, i m_k} \int_0^{2\pi} d\underline{\varphi}' e^{-i\underline{m}\cdot\underline{\varphi}'}\, \partial G(\underline{\varphi}', \underline{J}) / \partial \varphi'_k \\ &= \frac{1}{(2\pi)^n\, i m_k} \int_0^{2\pi} d\underline{\varphi}'\, e^{-i\underline{m}\cdot\underline{\varphi}'}\, R_k(\underline{\varphi}(\underline{\varphi}', \underline{J}), \underline{J}). \end{aligned}$$

The value of the subscript k is arbitrary. Since the map does not give $\underline{R}$ as a function of $\underline{\varphi}'$, one performs a change of coordinates in the integral to get an integral over $\underline{\varphi}$:

$$g_{\underline{m}}(\underline{J}) = \int_0^{2\pi} \frac{d\underline{\varphi}}{(2\pi)^n\, i m_k}\, e^{-i\underline{m}\cdot\underline{\varphi}}\, R_k(\underline{\varphi}, \underline{J})\, e^{-i\underline{m}\cdot\underline{\Theta}(\underline{\varphi}, \underline{J})} \det\left[\underline{1} + \partial \underline{\Theta}(\underline{\varphi}, \underline{J}) / \partial \underline{\varphi}\right] \tag{6.5}$$

[9] Much of the following is taken from Reference [38] and is based on the papers of Warnock, Ruth, *et al.* already cited.

This provides the connection between $\underline{\Theta}$, $\underline{R}$, and the coefficients, $g_{\underline{m}}$ and is the foundation of the numerical algorithm.

Once constructed, the canonical transformation replaces the original tracking program and can be used to find invariant tori, to calculate amplitude dependence of the tunes, or to do tracking itself. A significant application of this method is the estimation of long term bounds for the motion. If one finds that in some region, for some number of turns n, that the change in the new coordinate I is less than some value δI, then crossing a region of size ΔI will take at least $n\Delta I/\delta I$ turns. While δI cannot be determined rigorously over a region, it can be determined at many points in a region and at expected "bad" points (such as near resonance separatrices), to attempt to obtain some kind of "likely" bound.

The bounds so obtained have a connection with yet another important topic we omitted: the Nekhoroshev theorem. Unlike the KAM theorem, which establishes the existence of rigorously invariant tori, Nekhoroshev's theorem estimates long term bounds using the variation of an "almost invariant" action coordinate, bounds which tend to be more useful than those of the KAM theorem for establishing "stable" regions of phase space in practical applications.

Automatic differentiation and differential algebra We discussed in the last chapter the foundations of normal form algorithms for maps. The development of such algorithms and their numerical and symbolic implementations are among the important recent theoretical developments in accelerator physics. The numerical implementations hinge on the techniques of automatic differentiation [24] — first introduced into accelerator physics by Berz [15, 16, 17, 13] — and on the properties of Lie operators (of degree greater than one) as derivations, which assure the exact calculation of the normal form expansion to whatever order of interest [37, 80]. Although this text is not about computer implementations, a logical extension to Chapter 5 would have included discussion of the ideas behind these methods; all that can be done in a few sentences here is to provide a very brief introduction.[10]

We heuristically motivate automatic differentiation with the observation that basic theorems in differential calculus allow one to propagate information about derivatives through arithmetic manipulations. Associate with any smooth function, $f : U \subset R^n \longrightarrow R$, its "prolongation," or "jet,"[11]

$$\hat{f} \equiv \langle f, \underline{\partial} f, \underline{\partial}\,\underline{\partial} f, \underline{\partial}\,\underline{\partial}\,\underline{\partial} f, \ldots \rangle,$$

which is a structure explicitly carrying information on not only the function but also its derivatives. The first member of $\hat{f}$ is the function f itself, the second is its gradient, the third is its hessian, and so forth. Evaluating a jet at some value of its argument — say,

[10]Much of the following material is reproduced from Reference [69].

[11]This object may have a more acceptable name; if so, I do not know what it is. My calling it a prolongation is based on an abuse of terminology introduced on page 3 of Anderson and Ibragimov.[3] I am indebted to Gordon Pusch and Alex Dragt for introducing me to the term "jet."

$\hat{a} = \hat{f}(\underline{x})$—yields a numerical structure suitable for implementation in a computer.

$$\hat{a} = \hat{f}(\underline{x}) = \langle f(\underline{x}), \underline{\partial} f(\underline{x}), \underline{\partial}\,\underline{\partial} f(\underline{x}), \underline{\partial}\,\underline{\partial}\,\underline{\partial} f(\underline{x}), \ldots \rangle = \langle a, \underline{a}, \underline{\underline{a}}, \ldots \rangle \tag{6.6}$$

The first member of $\hat{a}$ is a real number, its second member is a singly indexed set of reals, its third, a doubly indexed set, and so forth. The number of indices associated with a member is its "order." A truncated jet represents an equivalence class of functions all of which have the same derivatives up to some finite order.

The arithmetic of jets is defined in a way that assures derivative values are propagated correctly. Rules for constructing members of sums, $\hat{s} = \hat{a} + \hat{b}$, products, $\hat{p} = \hat{a}\hat{b}$, and quotients, $\hat{q} = \hat{a}/\hat{b}$, can be obtained directly by applying theorems from differential calculus.

$$\begin{array}{lll} s = a + b & \underline{s} = \underline{a} + \underline{b} & \underline{\underline{s}} = \underline{\underline{a}} + \underline{\underline{b}} \\ p = ab & \underline{p} = a\underline{b} + b\underline{a} & \underline{\underline{p}} = b\underline{\underline{a}} + a\underline{\underline{b}} + \underline{a}\,\underline{b} + \underline{b}\,\underline{a} \\ q = a/b & \underline{q} = (1/b)\underline{a} - (a/b^2)\underline{b} & \underline{\underline{q}} = (1/b)\underline{\underline{a}} - (1/b^2)(\underline{a}\,\underline{b} + \underline{b}\,\underline{a}) \\ & & \qquad + (2a/b^3)\underline{b}\,\underline{b} - (a/b^2)\underline{\underline{b}} \end{array} \tag{6.7}$$

Moving from these low order examples to the general case is facilitated by defining the indexed family of operators,

$$D^{\underline{n}} = \prod_{k=0}^{r} \frac{1}{n_k!} \partial_k^{n_k},$$

where r is the number of coordinates, and $\underline{n}$ is an array of r non-negative integers (the "index" array). The evaluated numbers $D^{\underline{n}} f(\underline{x})$ are interpretable as coefficients in the multinomial expansion that belongs to the same jet as f. (Not surprisingly, this multinomial is typically taken as the representative of the equivalence class.) By carrying these values, rather than the derivatives themselves, multiplication is implemented easily using Leibniz's rule,

$$D^{\underline{n}}(uv) = \sum_{\underline{m}=\underline{0}}^{\underline{n}} (D^{\underline{m}} u)(D^{\underline{n}-\underline{m}} v) \tag{6.8}$$

Division is accomplished by a repeated partial multiplication. Notice that by combining $uv = w$ with Eq. (6.8) we can write a recursive procedure for defining the higher orders of $\hat{v}$ in terms of its lower orders,

$$D^{\underline{n}} v = \frac{1}{u}\left[D^{\underline{n}} w - \sum_{\underline{0} \leq \underline{m} < \underline{n}} (D^{\underline{m}} u)(D^{\underline{n}-\underline{m}} v) \right] \tag{6.9}$$

starting with $v = w/u$.

After arithmetic operations have been defined, we proceed to handle transcendental functions via Taylor series. For functions possessing summation formulas, this is expedited by splitting off the jet's "standard" part.

$$\hat{a} = \langle a, \underline{a}, \underline{\underline{a}}, \ldots \rangle = \langle a, \underline{0}, \underline{\underline{0}}, \ldots \rangle + \langle 0, \underline{a}, \underline{\underline{a}}, \ldots \rangle \equiv \tilde{a} + \varepsilon$$

To calculate $\exp \hat{a}$, for example, we then proceed as follows.

$$\exp \hat{a} = \exp[\tilde{a} + \varepsilon] = \exp \tilde{a} \cdot \exp \varepsilon = \exp a \cdot \sum_{n=0}^{\infty} \frac{\varepsilon^n}{n!}$$

The series is written as formally infinite, but it terminates when the jet is truncated — as it must be in real implementations — for then ε is a multiplicative nilpotent. (This is seen easily if you consider the polynomial representations. Let ε be a polynomial that satisfies $\varepsilon : \underline{0} \longmapsto \underline{0}$. Let the term of lowest degree be of degree k. Then the term of lowest degree in ε^n must be nk, and since $k \neq 0$ these degrees become arbitrarily large until the terms "fall off the edge" of the truncated jet.) Similar decompositions enable the implementation of trigonometric functions, logarithms, and so forth: for example, $\cos \hat{a} = \cos a \cos \varepsilon - \sin a \sin \varepsilon$.

A special set of jets come from projections. These are the jet equivalent of "coordinates," the "atoms" of the theory, and are used to *initialize* all jet computations. Effectively, they implement the identity function, $\underline{I}$, that appears throughout Chapter 5.

$$I_k(\underline{x}) \Rightarrow \hat{x}_k(\underline{x}) \equiv \langle x_k, \underline{e}_k, \underline{\underline{0}}, \ldots \rangle. \tag{6.10}$$

Recall that the components of $\underline{e}_k$ are $(\underline{e}_k)_i = \delta_{ik}$; all members of $\hat{x}_k$ vanish beyond the first order. Notice that the second member is indeed the gradient of the associated projector. All calculations are built starting from these functions, as they were in the final section of Chapter 5. Indeed, it is even possible to calculate derivatives of functions defined implicitly as limits of recursive or iterative procedures.

An addendum for those who know something of C++ programming: a library of C++ classes for doing automatic differentiation, based on a dlist representation, is available using anonymous ftp from the site calvin.fnal.gov in the directories pub/outgoing/michelotti/mxyzptlk and pub/outgoing/michelotti/beamline. Please contact me at the email address michelot@calvin.fnal.gov.

Catastrophe theory We have used the word "catastrophe" to refer to the local bifurcations that take place when critical orbits coalesce — fixed points, periodic orbits, resonances, attractors, repellors, and so forth. This was, perhaps, a slight extension of the term. Catastrophe theory, a well developed branch of mathematics in its own right, employs a more restricted definition that refers only to the fixed points of autonomous gradient vector fields. If the fixed point is simple (nondegenerate), the Hartman-Grobman theorem (Theorem 2.6) provides a chart in which the field is represented locally as a homogeneous, quadratic "potential" function of the coordinates.

$$U(\underline{x}) = \underline{x} \cdot \underline{\underline{A}} \cdot \underline{x}, \quad \underline{v}(\underline{x}) = \partial U(\underline{x}), \quad \text{and} \quad \underline{v}(0) = \underline{0}.$$

However, if the fixed point is degenerate, as happens at a point of bifurcation, such coordinates do not exist. Put another way, linearizing the flow does not provide enough information to classify its behavior, even in a small neighborhood of the fixed point. We must look beyond the linear terms, and write

$$U(\underline{x}) = F(\underline{x}) + \underline{x} \cdot \underline{\underline{A}} \cdot \underline{x},$$

where F is some polynomial of degree three or higher. Catastrophe theory deals with the equivalence classes of F and their representative normal forms. The situation is even more complicated than suggested here, however, for it expands these forms as functions not only of the dynamical coordinates but of the control coordinates as well. For example, in our treatment of the half-integer resonance in Section 4.3.2 the dynamical phase space, $\mathcal{P}$, is two-dimensional, the control space, C, is three-dimensional, and the combined space, $\mathcal{P} \times C$, would therefore be five-dimensional. The full bifurcation surface is the subset of $\mathcal{P} \times C$ on which the vector field over $\mathcal{P}$ vanishes. The catastrophe surface sketched in Figure 4.10 comprise the folds of its projection into the three-dimensional control space.

A splendid introduction to this field for scientists is the text by Gilmore. [40]

Quantum theory In studying the Van Hove theorem we very briefly made contact with one of the most fascinating problems in physics: how to understand the interface between classical and quantum mechanics. What is meant by the statement that classical mechanics arises from quantum mechanics "in the limit $\hbar \to 0$?" Following the wider understanding of chaos as a generic feature of classical mechanics, this question has received renewed interest, both theoretically and experimentally. We cannot go into the details here, but among the topics which should be studied are: (a) "scarring" of wave functions along underlying classical trajectories [34], (b) geometric quantization, which attempts to get around the impasse produced by Van Hove's theorem [92], (c) Maslov indices, in connection with semi-classical quantization [63, 62, 61], (d) distributions of energy levels in quantum mechanical systems, (e) the connections between highly excited states of atoms in a magnetic field and classical orbits of an electron spiraling in the field, (f) quantum suppression of chaos via the mechanism of the uncertainty principle [49], and, of course, (g) the foundations of quantum theory [50, 77], especially in connection with Bell's theorem [10] and the experiments that test it [8]. We have not treated any of these issues because they are too removed from the main thread of this book. They are, nonetheless, among the most interesting questions being pursued as we close the twentieth century. It is likely that the interface between classical and quantum mechanics will be an area of intense activity for some time to come.

APPENDIX A

Analysis of the Half-Integer Resonance

Fixed points of Eq. (4.36) are found by applying Hamilton's equations and setting derivatives to zero. This yields the following expressions (see Eq. (4.37)):

$$J^{(0)} = \frac{-\sin(\xi^{(0)} + \phi)}{\sin \xi^{(0)}},$$
$$\sin \xi^{(0)}[\delta + \cos(\xi^{(0)} + \phi)] = 2\sin(\xi^{(0)} + \phi)[\kappa + \cos \xi^{(0)}]. \tag{A.1}$$

The latter looks messy but is solved easily by writing it as a monic fourth order polynomial. We introduce the complex variables

$$z \equiv e^{i\xi^{(0)}}, \quad y \equiv e^{-i\phi} \tag{A.2}$$

and rewrite Eq. (A.1) in a form that we shall call the "fixed point polynomial," or "fpp" for short.

$$\mathrm{fpp}(z) = \sum_{k=0}^{4} a_k z^k = 0, \tag{A.3}$$

$$a_4 = 1,\ a_2 = 3(1 - y^2),\ a_0 = -y^2$$
$$a_3 = 4\kappa - 2y\delta,\ a_1 = 2y\delta - 4y^2\kappa. \tag{A.4}$$

A catastrophe occurs when fixed points coalesce. Thus on the catastrophe surface, the fixed point polynomial must have a unimodular solution of multiplicity two, or greater. Accordingly, it can be factorized

$$\text{fpp}(z) = (z - u)^2(z - v)(z - w), \tag{A.5}$$

where $|u| = 1$, in order that $\xi^{(0)}$ be real. The fpp coefficients, written in terms of these solutions, are

$$\begin{aligned} a_3 &= -(v + w + 2u), \\ a_2 &= u^2 + vw + 2u(v + w), \\ a_1 &= -u^2(v + w) - 2uvw, \\ a_0 &= u^2 vw. \end{aligned} \tag{A.6}$$

To map this as a surface in control space we must invert Eqs.(A.4), according to

$$y^2 = -a_0 \tag{A.7}$$

$$\kappa = \frac{3}{4}\frac{a_1 + a_3}{a_2}, \tag{A.8}$$

$$y\delta = 2\kappa - \frac{a_3}{2}. \tag{A.9}$$

It will be possible to do this and to interpret the answers as real, physical quantities only if the coefficients satisfy three "realizability conditions," which follow immediately from Eqs.(A.4).

$$\begin{aligned} |a_0| &= 1 \quad (\phi \text{ must be real}), \\ \tfrac{1}{3}a_2 - a_0 &= 1, \\ a_1 &= a_0 a_3^*. \end{aligned} \tag{A.10}$$

We shall show below that the first two conditions are sufficient as well as necessary by proving that they imply the third and that the control variables obtained are real. This statement is made formal in the following.

Key Proposition If a monic fourth order polynomial has a degenerate unimodular solution and satisfies the first two realizability conditions, then (a) it satisfies the third realizability condition, and (b) it is the fpp of a real-valued control point, (δ, κ, ϕ).

To prove this we begin with the following lemma, which assumes the same hypothesis as above.

Lemma $vw = (v + w)/(v + w)^*$ (v, w as defined in Eq. (A.5)).

Proof: Because of the first realizability condition, and $|u| = 1$,

$$|vw| = |u|^2|vw| = |u^2vw| = |a_0| = 1. \tag{A.11}$$

Now, multiply Eq.(A.17) by $3u^*$ to get

$$u(1 - 3vw) + 2(v + w) + u^*(vw - 3) = 0. \tag{A.12}$$

Take the complex conjugate, multiply by vw, and use Eq.(A.11) to get

$$u(1 - 3vw) + 2vw(v + w)^* + u^*(vw - 3) = 0. \tag{A.13}$$

Subtracting these two expressions yields the desired result. □

Proof of the Key Proposition: (a) Using Eqs.(A.6) and the lemma we calculate a_1.

$$\begin{aligned} a_1 &= -u^2(v + w) - 2uvw \\ &= -u^2vw[(vw)^*(v + w) + 2u^*] \\ &= y^2[(v + w)^* + 2u^*] \\ &= a_0a_3^*. \end{aligned} \tag{A.14}$$

(b) First, κ is real iff $(a_1 + a_3)(1 + a_0^*)$ is real, according to Eq.(A.8). But, using what we've proved in part (a),

$$\begin{aligned} (a_1 + a_3)(1 + a_0^*) &= (a_3 - y^2a_3^*)(1 - y^{*2}) \\ &= (y^*a_3 - ya_3^*)(y - y^*), \end{aligned} \tag{A.15}$$

which is manifestly a real quantity. On the other hand, using Eq.(A.9),

$$\begin{aligned} 2\delta|1 - y^2|^2 &= (a_3 + a_1y^{*2})(y - y^*) \\ &= (a_3 - a_3^*)(y - y^*) \end{aligned} \tag{A.16}$$

which again is manifestly real. □

Sections of the catastrophe surface are calculated in the following manner. Equation (A.10), in conjunction with Eqs.(A.6), implies that

$$u^2(\tfrac{1}{3} - vw) + \tfrac{2}{3}u(v + w) + (\tfrac{1}{3}vw + 1) = 0. \tag{A.17}$$

Now, ϕ = constant is equivalent to $a_0 = -y^2 = u^2vw$ = constant. Use this to eliminate w in Eq.(A.17); the resulting polynomial is quadratic in v.

$$(uy^*)v^2 + \tfrac{1}{2}[u^2y^* - u^{*2}y - 3(y^* - y)]v - u^*y = 0. \tag{A.18}$$

The skeleton of an algorithm can now be sketched, in pidgin algol, as follows.

```
for "all" φ ∈ (0, π)
begin  y ← e^{-iφ};
       for "all" u on the unit circle
       begin  solve Eq.(A.18) for v⁺ and v⁻;
              for v ∈ {v⁺, v⁻}
              begin  w ← −y²/u²v;
                     J^(0) ← (y² − u²)/y(u² − 1);
                     if J^(0) > 0 then evaluate and plot (δ, κ) using
                                 Eq.(A.8) and Eq.(A.9)
              end;
       end;
end.
```

The test $J^{(0)} > 0$ must be done explicitly; only if this inequality is satisfied does the control point produce a real non-Morse fixed point.

The algorithm does not handle the endpoints $\phi = 0$ and $\phi = \pi$, because κ evaluates to $0/0$. For $\phi = 0$, $J^{(0)}$ is always negative so nothing more need be done. At $\phi = \pi$ we have $y = -1$, $a_0 = 1$, $a_2 = 0$, and $a_3 = -a_1 = 4\kappa + 2\delta$. The fpp factorizes immediately.

$$\text{fpp}(z) = (z - 1)(z + 1)(z^2 + a_3 z + 1). \tag{A.19}$$

Now, a_3 must be chosen so that there is at least one degenerate solution. The possibilities are:

1. $z^2 + a_3 z + 1$ is a perfect square. This means that either
   (a) $a_3 = +2$, and $\text{fpp}(z) = (z - 1)(z + 1)^3$, or
   (b) $a_3 = -2$, and $\text{fpp}(z) = (z - 1)^3(z + 1)$.
2. $(z + 1) \mid (z^2 + a_3 z + 1)$ This reduces to case 1.
3. $(z - 1) \mid (z^2 + a_3 z + 1)$ This reduces to case 2.

In all cases we have $a_3 = \pm 2$, or, equivalently, $2\kappa + \delta = \pm 1$. Furthermore, the non-Morse solution is triply degenerate; that is, it corresponds to the merger of three fixed points.

APPENDIX B

What Are Nonlinear Dynamics?

The question in the title sounds facetious but is not meant to be so. A "dynamic," or "dynamical system," is a vector field possessing orbits on a manifold, usually of small dimension. The naive response to our question is that nonlinear dynamics are those that are not linear. Seen in this way, the appellation seems ill conceived, as though we identified the body of mathematics which does not specialize in studying prime numbers as "nonprime" mathematics. Loosely (and not quite correctly) stated, the category "nonlinear" covers everything except the harmonic oscillator. That the word is used so extensively may be a reaction to the subject matter having been underrepresented in curricula of the past, as though mathematicians had indeed spent 70% of their time studying and teaching properties of prime numbers while glossing over the rest of their field. Emphasizing the phrase, rather than simply using "dynamics," may be a way of stressing that it is time to stop counting on fingers and toes.

On the other hand, its continued use is assured. It has acheived the status of a buzz phrase, having entered into our popular culture as having something to do with such things as Mandelbrot (Julia-Fatou) sets or the standard map.

Sophistical tirades aside, the distinction between "linear" and "nonlinear" dynamics is fuzzier at its boundary than one would imagine. The fundamental property characterizing a *linear* vector field is

$$\forall \lambda \in R: \quad \underline{v}(\lambda \underline{z}; t) = \lambda \underline{v}(\underline{z}; t), \tag{B.1}$$

where $\underline{z}$ are the dynamical coordinates and t is the orbit parameter, "time." However, Eq. (B.1) is not a property of the vector field alone; its validity depends on the coordinate chart used to represent the field. On a "correct" chart, the components, $\underline{v}$, obey Eq. (B.1) and assume the representation

$$\underline{v}(\underline{z};t) = \underline{\underline{A}}(t) \cdot \underline{z},$$

the quintessential example of which is the simple harmonic oscillator in one degree of freedom:

$$\underline{\underline{A}} = \omega \begin{pmatrix} 0 & 1 \\ -1 & 0 \end{pmatrix},$$

$$\dot{x} = \omega p, \quad \text{and} \quad \dot{p} = -\omega x.$$

Consider, however, that the action-angle (polar) representation of this *same* vector field is written

$$\underline{v}((\underline{\varphi}, \underline{I}); t) = (\underline{\nu}, \underline{0}),$$

and this clearly does *not* satisfy Eq. (B.1). (One might object that the manifold of the simple harmonic oscillator is R^2 while action-angle coordinates exist on the cylinder, $C = S^1 \times R$, topologically equivalent to the punctured plane, $R^2 - \{(0,0)\}$. This is a red herring; we can use C, rather than R^2, to be the phase space of either representation by simply eliminating consideration of the origin throughout.)

As a condition defining linear dynamics, Eq. (B.1) must be preceded by the ominous phrase, "There exists a chart such that...." How can it be determined whether this is the case? It is easy to contrive tortured examples that emphasize the difficulty. Consider the complicated looking

$$\underline{z} = \begin{pmatrix} x \\ p \end{pmatrix}, \quad \underline{v} = \dot{\underline{z}} = \begin{pmatrix} p \\ f \end{pmatrix}, \tag{B.2}$$

where f is obtained by solving

$$t\,[\,f(\underline{z};t)(1 - x^2) + 3xp^2] - 2p(1 - x^2) + 9t^5 x(1 - x^2)^2 = 0.$$

This appears to describe a system that is neither autonomous nor linear. However, the substitutions

$$t \leftarrow t^{1/3}, \quad x \leftarrow x\sqrt{\frac{1}{1 + x^2}}$$

change Eq. (B.2) to

$$\dot{x} = p, \quad \dot{p} = -x.$$

Linearity was disguised by a particularly bad coordinate chart over extended phase space. It is not unusual for the solution of a problem to depend on an inspired choice of coordinates for its expression, but without advanced knowledge of the transformation to the "correct" coordinates, or even that one exists, how would one establish linearity? Put another way, *is there a coordinate-free way of characterizing linearity*?

To make this more clear, consider that a fixed point, z_0, of a vector field, $\underline{v}$, can be identified by the condition $\forall t : \underline{v}(\underline{z}_0; t) = \underline{0}$. This statement does *not* depend on the local chart, appearances notwithstanding, as $\underline{0} \longmapsto \underline{0}$ under any transformation; if a vector field vanishes at a point on one atlas, then it does so on them all. This would not be the case for a condition like "$\underline{v}(\underline{z}_0) = (3, 1, -9)$," which clearly depends on the chart used. It does not express an intrinsic, or "geometric," property but an accident arising from representation, much like saying that the atmospheric pressure at a point is 30, without mentioning the units.

The fixed point property is easy to formulate in a coordinate-free way. That of being a periodic orbit is more difficult, as it requires determining the period, constructing the appropriate map — which in its turn requires integrating the equations of motion somehow — and demonstrating that the orbit in question is indeed a fixed point of the *map*. If the system is autonomous, this only requires establishing that the orbit through $\underline{z}_0$ returns to $\underline{z}_0$ in some finite time, T; if not, then it must also be established that the vector field itself is periodic, $\forall \underline{z} : \underline{v}(\underline{z}, t + T) = \underline{v}(\underline{z}, t)$. The validity of all these conditions is clearly chart-independent, and so the periodic property is indeed "geometric."

For a yet more sophisticated example, consider the property of being a resonant (not necessarily periodic) orbit of a periodic, integrable (to keep things simple) Hamiltonian system on a $2N$-dimensional phase space. We have seen that resonant orbits can be defined in a geometric, chart-independent manner: (a) sample the orbit using forward and backward period-advance maps; (b) construct the closure, C, of this infinite, countable set; (c) C is a torus of dimension smaller than N. These are formidable tasks, but presuming one could carry them out, they would provide a chart-independent verdict; the "resonant" property is geometric also.

However, none of these geometric properties distinguish between linear and nonlinear dynamics: either can possess fixed points, periodic orbits, or resonant orbits. Returning to our original formulation of this question, the one property characterizing the simplest *nonlinear* systems, the so-called twist maps, is their possession of amplitude-dependent frequencies (tunes). Consider a very simple, decidedly nonlinear system: the first order averaged Hamiltonian associated with octupole-like terms, which represented in action-angle coordinates is

$$H = \nu J + \tfrac{1}{2} g J^2.$$

The equations of motion,

$$\begin{aligned}\dot{\varphi} &= \partial H/\partial J = \nu + gJ,\\ \dot{J} &= -\partial H/\partial \varphi = 0,\end{aligned}$$

reveal the amplitude-dependence of the tune via the term gJ. Consider, however, a new angle coordinate in extended phase space,

$$\psi \equiv \varphi - gJt.$$

Introducing ψ is just as valid as introducing resonance angle coordinates: writing $\eta = \underline{m} \cdot \underline{\varphi} + n\theta$, as was done in Eq. (4.10). The new components of the vector field are

$$\dot{\psi} = \nu \quad \text{and} \quad \dot{J} = 0,$$

which clearly correspond to the action-angle description of a harmonic oscillator, a linear system. This even can be done more formally via a Type 2 generating function

$$\begin{aligned} S(\varphi, I; t) &= \varphi I - \tfrac{1}{2} g I^2 t, \\ \psi &= \partial S / \partial I = \varphi - gIt, \\ J &= \partial S / \partial \varphi = I, \\ K = H + \frac{\partial S}{\partial t} &= \nu I. \end{aligned}$$

This example is easily extended to more general twist maps. Here we have an apparent instance of a whole class of nonlinear systems converted into a linear one with a judicious choice of coordinates.[1] Is anything awry?

The only point one may legitimately object to is that the transformation becomes somewhat ill-behaved for large $|t|$. Level sets of the angle coordinate, ψ, wind around phase space with greater rapidity as J increases until they become virtually indistinguishable from those of J itself. Furthermore, for large t, this behavior occurs for regions with arbitrarily small J. Perhaps we have to add a new condition that requires decent behavior at $t = \pm\infty$ as well as $J = \infty$, something like "regularity on compactified, extended phase space," as it were. (Notice that the transformation to resonance angle coordinates, Eq. (4.10), obeys this, so it is "safe.")

Indeed, no local set of conditions would be adequate. The straightening theorem (see Section 2.3.3) asserts that locally all vector fields can be made to look essentially identical, except in the vicinity of singular orbits. Therefore whatever final criteria are adopted must be global and thus difficult to implement via finite means.

An issue we have ignored is that of effectiveness or constructability. Phrased as yes-no questions — an orbit either does or does not possess a particular property — determining all of the above properties exactly would require infinite precision; some, infinite time. Even determining whether an orbit possesses the fixed point property requires one or the other, unless one has in hand an analytic proof. The Poincaré recurrence theorem, implies that nonperiodicity of a particular orbit on a compact

[1] Sundman's transformation, introduced in Problem 2.6, is similar but not as legitimate, as it depends on the orbit itself.

manifold is nondeterminable by finite means if arbitrarily large periods are allowed (unless there is some way of analytically proving that the orbit can never return to the same point under the same conditions). Yes-no questions tend to be nonconstructive, in general. It may be better to insist upon a requirement of providing finite algorithms that can determine these properties within definite limits: for limited times, within finite error.

In particular, the property of being linear is not "generic." Any linear vector field can be made nonlinear with an arbitrarily small perturbation whose effects would manifest themselves on arbitrarily small scales or only over arbitrarily long times. Linearity, like integrability, is an exact, analytic property, not a generic, observable one.

To bring this essay mercifully to an end, there is a geometric property that is possessed by all nonlinear systems *except* twist maps: the existence of *isolated* critical orbits, fixed points, periodic orbits, resonances, and the separatrices which are attached to those that are not elliptic. It may be impossible to establish that a system is linear; it is not impossible to establish the converse. A linear dynamic can possess at most one isolated fixed point, the origin, and separatrix. The existence of two or more is the unambiguous mark of nonlinearity.

We consider, therefore, that "nonlinear dynamics" should be taken to mean the study of separatrices. This declaration leaves the twist map in a class by itself, it being neither quite "linear" nor "nonlinear." Perhaps that is as it should be. Indeed, a more meaningful distinction would be between "integrable" and "nonintegrable," although this suffers from the same problem: while the property of "nonintegrability" is generic, any "integrable" vector field can be made "nonintegrable" by an arbitrarily small perturbation.

APPENDIX C

Resonance Widths

Island Width The most popular meaning assigned to the phrase "resonance width" seems to be associated with the width of a resonance island, measured in terms of the unperturbed action coordinate. In the general case this must be computed numerically, but an expression can be obtained in "quadratic approximation" for a simple model. Consider the isolated resonance Hamiltonian,

$$H = \delta\nu J + \frac{\alpha}{2}J^2 + gJ^{m/2}\cos m\eta. \tag{C.1}$$

The shear term could be thought of as arising from an averaged octupole distribution, while the resonance term can be associated with a (possibly) higher order multipole. The regular stable and unstable resonant orbits are found in the usual manner; the corresponding values of η and J obey the following relations:

$$\begin{aligned} \cos m\eta &= \pm 1, \\ \delta\nu + \alpha J \pm \tfrac{1}{2}mgJ^{m/2-1} &= 0. \end{aligned} \tag{C.2}$$

To estimate the island width, say, δJ, let us assume that g is small, in some sense, and work only to lowest order in this parameter. Let H_u and H_s symbolize the values of the Hamiltonian at the hyperbolic (unstable) and elliptic (stable) fixed points, respectively. Accordingly, H_u is its value on the separatrix, which marks the boundary of the resonance island. If g is small, the valley of the Hamiltonian surface within the island is sufficiently shallow that it can be adequately represented in quadratic approximation from the elliptic fixed point to the separatrix. Since the vector field vanishes at the elliptic fixed point, we have $(\partial H/\partial J)_s = 0$ and the expansion

$$H_u - H_s = \tfrac{1}{2}(\partial^2 H/\partial J^2)_s(\delta J/2)^2. \tag{C.3}$$

Now expand the left-hand side of this using Eq. (C.1) and Eq. (C.2).

$$H_u - H_s = (1 - m/2)g(J_u^{m/2} \cos m\eta_u - J_s^{m/2} \cos m\eta_s)$$
$$- \frac{\alpha}{2}(J_u^2 - J_s^2).$$

We expand everything only to lowest order in g,

$$J_u \approx J_s \equiv J_o \approx -\delta\nu/\alpha, \tag{C.4}$$
$$(\partial^2 H/\partial J^2)_s = \alpha,$$
$$-\frac{\alpha}{2}(J_u^2 - J_s^2) = \frac{1}{2} mg(J_u^{m/2} \cos m\eta_u - J_s^{m/2} \cos m\eta_s),$$

and substitute back into Eq. (C.3) to get our final expression for the island's width.

$$\frac{\alpha}{2}(\delta J/2)^2 \approx g J_o^{m/2}(\cos m\eta_u - \cos m\eta_s) \tag{C.5}$$
$$= 2\,\mathrm{sgn}(\alpha)|g|J_o^{m/2}$$
$$\delta J = 4\sqrt{|g/\alpha|}\, J_o^{m/4} = 4\sqrt{|g/\alpha|}\,(-\delta\nu/\alpha)^{m/4}. \tag{C.6}$$

Approximate 1.5 dof Central Width In 1.5 degrees of freedom, the resonant orbit is periodic, and its separatrix bounds a central region of orbits on KAM tori. The isolated resonance Hamiltonian has the following form:

$$H = \Delta J + (H_r(J)e^{i\eta} + \text{c.c.})$$
$$H_r(J) = (I/2)^{m/2}\zeta$$
$$\zeta = (-i)^m \oint \frac{B_o R}{|B\rho|} \frac{d\theta}{2\pi} b_{m-1}(\theta)\beta^{m/2}(\theta) e^{i[m\tilde{\psi}(\theta) - n\theta]}.$$

Recall that $\Delta = m\nu + n$, $\eta = m\varphi$, and $\tilde{\psi}(\theta) = \psi(\theta) - \nu\theta$. Written in terms of a cosine function,

$$H = \Delta J + 2gJ^{m/2} \cos(\eta + \xi),$$
$$g = (1/2)^{m/2}|\zeta|,$$
$$\xi = \arg \zeta.$$

The resonant orbits, corresponding to fixed points,

$$\eta_o = k\pi - \xi,$$
$$J_o = |\Delta/mg|^{2/(m-2)},$$

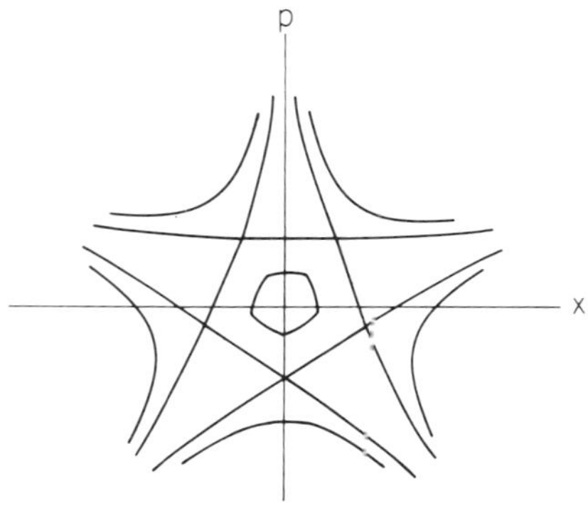

FIGURE C.1 The central stable region of an mth degree 1.5 dof resonance is approximately an m-sided polygon.

are arranged symmetrically about the origin, and the separatrix passing through the unstable ones can be approximated roughly as an m-sided polygon. The area of such a polygon (see Figure C.1) is given by

$$W = m \times \left[\sqrt{2J_o}\cos(\pi/m)\right] \times \left[\sqrt{2J_o}\sin(\pi/m)\right]$$
$$= |\Delta/mg|^{2/(m-2)} m \sin(2\pi/m). \tag{C.7}$$

The "central resonance width" is then defined as $\Delta\nu \equiv 2|\nu + n/m| = 2|\Delta/m|$.[1] It is the size of the tune interval within which no beam of emittance W can fit into the central stable region. Solving Eq. (C.7) for $\Delta\nu$ yields the desired result:

$$\Delta\nu = 2g\left(\frac{W}{m\sin(2\pi/m)}\right)^{m/2-1}.$$

Notice that the "resonance width" is not an absolute number but depends, as it should, on the emittance under consideration.

$\nu_1 + 2\nu_2$ *Central Width* Complications arise when we try extending this simple picture to 2.5 degrees of freedom. We shall examine one particular model: the $\nu_1 + 2\nu_2$ resonance considered earlier. There is indeed a central stable region surrounded by a three-dimensional separatrix, but the calculation of its volume is significantly more complicated.

Except for the irregular resonant orbits pinned on $I_1 = 0$ and $I_2 = 0$, the $(1, 2)$ resonance possesses no bounded orbits on the leaves for which $J_2 < -\frac{1}{30}\kappa^2$ or $\frac{1}{10}\kappa^2 < J_2$, whereas between these leaves bounded orbits fill some volume of phase space. This is the general behavior of all resonances, except the quadrupole resonances for which all orbits are either bounded or unbounded: the region of bounded orbits

[1] Apologies for giving the symbol Δ two different meanings here.

slowly shrinks as the resonance is approached. One quantitative measure of this approach to global instability is the "resonance width." We take this term to mean the size of the smallest strip in tune space that is centered on the resonance line, Eq.(2.34), and outside of which a beam is stable. This definition remains ambiguous, because it depends on the size and shape of the beam as well as on the experimental setup — for example, on whether the resonance is approached adiabatically or the beam is suddenly injected into the resonant situation. In order to avoid beam parameters entirely, we shall associate an "adiabatic resonance width" with each individual orbit. That is, we imagine initializing an orbit in phase space with control parameters set far from resonance, then approaching the resonance very slowly, and finally noting when the orbit becomes unbounded.

For the $(1, 2)$ resonance of our example this means beginning with $\kappa \approx \infty$ and letting $\kappa \longrightarrow 0$ on a time scale much greater than $\max(1/\nu_1, 1/\nu_2)$. At $\kappa = \infty$ all orbits are harmonic oscillator orbits, the variables I_1, I_2, J_1, and J_2 are conserved separately, and we can label an orbit with any two of the four initial values, I_1^{in}, I_2^{in}, J_1^{in}, and J_2^{in}.[2] According to the usual adiabatic theorems the variation of an orbit as κ approaches zero will be regulated by the adiabatic invariance of the action integrals. Because J_2 is a constant of motion for fixed κ, we can take $J_2 = \frac{1}{6\pi} \oint J_2 d\xi_2$ itself as the first adiabatic invariant. To the second we attach the symbol $\mathcal{A} \equiv \oint J_1 d\xi_1$, whose value is $\mathcal{A}^{in} = 6\pi J_1^{in}$.

What happens to an orbit as κ slowly decreases depends critically on the sign of J_2^{in}. For $J_2^{in} > 0$ the diagrams of Figure 4.16(e)–(g) are the relevant ones, and we now must think of them as *flow* diagrams for the projected Hamiltonian (see Eq.(4.40)) rather than *mapping* diagrams of the function F. As κ decreases the separatrix pushes downward. Each orbit remains on its leaf, $J_2 = J_2^{in}$, it maintains its value of $\mathcal{A}$, and it crosses the separatrix, thus becoming unbounded, when the area under the separatrix has decreased to $\mathcal{A}^{in}$.

For $J_2^{in} < 0$ the situation is much more interesting, as the separatrix contains two branches. Figures 4.16(a)–(e) are now the relevant ones, but they must be traversed in reverse order. As κ decreases from ∞ the upper branch pushes downward, as before, but simultaneously a bubble, representing the lower branch of the separatrix, forms and begins to grow. As these two branches grow closer, approaching their merger at the saddleswitch ($\kappa^2 = -40J_2^{in}$), orbits either are captured by the island or pass through the upper branch, depending on their values for $\mathcal{A}^{in}$. The total area under the saddleswitch is $\mathcal{A}_s = -(15 + 33\pi/4)J_2^{in}$. If $\mathcal{A}^{in} > \mathcal{A}_s$ the orbit passes through the upper branch of the separatrix; if $\mathcal{A}^{in} < \mathcal{A}_s$, then it is captured by and subsequently leaks through the lower branch. If the latter happens, $\mathcal{A}$ undergoes a discontinuous change upon passage through the separatrix, since only one of the three islands can capture the orbit. (Remember, the period 3 property refers to the phase space mapping, not the transformed flow.) As κ continues to decrease, the orbit will retain its new value for $\mathcal{A}$ as the island lifts and shrinks. Eventually — at some point before $\kappa^2 = -30J_2^{in}$ — the island becomes too small to contain the orbit.

[2]Because the system is linear for $\kappa = \infty$ we can legitimately associate I_1^{in} and I_2^{in} with the initial horizontal and vertical emittances divided by 2π [68].

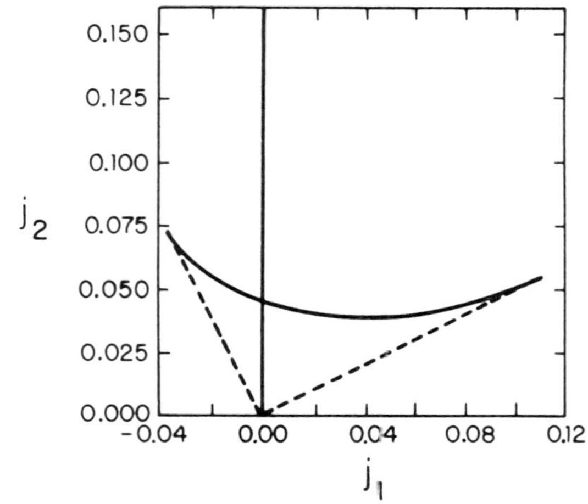

FIGURE C.2 Resonance width master curve.

Figure C.2 contains a "master curve," drawn in the normalized (j_1^{in}, j_2^{in}) coordinates of Eq. (4.42), which uses this scenario to assign resonance widths to individual orbits. The curve was computed by numerically integrating the area under the upper branch of the separatrix when $-1/40 < j_2 < 1/10$ and within the island when $-1/30 < j_2 < -1/40$. It is used in the following way. Suppose one starts an orbit at $\kappa \approx \infty$ with initial amplitude variables I_1^{in} and I_2^{in}. To find the value of κ at which the orbit becomes unbounded, first calculate J_1^{in} and J_2^{in}, using Eqs.(4.39), and take their ratio. The intersection of the ray $j_1^{in}/j_2^{in} = J_1^{in}/J_2^{in}$ with the "master curve" is now read off; call that point (j_1^{int}, j_2^{int}). The value of κ at which the orbit becomes unbounded is

$$\kappa \equiv \sqrt{J_1^{in}/j_1^{int}}.$$

For a given resonant coupling, the adiabatic resonance width of the orbit is then determined according to $2\Delta = 2g\kappa$.

A more dynamic picture is obtained by removing the $1/\kappa^2$ normalization: the curve of Figure C.2 would no longer be static but would sweep through the (I_1^{in}, I_2^{in}) space, converging on the origin as κ approaches zero and making orbits unbounded as it passes their initial conditions.

APPENDIX D

Resonance Overlap

An interesting application of the overlap criterion is made in the problem of resonance sidebands. Envision an isolated resonance model in which the parameters controlling the location of the separatrix are slowly varied, typically via a sinusoidal modulation. If this is done quasi-adiabatically, the intuitive picture one has is that of a separatrix moving to and fro in phase space. The dynamic is no longer integrable. The separatrix is no longer closed, and orbits can be temporarily captured, move with the island, and then escape in different regions of phase space. In this way an orbit can "diffuse" in an effectively random manner in regions of phase space not normally accessible to it.

Another way of describing what may take place involves recasting the modulation as a series of overlapping resonances. A practical application of this is found in the treatment of possible "diffusion" arising from synchrotron oscillations in the presence of an isolated transverse resonance. Orbit oscillations in longitudinal phase space produce a modulation of the small-amplitude transverse tunes via the mechanism of chromaticity. If the longitudinal coordinates are also small — that is, if the orbit is near the center of the bucket — then the resultant tune modulation is approximately sinusoidal. Consider the (one-dimensional, transverse) resonance model Hamiltonian,

$$H(\eta, J; \theta) = (\nu_0 + k/m + \nu_1 \cos \nu_s \theta)J + \frac{\alpha}{2}J^2 + \kappa J^{m/2} \cos m\eta,$$

where we keep the lowest order shear term, proportional to J^2, a higher order resonance term, and a sinusoidal tune modulation with amplitude ν_1 and frequency ν_s, the synchrotron oscillation frequency (see Eq. (3.87)). The modulational term prevents H from being a constant of the motion; the system is not integrable. We assume that $\nu_s << \nu_0$, as is the case in large synchrotrons. By neglecting all the other terms that must appear in a more realistic model we tacitly posit that they will at most distort the orbits in the phase space region of interest without changing their topology. It is therefore *necessary*, though not sufficient, that the *only* resonance of importance be the one singled out: $m\nu_0 + k = 0$. Treating the tune modulation as a single, pure sinusoid is also questionable except very near the center of the longitudinal bucket.

Having expressed these warnings, we proceed by absorbing the explicit θ dependence into the resonance term, something of a reversal of the transformation carried out in Eq. (4.10). This is accomplished using a canonical transformation of the second kind.

$$F(\eta, J) = \left(\eta - \frac{\nu_1}{\nu_s}\sin\nu_s\theta\right)J,$$

$$I = J \quad \text{and} \quad \eta = \phi + \frac{\nu_1}{\nu_s}\sin\nu_s\theta.$$

The new canonical coordinates are $(\phi, I) = (\phi, J)$; the new Hamiltonian is written as follows:

$$H \rightarrow H = (\nu_0 + k/m)J + \frac{\alpha}{2}J^2 + \kappa J^{m/2}\cos[m(\phi + \frac{\nu_1}{\nu_s}\sin\nu_s\theta)].$$

Now comes the purely mathematical step of expanding the cosine in a Fourier-Bessel series, exactly as was done in Section 4.3.1 when finding tune shifts arising from the beam-beam interaction (exponentials and trigonometric functions are the same thing, after all).

$$\cos(u + z\sin\gamma) = \sum_{n=-\infty}^{\infty} J_n(z)\cos(u + n\gamma).$$

We make the identifications, $u = m\phi$, $z = m\nu_1/\nu_s$, and $\gamma = \nu_s\theta$ to obtain the form

$$H = (\nu_0 + k/m)J + \frac{\alpha}{2}J^2 + \kappa J^{m/2}\sum_{n=-\infty}^{\infty} J_n(m\nu_1/\nu_s)\cos m(\phi + n\nu_s\theta/m).$$

This is now in the form of a multiple resonance model: the original isolated resonance is now surrounded by a series of "satellites." The overlap criterion warns that "diffusion" may occur when the island width of each resonance, treated individually, becomes comparable to the spacing between resonances. The former is estimated, in lowest approximation, in Eq. (C.6); the latter, within the same approximation, is estimated as given in Eq. (C.4). To make the connections, we begin by treating the satellite resonances one at a time. This is marginally valid only if they are widely separated. As that is the overlap criterion itself, the argument is at least self-consistent on this point. Formally, we simply erase the summation symbol, fixing n, and transform back to an appropriately rotating phase space via a by now familiar canonical transformation,

$$\eta = \phi + \frac{n}{m}\nu_s\theta.$$

Viewed from the rotating frame, the zero-amplitude tune has been shifted once again. The projected, isolated resonance Hamiltonians will look as in Eq. (C.1), provided we identify

$$\delta\nu = \nu_0 + k/m + n\nu_s/m,$$
$$g = \kappa J_n(m\nu_1/\nu_s).$$

We estimate the spacing between separate islands by using Eq. (C.4),

$$\Delta J_o = -\Delta(\delta\nu)/\alpha = |\nu_s/m\alpha|,$$

while the island widths, δJ, are estimated by using Eq. (C.6). A *nonoverlap* criterion is realized by demanding that $\delta J \ll \Delta J_o$, which in turn requires that

$$\nu_s \gg 4m|\alpha g|^{1/2} J_o^{m/4}.$$

Or, putting all the pieces together, we get the intimidating result

$$\nu_s \gg 4m|\alpha\kappa J_n(m\nu_1/\nu_s)|^{1/2}|(m\nu_0 + k + n\nu_s)/\alpha m|^{m/4},$$

as the conditions, one for each n, necessary to avoid the "diffusion" due to the overlapping of satellite resonances.

Bibliography

[1] Ralph Abraham and Jerrold E. Marsden. *Foundations of Mechanics*. Benjamin/Cummings, Reading, Massachusetts, 1982.

[2] Alex J. Dragt *et al.* Lie algebraic treatment of linear and nonlinear beam dynamics. In *Annual Review of Nuclear and Particle Science, Vol. 38*, pages 455–496. Annual Reviews, Inc., 1988.

[3] Robert L. Anderson and Nail H. Ibragimov. *Lie-Backlund Transformations in Applications*. Society for Industrial and Applied Mathematics, Philadelphia, Pennsylvania, 1979. SIAM Studies in Applied Mathematics.

[4] V. I. Arnold. Proof of A. N. Kolmogorov's theorem on the preservation of quasi-periodic motions under small perturbations of the hamiltonian. *Russian Mathematical Surveys*, 18:9–36, 1963.

[5] V. I. Arnold. *Mathematical Methods of Classical Mechanics*. Springer-Verlag, New York, 1978. Translated by K. Vogtmann and A. Weinstein.

[6] V. I. Arnold. *Ordinary Differential Equations*. The MIT Press, Cambridge, Massachusetts, 1980.

[7] V. I. Arnold and A. Avez. *Ergodic Problems of Classical Mechanics*. W. A. Benjamin, New York, 1968.

[8] Alain Aspect, Philippe Grangier, and Gérard Roger. Experimental tests of realistic local theories via bell's theorem. *Physical Review Letters*, 47(7):460–3, 17 August 1981.

[9] P. Bambade. *Effets faisceau-faisceau dans les anneaux de stockage e^+e^- a haute energie: Grossissement resonant des dimensions verticales dans le cas de faisceaux plats*. PhD thesis, Universite de Paris sud, Centre d'Orsay, 1984. LAL 84/21.

[10] J. S. Bell. *Physics (N.Y.)*, 1:195, 1965.

[11] Richard Bellman. *Introduction to Matrix Analysis*. McGraw-Hill, New York, 1970.

[12] J. S. Berg and R. L. Warnock. Symplectic full-turn maps in a fourier representation. In *Proceedings of the 1991 IEEE Particle Accelerator Conference*. IEEE, 1991. IEEE Catalog Number 91CH3038-7.

[13] M. Berz. *COSY INFINITY: User's Guide and Reference Manual*. Department of Physics and Astronomy. Michigan State University, East Lansing, MI 48824.

[14] M. Berz, W. M. Fawley, and K. Hahn. High order calculation of the multipole content of three dimensional electrostatic geometries. In L. Lizama and J. Chew, editors, *Proceedings of the 1991 IEEE Particle Accelerator Conference*, volume 5, pages 3020–2. IEEE, 1991. IEEE Catalog Number 91CH3038-7.

[15] M. Berz and H. Wollnik. *Nuclear Instruments and Methods*, A258:364, 1987.

[16] Martin Berz. *Nuclear Instruments and Methods*, A258:431, 1987.

[17] Martin Berz. Differential algebraic description of beam dynamics to very high orders. *Particle Accelerators*, 24(2):109, March 1989.

[18] Patrick Billingsley. *Ergodic Theory and Information*. John Wiley & Sons, New York, 1965.

[19] James D. Bjorken. Particle physics — where do we go from here? *Beam Line*, 22(4):8–14, Winter, 1992. Published by the Stanford Linear Accelerator Center.

[20] P. J. Channell and C. Scovel. Symplectic integration of hamiltonian systems. *Nonlinearity*, 3(2):231–59, May 1990.

[21] Boris V. Chirikov. A universal instability of many-dimensional oscillator systems. *Physics Reports*, 52(5):263–379, 1979.

[22] Yvonne Choquet-Bruhat and Cécile DeWitt-Morette. *Analysis, Manifolds and Physics*. North-Holland, New York, 1982.

[23] S. Coffey, A. Deprit, E. Deprit, and L. Healy. Painting the phase space portrait of an integrable dynamical system. *Science*, 247:833–6, February 16, 1990.

[24] G. Corliss and A. Griewank, editors. *Automatic differentiation of algorithms: theory, implementation, and application*. SIAM, Philadelphia, PA, 1991.

[25] E. D. Courant and H. S. Snyder. Theory of the alternating-gradient synchrotron. *Annals of Physics*, 3(1):1, 1958.

[26] H. S. M. Coxeter. *Introduction to Geometry*. John Wiley & Sons, New York, 1969.

[27] Harold T. Davis. *Introduction to Nonlinear Differential and Integral Equations*. Dover Publications, Inc., New York, 1962.

[28] A. Deprit. Canonical transformations depending on a small parameter. *Celestial Mechanics*, 1:12–30, 1969.

[29] P. A. M. Dirac. *The Principles of Quantum Mechanics*. Oxford at the Clarendon Press, London, fourth edition, 1966.

[30] A. J. Dragt. Lectures on nonlinear orbit dynamics. In *Physics of High Energy Particle Accelerators*, pages 147–313. American Institute of Physics, New York, 1982.

[31] A. J. Dragt and J. M. Finn. *Journal of Mathematical Physics*, 17:2215, 1976.

[32] D. A. Edwards and L. C. Teng. Parametrization of linear coupled motion in periodic systems. *IEEE Transactions on Nuclear Science*, NS-20(3):885, 1973. Proceedings of the 1973 IEEE Particle Accelerator Conference. (March 5–7).

[33] George M. Ewing. *Calculus of Variations with Applications*. W. W. Norton & Company, Inc., New York, 1969.

[34] M. Feingold, R. G. Littlejohn, S. B. Solina, J. S. Pehling, and O. Piro. Scars in billiards: the phase space approach. *Physics Letters A*, 146(4):199–203, 21 May 1990.

[35] Harley Flanders. *Differential Forms*. Academic Press, New York, 1963.

[36] Etienne Forest. A hamiltonian-free description of single particle dynamics for hopelessly complex periodic systems. *Journal of Mathematical Physics*, 31:1133, 1990.

[37] Etienne Forest, Martin Berz, and John Irwin. Normal form methods for complicated periodic systems: A complete solution using differential algebra and lie operators. *Particle Accelerators*, 24:91, 1989.

[38] Etienne Forest, Leo Michelotti, Alex J. Dragt, Scott J. Berg, and Johan Bengtsson. The hybrid approach to single particle dynamics for circular rings. To be published in the Proceedings of the Workshop on Instabilities in Storage Rings. Brookhaven National Laboratory. October, 1992.

[39] W. E. Gabella, Ron D. Ruth, and Robert L. Warnock. Iterative determination of invariant tori for a time-periodic hamiltonian with two degrees of freedom. *Physical Review A*, 46(6):3493–512, 1992.

[40] Robert Gilmore. *Catastrophe Theory for Scientists and Engineers*. John Wiley & Sons, New York, 1981.

[41] Herbert Goldstein. *Classical Mechanics*. Addison-Wesley, Reading, Massachusetts, 1959.

[42] H. J. Groenwold. On the principles of elementary quantum mechanics. *Physica*, 12:405–460, 1946.

[43] John Guckenheimer and Philip Holmes. *Nonlinear Oscillations, Dynamical Systems, and Bifurcations of Vector Fields*. Springer-Verlag, New York, 1983.

[44] Victor Guillemin and Shlomo Sternberg. *Symplectic Techniques in Physics*. Cambridge University Press, New York, 1984.

[45] Enrique Henestroza. Pertubative Analysis of Nonlinear Betatron Oscillations. Master's Thesis. Massachusetts Institute of Technology, 1988.

[46] G. W. Hill. On the part of the motion of the lunar perigee which is a function of the mean motions of the sun and the moon. *Acta Mathematica*, 8:1–36, 1886. Expanded from a paper written in 1877 at Cambridge, USA.

[47] P. J. Hilton and S. Wylie. *Homology Theory: An Introduction to Algebraic Topology*. Cambridge University Press, New York, 1967.

[48] L. Van Hove. Sur certaines representations unitaires d'un groupe infini de transformations. *Mem. de l'Acad. Roy. de Belgique (Classe des Sci.)*, XXVI:61–102, 1951.

[49] F. M. Izrailev. Simple models of quantum chaos: Spectrum and eigenfunctions. *Physics Reports*, 196:299–392, November, 1990.

[50] Josef M. Jauch. *Foundations of Quantum Mechanics*. Addison Wesley, Reading, Massachusetts, 1968.

[51] A. A. Kamel. Expansion formulae in canonical transformations depending on a small parameter. *Celestial Mechanics*, 1:190–199, 1969.

[52] A. A. Kamel. Perturbation method in the theory of nonlinear oscillations. *Celestial Mechanics*, 3:90–106, 1970.

[53] Feng Kang. On difference schemes and symplectic geometry. In *Proceedings of the 1984 Beijing Symposium on Differential Geometry and Differential Equations*. Science Press, Beijing, China, 1985.

[54] Feng Kang. Symplectic geometry and numerical methods in fluid dynamics. In F. G. Zhuang and Y. L. Zhu, editors, *Tenth International Conference on Numerical Methods in Fluid Dynamics. June 23-27, 1986*, New York, 1986. Springer-Verlag.

[55] Feng Kang. Difference schemes for hamiltonian formalism and symplectic geometry. *Journal of Computational Mathematics*, 4(3):279–289, July 1986.

[56] D. W. Kerst and R. Serber. *Physical Review*, 60:53, 1941.

[57] A. N. Kolmogorov. *Akad. Nauk. SSSR Dokl.*, 98:527, 1954.

[58] Serge Lang. *Differential Manifolds*. Addison Wesley, Reading, Massachusetts, 1972.

[59] A. J. Lichtenberg and M. A. Lieberman. *Regular and Stochastic Motion*. Springer-Verlag, New York, 1983.

[60] B. A. Lippmann and J. Schwinger. *Physical Review*, 79:469, 1950.

[61] Robert G. Littlejohn. Maslov indices of resonant tori. *Physical Review Letters*, 58(14):1388–91, 1987.

[62] Robert G. Littlejohn. New way to compute maslov indices. *Physical Review A*, 36(6):2953–61, 1987.

[63] Robert G. Littlejohn. Cyclic evolution in quantum mechanics and the phases of bohr-sommerfeld and maslov. *Physical Review Letters*, 61(19):2159–62, 1988.

[64] Wilhelm Magnus and Stanley Winkler. *Hill's Equation*. Interscience Publishers (John Wiley and Sons), New York, 1966.

[65] Benoit B. Mandelbrot. *Fractals: Form, Chance, and Dimension*. W. H. Freeman and Company, San Francisco, 1977.

[66] Benoit B. Mandelbrot. *The Fractal Geometry of Nature*. W. H. Freeman and Company, San Francisco, 1983.

[67] S. R. Mane. Solutions of laplace's equations in two dimensions with a curved longitudinal axis. *Nuclear Instruments and Methods in Physics Research*, A321:365–75, 15 September 1992.

[68] Leo Michelotti. Introduction to the nonlinear dynamics arising from magnetic multipoles. In Melvin Month and Margaret Dienes, editors, *Physics of Particle Accelerators*. American Institute of Physics, 1987. AIP Conference Proceedings No. 153.

[69] Leo Michelotti. MXYZPTLK: A C++ hacker's implementation of automatic differentiation. In Reference [24].

[70] Leo Michelotti. Phase space concepts. In Melvin Month and Margaret Dienes, editors, *Physics of Particle Accelerators*. American Institute of Physics, 1989. AIP Conference Proceedings No. 184.

[71] Leo Michelotti. Perturbation theory and the single sextupole. In Y.S. Kim and W.W. Zachary, editors, *The Physics of Phase Space*. Springer-Verlag, 1987.

[72] B. W. Montague. Fourth-order coupling resonance excited by space-charge forces in a synchrotron., October 1968. CERN ISR/68-38.

[73] J. Moser. *Nachr. Akad. Wiss. Gottingen Math. Phys.*, 2, 1962.

[74] Charles Nash and Siddhartha Sen. *Topology and Geometry for Physicists*. Academic Press, New York, 1983.

[75] Roger G. Newton. *Scattering Theory of Waves and Particles*. McGraw-Hill Book Company, New York, 1966.

[76] G. Nicolis and I. Prigogine. *Self-Organization in Non-Equilibrium Systems*. John Wiley & Sons, Inc., New York, 1977.

[77] C. Piron. *Foundations of Quantum Physics*. W. A. Benjamin, Inc., Reading, Massachusetts, 1976.

[78] M. B. Pour-El and I. Richards. Noncomputability in models of physical phenomena. *International Journal of Theoretical Physics*, 21(6-7):553–5, 1982.

[79] S. Neil Rasband. *Dynamics*. John Wiley & Sons, New York, 1983.

[80] Joseph Fels Ritt. *Differential Algebra*. American Mathematical Society, New York, 1950.

[81] Ronald D. Ruth. A canonical integration technique. *IEEE Transactions on Nuclear Science*, NS-30(4):2669–71, August 1983.

[82] F. Schmidt, F. Willeke, and F. Zimmerman. Comparison of methods to determine long-term stability in proton storage rings. *Particle Accelerators*, 35(4):249–56, 1991.

[83] C. E. Shannon. A mathematical theory of communication. *Bell System Technical Journal*, 27:379–423,623–656, 1948.

[84] Carl Ludwig Siegel. *Symplectic Geometry*. Academic Press, New York, NY, 1964.

[85] Y. G. Sinai. *Introduction to Ergodic Theory*. Princeton University Press, Princeton, New Jersey, 1976.

[86] S. C. Snowdon. Transformations useful in linear betatron theory. Fermi Note 185, Fermilab, May 8, 1969.

[87] Keith R. Symon. *Mechanics*. Addison-Wesley, Reading, Massachusetts, 1961.

[88] Walter Thirring. *Classical Dynamical Systems*. Springer-Verlag, New York, 1978.

[89] R. Thom. *Structural Stability and Morphogenesis: an Outline of a General Theory of Models*. W. A. Benjamin, Reading, Massachusetts, 1975.

[90] R. L. Warnock and R. D. Ruth. Invariant tori through direct solution of the hamilton-jacobi equation. *Physica*, 26D:1–36, 1987.

[91] Robert L. Warnock and Ron D. Ruth. Long-term bounds on nonlinear hamiltonian motion. *Physica D*, 56:188–215, 1992.

[92] Nicholas Woodhouse. *Geometric Quantization*. Clarendon Press, Oxford, 1980.

[93] H. Yoshida. *Physics Letters A*, 47:190, 1990.

Index